Annual Conference and Expo of the Association of Avian Veterinarians (AAV 2024)

Kansas City, Missouri, USA
28 June – 1 July 2024

ISBN: 979-8-3313-0827-8

Printed from e-media with permission by:

Curran Associates, Inc.
57 Morehouse Lane
Red Hook, NY 12571

CURRAN ASSOCIATES INC. proceedings.com

Some format issues inherent in the e-media version may also appear in this print version.

Association of Avian Veterinarians
PO Box 9
Teaneck, NJ 07666
USA

Phone: 720-458-4111
Fax: 720-398-3496

office@aav.org

Additional copies of this publication are available from:

Curran Associates, Inc.
57 Morehouse Lane
Red Hook, NY 12571 USA
Phone: 845-758-0400
Fax: 845-758-2633
Email: curran@proceedings.com
Web: www.proceedings.com

ASSOCIATION OF AVIAN VETERINARIANS

AAV ANNUAL CONFERENCE & EXPO

2024 PROCEEDINGS

KANSAS CITY MARRIOTT DOWNTOWN

FRIDAY, JUNE 28 - MONDAY, JULY 1, 2024

ASSOCIATION OF AVIAN VETERINARIANS AAV AVIAN MEDICINE VIRTUAL

INCLUDES BONUS SECTION

AAV.ORG

Introduction

We invite you to enjoy the 45th Annual AAV Conference Proceedings. Within the proceedings, you will find a wealth of information related to the variety of lectures presented at this year's conference.

AAV recognizes and thanks the following committees and staff for their time and hard work in assuring a quality program and publication:

AAV Executive and Conference Committees

Stephanie Lamb, DVM, DABVP (Avian Practice), *President, Conference Committee Member*
Jennifer Graham, DVM, DABVP (Avian Practice), DABVP (Exotic Companion Mammal Practice), Dipl ACZM, *President-Elect, Conference Committee Member*
Anna Osofsky, DVM, DABVP (Avian Practice), *Immediate Past President, Conference Committee Chair*
Tia Greenberg, DVM, *Conference Committee Member*
Susie Baley, DVM, DABVP (Avian Practice), *AAV Hands-On Workshop Coordinator*
Salina Locke, DVM, DABVP (Avian Practice), Cert AqV Len Donato, VMD, DABVP (Avian Practice), *Treasurer*
Jen Riley, DVM, *Advisory Council Representative to the AAV Executive Board*
Elizabeth Mackey, DVM, *Executive Director*

Veterinary Review and Editing Committee

Natalie Antinoff, DVM, DABVP (Avian Practice)
Heather Barron, DVM, DABVP (Avian Practice)
Victoria Bender, DVM
Kyra Berg, DVM, Dipl ACZM
Joao Brandao, LMV, MS, DipECZM (Avian)
Erica Cervasio, DVM
Sue Chen, DVM, DABVP (Avian Practice)
John Chitty, BVetMed, CertZooMed, CBiol, MRSB, MRCVS
Bob Dahlhausen, DVM
M. Scott Echols, DVM, DABVP (Avian Practice)
Jesse Fallon, DVM, PhD
Jennifer Graham, DVM, DABVP (Avian Practice), DABVP (Exotic Companion Mammal Practice), Dipl ACZM
Cheryl Greenacre, DVM, DABVP (Avian Practice), DABVP (Exotic Companion Mammal Practice)
Michelle Hawkins, VMD, DABVP (Avian Practice)
Stephanie Lamb, DVM, DABVP (Avian Practice)
Isabelle Langlois, DVM, DABVP (Avian Practice)
Stephany Lewis, DVM, CWR, DABVP (Avian Practice)
Salina Locke, DVM, DABVP (Avian Practice), Cert AqV
Brynn McCleery, DVM, DABVP (Avian Practice)
Alicia McLaughlin, DVM
Mark Mitchell, DVM, MS, PhD, DipECZM (Herpetology)
Bianca Murphy, DVM, DABVP (Avian Practice)
Brendan Noonan, DVM, DABVP (Avian Practice)
Anna Osofsky, DVM, DABVP (Avian Practice)
Olivia Petritz, Olivia Petritz, DVM, Dipl ACZM
Christal Pollock, DVM, DABVP (Avian Practice)
Julianne Richard, DVM, MPH, CertAqV
Jaime Samour, MVZ (Hons), PhD, Dip ECZM (Avian), FRSB
David Sanchez-Migallon Guzman, LV, MS, DipECZM (Avian), DipECZM (Small Mammal), Dipl ACZM
Marcy Souza, DVM, MPH, DABVP (Avian), DACVPM
Brian Speer, DVM, DABVP (Avian Practice), DipECZM
Lauren Thielen, DVM, DABVP (Avian Practice)
Thomas N. Tully, Jr., DVM, MS, DABVP (Avian), DipECZM (Avian)
Shangzhe Xie, BSc/BVMS, MVS (Conservation Medicine), PhD, DABVP (Avian Practice), DACC

Technical Editor
Charlotte Kirk Baer, MS

Proceedings Layout and Design — BrightLizard, Inc.
Elizabeth Smith

Conference Sponsors

Continuing Education Accreditation

The American Association of Veterinary State Boards (AAVSB) Registry of Approved Continuing Education (RACE) committee has reviewed and approved the 45th Annual Association of Avian Veterinarians Conference & Expo program as meeting the standards adopted by the AAVSB. Please contact the AAVSB RACE program at race@aavsb.org or 877-698-8482 if you have any questions regarding Continuing Education for this program.

Disclaimer

The material appearing in this publication comes exclusively from the authors and contributors identified in each manuscript. The techniques and procedures discussed reflect the personal knowledge and experience of the authors and contributors, and demonstrate their views of the methods that may be used for these medical procedures. The procedures demonstrated do not incorporate all known techniques, are not exclusive, and other techniques and technology may also be available. Any questions or requests for additional information concerning any of the manuscripts should be addressed directly to the authors.

AAV does not research, review, or otherwise verify any of the information contained in this publication. Opinions expressed in this publication are those of the authors and contributors and not necessarily those of AAV. AAV is not responsible for errors or for opinions expressed in this publication.

AAV expressly disclaims any warranties or guarantees, expressed or implied, and shall not be liable for damages of any kind in connection with the material, information, techniques, or procedures set forth in this publication.

Publication Terms

Conference proceedings manuscripts are accepted as full manuscripts or as scientific abstracts. Scientific abstracts are accepted if the author(s) express(es) intent to submit their work for consideration for future publication in a peer-reviewed scientific journal. Several distinguished veterinary journals, including the *Journal of the American Veterinary Medical Association* and the *American Journal of Veterinary Research* may reject work that has been previously published of 250 words or more and included additional content such as figures, tables, illustrations, and references. Scientific abstracts published in the AAV proceedings are still required to contain data and meaningful information to the reader.

For permission to reproduce or order copies of this publication, contact the AAV association office: www.aav.org, 1802 Antioch Church Road, Watkinsville, GA 30677, USA, email pubs@aav.org.

Visit www.aav.org for information on AAV educational conferences, membership, and other association programs.

Table of Contents

Section 1: Friday Fundamentals Session

Section 2: Plenary Session

Section 3: Hot Topics Session

Section 4: Wildlife and Conservation Session

Section 5: Potpourri Session

Section 6: Technician Session

Section 7: Fundamentals Masterclasses

Section 8: Clinical Practice Masterclasses

Section 9: Wildlife and Conservation Masterclasses

Section 10: Monday Masterclasses

Author Index

Author Index Continued

Section 1

Friday Fundamentals Session

Jennifer Graham, DVM, DABVP (Avian Practice), DABVP (Exotic Companion Mammal Practice), Dipl ACZM;
Cheryl Greenacre, DVM, DABVP (Avian Practice), DABVP (Exotic Companion Mammal Practice)
Session Chairs

Avian Top Ten Emergencies

Jennifer Graham, DVM, DABVP (Avian Practice), DABVP (Exotic Companion Mammal Practice), Dipl ACZM

Session #121

Affiliation: From Graham Veterinary Consulting, LLC, Madison, AL 35758, USA.

Note: Adapted from Graham J. Avian top 10 emergencies. *Proc Annu Conf Int Vet Emerg Crit Care Soc.* 2019.

Abstract: Avian emergency presentations can be challenging. Because birds are prey species, they hide their signs of illness to avoid predation. As such, most birds presenting in an emergency have significant underlying disease. Rapid assessment and appropriate supportive care measures are integral to a good case outcome. This presentation provides an overview of common avian presentations including ain't doing right, respiratory, gastrointestinal, reproductive, heavy metal toxicosis, traumatic, neurologic, feather-damaging, renal, and ophthalmic emergency presentations.

Ain't Doing Right

My bird ain't doing right (ADR) is a common avian emergency presentation that can include a combination of non-specific signs of illness: fluffed on bottom of the cage, not eating, not vocalizing, among others. There are numerous potential underlying disease processes and organ system failures that could cause a client to finally notice that their bird is sick, but the reason that the patient is being presented is because it is too weak to continue hiding more obvious signs of illness. Underlying pathophysiologic mechanisms that could lead to this include septicemia, toxemia, cardiovascular collapse, anemia, dehydration, hypoxemia, electrolyte abnormalities, acid-base imbalances and caloric deficits or hypoglycemia. Essentially all birds presented in critical condition should be hospitalized and treated aggressively. Diagnostic and therapeutic plans need to be determined at the outset, and it is important to understand that sometimes it is best to delay any invasive or hands-on diagnostic tests until the patient is more stable.

Respiratory Emergencies

Aspiration pneumonia

Aspiration pneumonia should be considered as a differential when handfed babies present in respiratory distress. Aspiration of foreign material can occur iatrogenically, especially with improper feeding techniques with baby or adult birds, or with loss of normal protective mechanisms, such as in a very weak bird. Initial radiographs may be normal, but evidence of local pneumonia or focal pulmonary abscesses can develop with time. Treatment may include hospitalization with oxygen therapy, nebulization, fluid therapy, and broad-spectrum antibiotics and antifungal medications.

Airway obstruction

Airway obstruction is characterized by stridor and dyspnea that is exacerbated by exertion, and intervention is necessary to relieve obstruction. In severe cases, immediate placement of an air sac cannula is indicated. Following cannula placement, endoscopic examination of the trachea can allow removal of foreign bodies or identification of granulomas in larger birds. In smaller birds, a 1.1-mm endoscope can be passed, but foreign bodies cannot be retrieved in this manner. A suction tube (e.g., urinary catheter) smaller than the diameter of the trachea may be passed to suction out foreign bodies. Tracheal surgery has been described to remove tracheal foreign bodies in birds. Membranous stenosis has been seen by the author and others in birds following suspected tracheal damage caused by intubation and ventilation; this should be kept in mind when a bird presents with evidence of airway obstruction with a history of recent anesthesia.

Airborne toxins

Birds are sensitive to the effects of inhaled toxins, and clients should be educated about the risks of inhaled fumes such as nonstick cookware, self-cleaning ovens, bleach, ammonia, cigarette smoke, and others. Nonstick cookware can release polytetrafluoroethylene gas (PTFE) when cooking surfaces overheat. PTFE inhalation can cause pulmonary congestion, hemorrhage, and death. If a bird presents with a history of an inhaled toxin, immediate oxygen support is ideal. Saline nebulization can be beneficial. Bronchodilators such as terbutaline (Brethine, Novartis Pharmaceuticals Corporation, East Hanover, NJ) have been used parenterally and via nebulization in dyspneic birds suffering from inhaled toxins.

Air sac rupture

Air sac rupture can occur spontaneously or in association with trauma, such as bite wounds. Air sac rupture may cause local subcutaneous emphysema, particularly in Amazon parrots, macaws, and cockatiels. Needle aspiration can aid in a diagnosis and reduce the swelling but creating a draining site that is open for several weeks may be necessary to effectively treat this condition. Indwelling stents have been described as one treatment option. When possible, a complete medical workup should be performed on birds presenting with air sac rupture to rule out the possibility of systemic disease, such as air sac infections.

Sinusitis

Sinusitis and rhinitis may be caused by a variety of etiologies, including infection, irritation, hypovitaminosis A, neoplasia, and others. If a bird presents in an emergency with clinical signs suggestive of upper airway disease localized to the sinuses, supportive care including oxygen, warmth, nebulization, fluid therapy, and nutrition may be indicated depending on the degree of disease. A thorough history and physical examination are required to determine the best approach for diagnosing and treating the underlying problem.

Fungal disease

Aspergillus sp. is a common fungal disease affecting psittacine birds, especially African grey parrots. *Aspergillus* may affect the sinuses, trachea, lungs, and air sacs; the syringeal area is a site typically prone to infection. A change of voice, dyspnea, and sudden death are characteristic of this disease. Diagnosis is best made with direct visualization of lesions with cytology, histopathology, culture, or genetic sequencing. A complete blood count (CBC) usually shows a marked leukocytosis with a monocytosis and heterophilia. *Aspergillus* antibody titers have only moderately predictive value.

Coelomic distention

Coelomic distention is a common cause of respiratory distress. See reproductive emergencies for more information.

Gastrointestinal Emergencies

Diarrhea

Diarrhea is characterized by abnormally increased frequency, liquidity, and volume of fecal discharge. Birds with this condition may have abnormal stool (e.g., color, character), undigested food in the droppings, and soiled vent feathers. Causes of diarrhea include bacterial infection (e.g., *E. coli*, *C. psittaci*, *Salmonella* sp.), viral infection (e.g., bornavirus, paramyxovirus, adenovirus), parasitic infection (*Cryptosporidium* sp., *Hexamita* sp.), obstruction, metabolic disorders, dietary causes, maldigestion, and toxins. Diagnostics that can be helpful to determine the cause of diarrhea include CBC and chemistries (may be nonspecific or show signs of infection and dehydration), fecal or gastrointestinal (GI) cytology, and radiographs, with or without additional imaging. Treatment is dependent on the cause and nursing care including fluids, warmth, and nutritional support can be beneficial.

Crop stasis and ileus

Ingluvial hypomotility and ileus are relatively common GI emergency presentations and are generally secondary to an underlying disease. Historical findings may include crop distention, decreased fecal production, and sudden death. Crop distention, dehydration, respiratory distress, and weakness may be noted on examination. Causes of crop stasis and ileus include infectious (e.g., *Candida* sp., bacterial, viral, parasitic), atonic ingluvies (e.g., overstretched), trauma, stricture, foreign bodies, metabolic disease, toxicosis (e.g., lead, zinc), and neoplasia. Helpful diagnostics can include CBC and chemistries, fecal or crop cytology, heavy metal levels, and radiographs (empty crop if full before positioning), with or without advanced imaging. Treatment depends on etiology and supportive care measures are warranted.

Regurgitation and vomiting

Regurgitation is the retrograde expulsion of ingesta from the esophagus or crop, and vomiting is the retrograde expulsion of ingesta from the proventriculus, ventriculus, or intestine. Physical examination findings can include foul odor to breath, dried food or mucus matted to head feathers, and crop distention. A thorough history is required as this condition can be a behavioral condition or due to motion sickness. Crop or esophageal disorders can include infections (e.g., bacterial, yeast, trichomonas), foreign bodies or masses, and trauma including crop burns. Proventricular and ventricular disorders include infectious (e.g., bacterial, fungal, viral including bornavirus, *M. ornithogaster*, *Cryptosporidium* sp.), toxins, and neoplasia. Recommended diagnostics include CBC and chemistries (may have nonspecific findings); cytology of fecal, crop, or gastric samples; fecal flotation; and radiographs, with or without additional testing as indicated. Treatment depends on etiology and supportive care measures should be given if warranted.

Fecal discoloration

Fecal color can be influenced by bile pigments (i.e., dark green color), diet (i.e., food can discolor feces), substrate (i.e., ink from newsprint), and blood. Physical examination findings may vary depending on cause: lead toxicosis (e.g., hemoglobinuria, weakness, seizures, polyuria and polydipsia, anemia); liver disease (e.g., regurgitation or vomiting, diarrhea, biliverdinuria, hepatomegaly); intestinal disease (e.g., diarrhea, undigested food in droppings, melena or frank blood). Brown color of the feces is normal in most situations, but coliform infections may cause liquidity and distinctly brown feces. Green color of the feces can be normal (light) or can be a sign of inadequate food intake (i.e., dark). Clay colored feces can indicate maldigestion or malabsorption, or intestinal, pancreatic or liver disease. Black color indicates melena or severe anorexia; budgie feces often looks dark when dry but appears green when fresh stool is examined. Red feces can indicate lower GI disease (e.g., cloacal pathology including bleeding papilloma) but could also be coming from the urinary or reproductive tract. Ingestion of certain fruits and vegetables and artificial dyes or food coloring in food and toys can cause feces to appear orange, yellow, blue, red, orange, and purple. Diagnostics should be based on the suspected organ system affected and can include bloodwork and fecal cytology (including direct wet mount, Gram's stain), flotation, fecal occult blood testing, acid-fast stain, or culture and sensitivity. Heavy metal testing may also be warranted. Radiographs and additional imaging (e.g., cloacal endoscopy) may also be helpful. Treatment includes supportive care and therapeutics based on cause.

Undigested food in droppings

Undigested food in the dropping is most obvious when a bird is eating seeds. Bulky or loose stools may be noted if a bird is eating a formulated diet. Proper fecal formation is based on intestinal, hepatic, and pancreatic functions along with proper mechanical functions of the GI tract. Causes of undigested food in the droppings include GI foreign bodies or impaction; infectious (e.g., bacterial, viral, fungal, parasitic); neoplasia; *M. ornithogaster*; toxins; and hepatic, pancreatic, or coelomic disease. Diagnostics helpful in working up these cases include bloodwork (e.g., anemia with lead poisoning, leukocytosis with inflammation or infection), cytology (e.g., direct fecal examination, Gram's stain), fecal flotation, fecal occult blood testing, imaging (e.g., radiographs, contrast study, fluoroscopy), heavy metal testing, PCR screening (e.g., *Chlamydia*, *M. ornithogaster*, bornavirus, *Cryptosporidium* sp.). Treatment is based on underlying etiology.

Reproductive Emergencies

Egg binding

Chronic egg laying can result in weakness, osteopenia, egg binging, yolk peritonitis, and pathologic fractures. Initial stabilization measures involve the administration of fluids, parenteral calcium, and possibly vitamin D; injectable vitamin A, D, and E formulations are most often used. Diagnostics including bloodwork (including iCa2+), radiographs, ultrasound, and coelomocentesis can be helpful to determine the extent and type of disease. Care should be taken to remove any coelomic fluid before putting the bird in a non-vertical position. Prostaglandin and hormonal therapy may be warranted to induce oviductal contractions once the patient has been stabilized if non-obstructive egg binding is diagnosed. Percloacal ovocentesis may be performed to facilitate passage of an egg but runs the risk of inducing coelomitis and is usually reserved for cases in which conservative therapy is unsuccessful. Salpingohysterectomy is recommended for birds with chronic recurring disease or when egg removal is otherwise unsuccessful.

Coelomic disease

Ascites, masses, and dystocia may cause pronounced respiratory signs due to compression of the air sacs. If coelomic palpation reveals distention or the presence of a mass, ultrasound is useful to determine if free fluid is present. Ultrasound-guided coelomic aspiration can be performed on an emergency basis to remove fluid and provide immediate relief to the patient. Care should be taken to maintain the patient in an upright position during this procedure to prevent fluid from entering the lungs; oxygen provided via a facemask and mild sedation may be helpful for the procedure. Ultrasound can also help characterize any masses in the coelom. Dystocia can initially be treated with warmth, subcutaneous fluids, and calcium supplementation. Additional therapies such as oxytocin, topical prostaglandin gel, ovocentesis, or coelomic surgery may be indicated depending on the severity of disease. Consult avian texts for more complete information.

Cloacal prolapse

Cloacal and oviductal prolapse can occur secondary to straining from a variety of conditions including chronic masturbatory behavior, space-occupying masses, papillomatous lesions, egg laying, or dystocia. Prolapsed tissues should be cleaned and kept moistened to prevent desiccation. Temporary stay sutures (avoid purse-string suture) may be warranted to keep prolapsed tissue in place, but this is contraindicated if an egg is still within the reproductive tract. Phallus prolapse is common in waterfowl. Treatment as described for cloacal prolapse is warranted and phallus amputation can be performed if medical management is not effective.

Heavy Metal Toxicosis

Ingestion of particles of heavy metals, either directly (e.g., parts of toys, galvanized cages, lead-based paints, fishing weights, lead-coated seeds, pennies > 1982) or indirectly (e.g., ingestion of a prey shot with lead pellets) can cause heavy metal toxicosis. Heavy metals are absorbed mainly by the GI tract and rarely by the lungs after inhalation. Due to the poor rate of absorption, lead shot embedded in muscles does not usually induce toxicity. Lead is widely distributed in soft tissues, and bones serve as the long-term storage site for the metal. Lead is toxic to multiple enzymatic systems and interferes with the numerous cellular functions that require calcium. Lead will affect the red blood cells (i.e., microcytic anemia and/or porphyrinuria), the renal tubules (i.e., Fanconi syndrome nephropathy), and the central and peripheral nervous system (i.e., encephalopathy and polyneuropathy by demyelination and neuronal necrosis). Both the immune and reproductive systems can also be impaired (e.g., infertility, embryonic mortality, and teratogenicity). Although the exact mechanism of zinc toxicity is not completely understood, zinc has a high affinity for the pancreas, liver, and kidneys. Through both its direct and indirect effects, zinc can cause anemia, pancreatitis, hepatic failure, and renal failure. Zinc salts have direct irritant and corrosive effects on tissue, interfere with the metabolism of other ions such as copper, calcium, and iron, and inhibit erythrocyte production and function.

Birds with heavy metal toxicosis can present with nonspecific signs such as lethargy, anorexia, and depression. GI signs including regurgitation and diarrhea are common. Neurologic signs may include seizures and weakness. Polyuria and polydipsia, red urine (particularly in Amazon parrots), and sudden death have been reported. CBC changes can include a regenerative microcytic hypochromic anemia with increased immature RBCs, poikilocytosis, and nuclear abnormalities (i.e., fusiform, elongated, and irregular nuclei are seen with zinc toxicosis in birds). Chemistry changes such as elevation of uric acid, AST, LDH and CK have been reported. Whole blood on EDTA or heparin should be used to measure lead levels; do not submit serum or plasma. Generally, intoxication is confirmed if the lead concentration in whole blood is > 0.2 ppm (20 µg/dl). Subclinical toxicosis occurs at lower levels. Some species (e.g., pigeons, poultry) seem to be able to tolerate higher levels before showing clinical signs of acute toxicity. Decreased ALAD activity is also reported. Blood should be collected in serum tubes without rubber components for measuring zinc levels. Most pet birds have normal zinc levels below 3.5 ppm (0.35 mg/dl).

Aggressive fluid therapy including intravenous (IV) or intraosseous (IO) fluids is ideal, particularly when renal disease is present. Nutritional support and keeping the bird in an environment appropriate to the neurologic conditions are recommended (e.g., padded cage, if having seizures; comfortable bedding, if sternally recumbent). Edetate calcium disodium ($CaNa_2EDTA$, injectable sterile solution) 35 mg/kg q12h IV or intramuscularly (IM) is administered for 5 days, followed by a rest period of 2 to 3 days to allow a redistribution of tissue and fluid lead concentrations and to prevent excessive chelation of endogenous minerals. Assessment of blood concentrations after the rest period will indicate if the protocol needs to be repeated. If seizures are present, midazolam 0.5-2 mg/kg IM can be administered (see neurologic emergencies for more information about seizure management). Antibiotics (e.g., amoxicillin/clavulanic acid 125 mg/kg q12h PO, enrofloxacin 15 mg/kg q12h PO) might be indicated due to the immunosuppressive effects of lead and in cases with severe enteritis.

Trauma

Hemorrhage

Common bleeding emergencies include broken blood feathers, injured nails or beak, lacerations, and traumatized wing tips. Hemostasis is critical and fluid therapy is beneficial in most cases of hemorrhage. Although hypertonic saline, colloidal fluids, plasma, or blood transfusion can be administered in cases of severe hemorrhage, colloid or crystalloid solution administration alone may be sufficient.

Broken blood feather

Broken blood feathers should be pulled with pressure applied to the follicle for 60-90 seconds. Avoid applying styptic powders or other irritating substances into the feather follicles or skin. If hemorrhage continues, tissue glue or a small absorbable suture can be placed to close the follicle.

Nail injury

Toes may become traumatized in toys, on cages, or by conspecifics. Styptic powder or cautery can be used on bleeding nail tips. In the case of complete nail avulsion, the exposed tissue should be cleaned and covered with a light bandage or tissue glue.

Beak trauma

Beak trauma can be seen in cases of head trauma or falls, toy entrapment, and bites from conspecifics. Simple wounds can be managed with manual pressure with styptic powder or cautery applied to maxillary tip fractures. More complex fractures or avulsions may require surgical or acrylic stabilization. Analgesia is important, and some birds may benefit from temporary gavage feeding or being fed a softer diet.

Laceration and other wounds

Skin wounds can be self-inflicted or caused by conspecifics, falls, and predators. Wing tip trauma can occur with episodes of night fright and falls. Open wounds should be cleaned and closed if fresh and non-contaminated. Some wounds cannot be managed by primary closure and collars or bandages may be required to protect the wound while healing.

Constriction injury

Constriction injuries can occur to an extremity from a leg band or trauma from toys, strings or nesting material, injuries, and skin hyperplasia secondary to malnutrition. Magnification can be helpful when removing constricting material and great care should be taken to avoid causing further injury. Hemostats, wire cutters, or bolt cutters may be used to remove metal bands. Appropriate wound care should follow removal of the constricting material. In some cases, constriction may be severe enough to cause loss of extremities.

Fractures

Fractures may occur in association with flying accidents or crushing trauma and when pathology of the bone is present. Fractures should be immobilized in stable patients to prevent further trauma. Open fractures should be cleaned and reduced, taking care not to flood air sacs in the case of a pneumatic bone, and antibiotic therapy should be initiated. Appropriate analgesia is vital when managing these cases. The figure-of-eight bandage can be useful to manage wing fractures distal to the humerus, soft tissue injuries, and luxations. If there is a humeral or pectoral girdle fracture, a body wrap should be placed. A modified Robert-Jones bandage for birds > 200 g or Altman tape splint for birds < 200 g can be used to stabilize fractures distal to the femur. Once the bandage is in place the bird should be monitored to make sure there is no evidence of increased stress, dyspnea, or chewing/destroying the bandage. Appropriate fracture repair should follow when appropriate.

Burns

The most common burns seen on an emergency basis occur from birds flying into pots of hot liquid. Less commonly, birds may present for care following electrical burns, burns secondary to fires, or with crop burns. Burns can be superficial, partial thickness, or full thickness. It can be difficult to assess the severity of the burn on initial presentation as damage may become more apparent in the days following the initial injury. Cold compresses can be applied for the first 20-30 min after the burn with careful monitoring of the patient to avoid hypothermia. Analgesia and supportive care are critical before wound management. Oxygen and fluid therapy may be beneficial, particularly if smoke inhalation has occurred or if the patient is dehydrated. An air sac cannula can be helpful in the case of laryngeal edema. Systemic antibiotics and antifungals are recommended in cases of burns to prevent sepsis. Superficial birds may result in transient erythema and epidermal desquamation; saline and 5% povidone iodine or chlorhexidine can be used to gently cleanse wounds. Partial-thickness burns extend to mid-dermis with associated exudation and pain. Full-thickness burns cause coagulation of the epidermis and dermis with subsequent necrosis of the damaged tissue. Partial and full-thickness burns can be gently cleansed and debrided daily followed by topical treatment with a water-soluble antibiotic (e.g., silver sulfadiazine). Anesthesia may be needed for wound debridement. Wounds can be covered if likely to get contaminated or injured by the bird. The first 24-48 hours following the burn is the most critical time and care should be taken to monitor hydration and renal function. Crop burns may be identified initially by edema and then eschar leading to fistula over a period of days to a week after the injury. Devitalized tissue can be resected followed by two-layer closure.

Neurologic Emergencies

Head trauma

Flight injuries are a common cause of head trauma in pet birds. Patients with head trauma can present with evidence of bruising or bleeding, lethargy, and altered consciousness. Patients with head trauma should be placed in an oxygen-enriched, quiet, dark, well-padded environment. Additional thermal support is not recommended if intracranial hemorrhage is suspected. Vascular access is ideal with placement of an IV or IO catheter. Hypovolemia can be managed with low-volume resuscitation (3 ml/kg) over 10 min with hypertonic saline with or without colloids. Maintenance fluid support can be reduced by 35-50% if there are concerns about increased intracranial pressure. Mannitol (0.25-1.0 g/kg) IV or IO as a slow bolus can be given every 4-6 hours for a total of three doses in 24 hours. Head elevation by 30% is recommended for recumbent patients. Analgesia, including opiods, can be given, but should be used cautiously to avoid respiratory depression. If hydrated, meloxicam (0.5-1 mg/kg q12h) may decrease cerebral edema and intracranial pressure. Propofol CRI to control seizures may be preferable over benzodiazepines to avoid neurologic depression. Imaging including computed tomography scans can be performed once the patient is stable to evaluate for beak or skull fractures.

Seizures

Seizures are a relatively common presentation in emergency settings. Seizures can occur secondary to trauma, heavy metal ingestion, hypoglycemia, cardiovascular disease (e.g., claudication syndrome or egg yolk embolism), hypocalcemia, idiopathic epilepsy, and infectious disease including viral (e.g., bornavirus), bacterial, fungal, and parasitic causes. Once the patient's airway is ensured, oxygen therapy should be provided while obtaining IV or IO access. Benzodiazepines including diazepam (0.5-2 mg/kg IV/IO/cloacal) or midazolam (0.1-2 mg/kg IV/IO/IM/IN) can be administered 2-3 times or by CRI. In cases of refractory seizures, levetiracetam (20-60 mg/kg IV) or phenobarbital can be attempted, followed by propofol or ketamine boluses if unsuccessful. Additional therapies, including calcium, glucose, chelating agents, among others should be used as warranted. Levetiracetam and zonisamide may be more helpful than phenobarbital in some avian species for long-term seizure management (i.e., African grey parrots).

Bite wounds

Bite and scratch wounds can lead to fatal septicemia if not managed in a timely fashion, with cat bites being particularly problematic. Wounds can range from superficial puncture wounds to lacerations, coelomic cavity penetration, and traumatic limb amputation. Patients should be stabilized with oxygen therapy, fluids, and analgesia. Broad-spectrum antibiotic therapy should be initiated as soon as possible, with attention paid to gram-negative and anaerobic coverage in the case of predator attacks. Wound management may necessitate flushing, bandaging, suturing, or fracture repair; care should be taken to avoid flushing wounds that could connect to the air sacs or lungs.

Glue traps

Birds can become entrapped in glue traps intended for rodent control if allowed unsupervised access. If a small part of the feathers are adhered to the trap they can be cut to release the bird. Mineral oil can be applied to detach larger areas of feathers and skin from the glue trap. Once released, warm water (100-105 °F) and commercial dish detergent can be used to remove the glue or oil from the feathers.

Feather Damaging

Feather-damaging behavior (FDB) is a condition in which the parrot mutilates its feathers with the beak and may involve chewing, biting, plucking and/or fraying. Typically, the feathers of the head and crest remain unchanged as these are inaccessible to the bird's beak. In case of self-injurious behavior or auto-mutilation, skin damage is also present. In general, three underlying pathophysiological mechanisms should be taken into consideration for this condition: 1) maladaptive behavior, resulting from attempts of the animal to cope with an abnormal or inadequate environment that is lacking the appropriate stimuli needed and/or in which stressors are present; 2) malfunctional behavior, resulting from an abnormal psychology, brain development, or neurochemistry that may have developed as a result of the bird's living conditions, particularly in early life; and 3) abnormal behavior resulting from underlying medical (physical) problems. Any disease causing pain, discomfort, irritation and/or pruritus may result in the bird damaging its feathers. Although the condition may occur in any parrot species, grey parrots (*Psittacus* spp.), Eclectus parrots (*Eclectus roratus*) and cockatoos (*Cacatua* spp.) appear particularly prone. The condition is less common in budgerigars (*Melopsittacus* spp.). Automutilation appears most common in cockatoos, particularly umbrella (*Cacatua alba*) and Moluccan cockatoos (*Cacatua moluccencis*).

Feather-damaging behavior should be distinguished from other causes of alopecia or feather loss including normal molt and/or apteria (not recognized by an inexperienced owner), cage mate plucking (excessive allopreening), (iatrogenic) trauma, feather loss due to parasitism or bacterial, mycotic or viral infections, and plucking related to normal brooding behavior (bird preparing a brood patch). In addition, causes for lack of feather growth (e.g., hypothyroidism, malnutrition, PBFD) should be ruled out as well. A thorough history and physical examination (including a thorough dermatologic examination of skin and feathers) are needed to identify potential underlying causes for feather-damaging or self-injurious behavior. Before a (definite) diagnosis of feather damaging as a primary behavioral disorder can be made, the presence of underlying medical conditions disease should be ruled out. In case a psychological or behavioral origin of the disorder is likely, effort should be made to identify the potential underlying triggers (antecedents) and reinforcing factors (consequences) that may have contributed to the onset and maintenance of FDB.

The diet should be corrected and/or the bird's housing and living conditions should be modified to address any environmental factors that may be involved. A more stimulating environment through increasing the size of the enclosure, taking the bird outdoors, providing social contact, (chewing) toys, puzzle feeders and other forms of environmental enrichment should be promoted. Foraging enrichment has been shown to effectively reduce feather-damaging behavior. Training and behavior modification techniques such as desensitization, counter-conditioning and differential reinforcement of alternate (desired) behaviors may be employed to alter the behavior of the bird and provide the bird with a mentally stimulating challenge or task. Medical conditions should be addressed appropriately if present. In case of severe auto-mutilation, mechanical barriers such as Elizabethan collars, neck braces, ponchos, jackets and vests, can be used. These merely provide symptomatic treatment and may cause additional stress. Local application of foul-tasting substances is controversial as these may sometimes result in deterioration. In case of secondary skin infections, the use of appropriate antibiosis, preferably based on results of a culture and sensitivity testing, and/or NSAIDs may be considered. Drug therapy may be warranted in cases that are refractory to treatment with enrichment and/or behavior modification.

Renal Disease

Renal failure can manifest as polyuria and polydipsia as in other species. Persistently elevated uric acid levels in a hydrated bird may be the most consistent finding associated with renal disease in avian species. Visceral and articular gout can be seen with renal disease. Bloodwork, radiographs, and renal biopsy may be required in some cases. Some common causes of renal disease include nephritis, glomerulopathies (caused by bacterial, fungal, or viral disease), diet-induced renal disease of color variety psittacine birds, vitamin D toxicosis, nutritional disease, urolithiasis, neoplasia, trauma, and others. Treatment involves correction of dehydration, thermal support, nutritional support, and other therapies as indicated based on etiology.

Ophthalmic Emergencies

The most common ophthalmologic lesions include cataracts, conjunctivitis, periocular swelling, uveitis and retinal detachment. A cataract is the opacity of any part of the lens or its capsule. Cataracts may be associated with malformation, genetic disorders, nutritional deficiency, trauma, senescence, toxicity, or other ocular disorders. Conjunctivitis is inflammation of the conjunctiva and is one of the most common ocular lesions in pet birds. Common causes of conjunctivitis include vitamin A deficiency, infection, and toxins. Inflammation can be secondary to trauma, such as that experienced during handling, shipping, self-induced, or from a conspecific. Periocular swelling involves focal or diffuse disorders of the eyelids, infraorbital sinus, or nasal gland. Common causes of periocular swelling include trauma, infections, neoplasia, and nutritional deficiency. Uveitis is the term for inflammation of the uvea and can be divided into categories based on location: anterior, intermediate, posterior, and panuveitic. Causes of uveitis can include trauma, infection, immune-mediated, and neoplasia. Retinal detachment can be either retinal separation or a tear. Retinal separation occurs between the retinal pigment epithelium and neural retinal, resulting in subretinal fluid and exudate. A retinal tear is most often due to trauma. Supportive care should be given as indicated, including fluid therapy and gavage feeding in anorexic patients. Warm compresses and ocular flushing may help loosen up caked debris resulting in eyelid matting. In cases of suspected infectious disease, birds should be isolated. Protective gear should be worn by caretakers if zoonotic disease is suspected. Elizabethan or protective collars may be helpful in cases of self-trauma. Supportive care and management of secondary bacterial and fungal infections may be the only treatment option for some viral diseases. Systemic antibiotic, antifungal, antiparasitic, or antiviral therapies should be administered as warranted. Broad-spectrum topical ophthalmic antibiotics (e.g., bacitracin-polymyxin B-neomycin, tetracycline, chloramphenicol, gentamycin, fluoroquinolone) are indicated in some cases of infectious conjunctivitis, keratoconjunctivitis, and uveitis. Systemic NSAIDs are warranted in cases of severe inflammation or uveitis. Topical NSAIDs including flurbiprofen or diclofenac can help control local inflammation. Mydriasis cannot realistically be obtained with topical therapy in birds, so pupil preservation is best controlled by reducing inflammation. Lubricating therapies should be instituted if lagophthalmos is present or associated with eyelid swelling. Parenteral vitamin A (10,000-25,000 IU/300 g body weight) is helpful in vitamin A deficiency.

References

1. Carpenter J. *Exotic Animal Formulary.* 5th Ed. St. Louis, MO: Elsevier; 2018.
2. Speer B. *Current Therapy in Avian Medicine and Surgery.* St. Louis, MO: Elsevier; 2016.
3. Graham J. *Blackwell's Five-Minute Veterinary Consult: Avian.* Ames, IA: Wiley-Blackwell; 2016.
4. Jenkins J. Critical care of pet birds. *Vet Clin North Am Exot Anim Pract.* 2016;19(2):501-512.
5. Stout J. Common emergencies in pet birds. *Vet Clin North Am Exot Anim Pract.* 2016;19(2):513-433.
6. Bowles H, Lichtenberger M, Lennox A. Emergency and critical care of pet birds. *Vet Clin North Am Exot Anim Pract.* 2007; 10(2): 345-394.
7. Jenkins J. Avian critical care and emergency medicine. In: Altmann RB, Clubb SL, Dorrestein GM, Quesenberry K, eds. *Avian Medicine and Surgery.* Philadelphia, PA: WB Saunders; 1997:839-864.
8. Harrison G, Lightfoot T, eds. *Clinical Avian Medicine.* Palm Beach, FL: Spix Publishing; 2006.
9. Ritzman T. Wound healing and management in psittacine birds. *Vet Clin North Am Exot Anim Pract.* 2004;7(1):87-104.

Interpreting Avian Radiographs in Health and Disease

Cheryl B. Greenacre, DVM, DABVP (Avian Practice),
DABVP (Exotic Companion Mammal Practice)

Session #109

Affiliation: From College of Veterinary Medicine, University of Tennessee, Knoxville, TN 37996, USA (Emeritus).

Abstract: Radiography can be a very useful diagnostic tool, and, in many instances, it can be considered part of a minimum database in birds. Various common and not-so-common diseases of birds can be diagnosed, or at least be a part of a definitive diagnosis, by using radiography. Understanding proper technique, positioning, anatomy and diseases of birds (including poultry, parrots, and wildlife), as well as thoroughly examining the radiograph in a systematic manner, are all part of arriving at a diagnosis.

Introduction

The goals of this masterclass are to provide the participant with the tools to be able to confidently evaluate radiographs from a variety of species of birds using a systems approach. In addition, the class allows the participant to be able to find reference materials quickly, as needed. Questions are encouraged during the presentation.

Taking and interpreting radiographs of birds is different than it is for mammals for various reasons: birds are usually smaller, avian anatomy is different, and usually, birds need to be sedated or anesthetized for ideal positioning and interpretation of radiographs. First, determine the goals of your imaging study. To do this, it is important to obtain a thorough history and physical examination, develop a problem list and differential list for each problem, and develop a diagnostic and treatment plan.

Keep in mind the spectrum of care. Can the client afford both radiographs and potential treatment, or just treatment? Sometimes radiographs just confirm what was already known (e.g., transverse mid to distal tibiotarsal fracture that is not only easily palpable but visible with our own eyes from several feet away). It would be ideal to determine if there is concurrent trauma to other structures in this case, but explain (and document) to the client who can only afford the bandage, the risks rather than euthanizing the animal after spending their money on a radiograph with low yield potential.

Radiographs alone can help to make a diagnosis in some cases, but this is not always the case. Other diagnostics will most likely be necessary for a complete diagnosis. Obtain high-quality images. This is especially important with avian patients since they are so small and fine detail is necessary. Consider if you want to sedate, anesthetize, or restrain your patient for radiographs and how positioning may affect your ability to determine if a radiographic lesion is present or not. Can the bird withstand the diagnostic test you have planned? And lastly, interpret your radiographic images. There are many texts and resources available to help in evaluating avian radiographs (Table 1).

Table 1. Resources and reference materials for evaluating avian radiographs.

Books
Silverman S, Tell L. *Radiology of Birds: An Atlas of Normal Anatomy and Positioning.* St. Louis, MO: Elsevier Mosby; 2010.
Smith BJ, Smith SA, Spaulding KA, Flammer K, Smallwood JE. *Atlas of Avian Radiographic Anatomy.* Philadelphia, PA: WB Saunders; 1990. Note: Species-specific radiographs with matching xerographs of a cockatiel, orange-winged Amazon parrot, budgerigar, Java rice finch, toco toucan, domestic pigeon, red-tailed hawk, great horned owl, mallard duck, bobwhite quail, and ring-billed gull.
Smith BJ, Smith SA. Radiology. In: Altman RB, Clubb SL, Dorrestein GM, Quesenberry K, eds. *Avian Medicine and Surgery.* Philadelphia, PA: WB Saunders; 1997:170-199.
Samour JH, Naldo JL. *Anatomical and Clinical Radiology of Birds of Prey.* Philadelphia, PA: WB Saunders; 2007.
McMillan MC. Imaging techniques. In: Ritchie BW, Harrison GJ, Harrison LR, eds. *Avian Medicine: Principles and Application.* Lake Worth, FL: Wingers Publishing; 1994:246-326.
Hanna AL, Logsdon ML, Mattoon JS. Radiographic evaluation of normal and common diseases. In: Greenacre CB, Morishita TM, eds. *Backyard Poultry Medicine and Surgery: A Guide for the Veterinary Practitioner.* 2nd Ed. Hoboken, NJ: Wiley-Blackwell; 2001:173-191.
Krautwald-Junghans ME, Schroff S, Bartels T. Birds. In: Krautwald-Junghans ME, Pees M, Reese S, Tully T, eds. *Diagnostic Imaging of Exotic Pets.* Hanover, GM: Schlutersche Veragsgesellschraft mbH & Co, KG; 2011:1-26.
Krautwald ME, Tellhelm B, Hummel GH, et al. *Atlas of Radiographic Anatomy and Diagnosis of Cage Birds.* Berlin, GM: Verlag Paul Bracey; 1992.
Konig HE, Korbel R, Liebich HG. *Avian Anatomy: Textbook and Colour Atlas.* Essex UK: 5m Books Ltd; 2016.
Speer B. *Current Therapy in Avian Medicine and Surgery.* St. Louis, MO: Elsevier; 2016.
Farrow CS. *Veterinary Diagnostic Imaging: Birds, Exotic Pets and Wildlife.* St. Louis, MO: Elsevier Mosby; 2009.
Crosta L, Melillo A, Schnitzer P. Basic radiography. In: Chitty J, Monks D eds. *BSAVA Manual of Avian Practice: A Foundation Manual.* Gloucester, UK: British Small Animal Veterinary Association; 2018:269-285.
Harcourt-Brown HN. Diagnostic imaging, In: Harcourt-Brown N, Chitty J, eds. *BSAVA Manual of Psittacine Birds.* 2nd Ed. Gloucester, UK: British Small Animal Veterinary Association; 2005:97-106.
Orosz SE, Ensley PK, Haynes CJ. *Avian Surgical Anatomy Thoracic and Pelvic Limbs.* Philadelphia, PA: WB Saunders; 1992.
Orosz S, Echols S, Redig P. *Avian Surgical Anatomy and Orthopedic Management.* Florence, OR: Teton NewMedia; 2024.
Rübel GA, Isenbügel E, Wolverkamp P. *Atlas of Diagnostic Radiology of Exotic Pets.* Philadelphia, PA: WB Saunders; 1993.
Baumel JJ, King AS, Breazile JE, et al. *Handbook of Avian Anatomy: Nomina Anatomica Avium.* 2nd Ed. Cambridge, MA: Nuttall Ornithological Club; 1993.
Articles
Naguib M. Avian radiography and radiology, Part 1. *Companion Anim.* 2017;22(8):486-493.
Naguib M. Avian radiography and radiology, Part 2. *Companion Anim.* 2017;22(10):614-621.
Visser M, Hespel AM, de Swarte M, Bellah JR. Use of a caudoventral-craniodorsal oblique radiographic view made at 45° to the frontal plane to evaluate the pectoral girdle in raptors. *J Am Vet Med Assoc.* 2015;247(9):1037-1041. doi: 10.2460/javma.247.9.1037.
Dennison SE, Paul-Murphy JR, Adams WM. Radiographic determination of proventricular diameter in psittacine birds. *J Am Vet Med Assoc.* 2008;232(5):709-714. doi: 10.2460/javma.232.5.709.

Table 1 Continued.

Books
Matt CL, Di Girolamo N, Hallman RM, et al. Diagnostic accuracy of seven radiographic views, alone and in combination, for diagnosis of pectoral girdle fractures in wild passerines after window collisions. *J Am Vet Med Assoc.* 2022;1-6. doi: 10.2460/javma.20.11.0642.
Straub J, Pees M, Krautwald-Junghanns ME. Measurement of the cardiac silhouette in psittacines. *J Am Vet Med Assoc.* 2002;221(1):76-79. doi: 10.2460/javma.2002.221.76.
Woo KMT, Barron GH, Daugherty AL, et al. Measurements of the radiographic cardiac silhouette of ospreys (*Pandion haliaetus*). *Am J Vet Res.* 2019;80(9):840-845. doi: 10.2460/ajvr.80.9.840.
Zoller G, Chassang L, Steyvoort OV, Huynh M. Comparison of dorsoventral erect and ventrodorsal supine radiographic views for the evaluation of intracoelomic organs in clinically normal African grey parrots (*Psittacus erithacus*). *J Avian Med Surg.* 2019;33(3):218-228. doi: 10.1647/2018-361. Note: Conclusions "…the DVE [dorsoventral erect] view could be adequate to assess the heart and the upper digestive tract. This positioning is likely to provide clinically relevant information for cases in which general anesthesia or dorsal recumbency is contraindicated."
Malka S, Hawkins MG, Jones JH, et al. Effect of body position on respiratory system volumes in anesthetized red-tailed hawks (*Buteo jamaicensis*) as measured via computed tomography. *J Am Vet Med Assoc.* 2009;70(9). doi: 10.2460/ajvr.70.9.1155 Note: Conclusions and Clinical Relevance "In anesthetized red-tailed hawks, positioning in sternal recumbency resulted in the greatest lung and air-sac volumes and lowest lung density, compared with positioning in right lateral and dorsal recumbency. Additional studies are necessary to determine the physiologic effects of body position on the avian respiratory system."
McCready JE, Gardhouse SM, Appleby R, Beaufrere H. Mortality rate of birds following intravenous administration of iodinated contrast medium for computed tomography. *J Am Vet Med Assoc.* 2021;259(1). doi: 10.2460/javma.259.1.77 Note: Results "Following 131 CT scans with contrast medium administration (3 scans were excluded because of perivascular or subcutaneous leakage of contract medium), small birds had a mortality rate of 45.4% (5/11), compared with a mortality rate of 0.8% (1/120) for large (> 150 g) birds."
Freitas FP, Koehncke NK, Waldner CL, et al. Self-reported radiation safety behaviors among veterinary specialists and residents performing fluoroscopic procedures on small animals. *J Am Vet Med Assoc.* 2021;259(5). doi: 10.2460/javma.259.5.518 Note: Results "More than 95% of respondents believed that radiation causes cancer, yet approximately 60% of respondents never wore hand or eye protection during fluoroscopic procedures…"
International Atomic Energy Agency. *Radiation Protection and Safety in Veterinary Medicine.* Safety Reports Series No. 104. Vienna, AUT: IAEA; 2021. https://www.iaea.org/resources/safety-standards.
Online (some books and articles above are online as well)
Avian Studios. The Grey Parrot Anatomy Research Project. Available at: https://www.avianstudios.com/the-grey-parrot-anatomy-project/. Accessed March 1, 2024.
The Wildlife Center of Virginia. Education and Outreach, Radiographs. Available at: http://www.wildlifecenter.org. Accessed March 1, 2024.
Huynh M. Avian radiology: Basic principles, reference values and clinical cases. Jan 25, 2022. LafeberVet web site. Available at: https://lafeber.com/vet/avian-radiology-webinar/. Accessed March 1, 2024.
Association of Avian Veterinarians. AAV.org
Hallwarth A. How to take and interpret avian radiographs. Veterinary Practice. 2019. Available at: https://www.veterinary-practice.com/article/how-to-take-and-interpret-aviian-radiographs. Accessed March 1, 2024

Fundamentals

Anatomy

There is no such thing as a generic bird species, so learning avian anatomy and the various species differences will help you better evaluate radiographs. Refer to anatomy references above and the specific systems outlined below.[1-11] On the ventrodorsal (VD) or dorsoventral (DV) view, most parrots have a well-defined cardiac and hepatic silhouette, whereas raptors and poultry do not. Examples of some unusual common findings include the Silkie chicken breed having an additional one or two toes per foot, and skull malformations in crested ducks.[1]

Restraint

To obtain the best quality radiographs without artifacts from movement of fast respiratory rates or struggling it is best to sedate or anesthetize the bird.[2] When considering the method of restraint, think about stress to the patient, suspected disease process, pain level and stability of the patient, the degree of correct positioning needed, and staff exposure to radiation and anesthetic gas. The author prefers to briefly anesthetize to a light plane of anesthesia with either isoflurane or sevoflurane ± the use of midazolam, ± butorphanol, ± intubating. It is the author's opinion that physical restraint is very stressful to the bird and that positioning is less than ideal. There are plexiglass restraint boards available to tape or Velcro the bird to the board. There are times when the bird is too debilitated for restraint or anesthesia, and if the goal is to ascertain whether there is an egg present or whether there is heavy metal present, then a bird-in-the-box radiograph can be taken where positioning doesn't matter. Bird-in-the-box radiography refers to simply placing the bird in a cardboard box and taking radiographs through the box. Always be prepared to monitor the bird via multiple methods (e.g., stethoscope, pulse oximeter, Doppler, ECG), provide thermal support (e.g., forced heated air blanket), be ready in case of an emergency (i.e., have ET tubes there with you that will fit the bird, epinephrine and atropine), and have all equipment available that you think you will need, especially if you are going to combine radiography with another procedure, even if it is brief.

Positioning

Typically, whole-body orthogonal views are taken (ventrodorsal and right lateral). Recently a study showed that a dorsoventral erect (DVE) view in an awake bird "could be adequate to assess the heart and the upper digestive tract. This positioning is likely to provide clinically relevant information for cases in which general anesthesia or dorsal recumbency is contraindicated."[3]

For the VD view, the bird is laid in dorsal recumbency with both wings extended equally laterally, and both legs extended equally caudally. The keel should be superimposed over the vertebral column, and the femurs, scapula and acetabula should be parallel. The focal point of the beam should be at the caudal point of the sternum.[1] On the VD view most normal parrot species will demonstrate a distinct cardiohepatic waist with a cardiac silhouette within the boundaries of the scapulae and the hepatic silhouette (representing not only the liver and proventriculus but other nearby structures) within the boundaries of the acetabulae.

For the lateral view, the bird is laid in right lateral recumbency, with the wings extended equally dorsally and the legs extended equally caudally (so the knee is caudal to ventriculus). If views of the wings or legs are desired along with the coelomic view, some prefer to stagger the positions of the wings and legs such that the dependent limb is slightly cranial to the other limb. If the limb is the structure of interest, then a focused radiograph of the individual limb should be obtained, allowing for more appropriate positioning and interpretation. The acetabula, ribs, shoulder joints, and kidneys should be superimposed. The focal spot for the beam is just cranial to the caudal tip of the sternum.[1] On the lateral view most normal parrot species will demonstrate a triangular area of no organs just dorsal to the liver/proventriculus.

If the wing is to be evaluated, an additional view is needed, specifically the craniocaudal view, to obtain orthogonal views. The bird is placed in dorsal recumbency and a horizontal beam is used. If a horizontal beam is not available, an alternative method is to use a 45-degree wedge placed under the body of the bird to allow the wing to be extended and achieve the orthogonal view. If the foot is to be evaluated, then a caudoplantar view of the foot is also needed. For skull radiographs, the views include both lateral views, obliques, DV and rostrocaudal, but CT scans are best for evaluating the intricate structures of the skull in birds. Care must be taken when using IV contrast during CT of small birds (< 150 g) as a recent study found an associated 45.4% mortality rate.[4]

Radiographic technique and interpretation

A standard radiograph machine and film, or mammography film, or digital can be used. The ideal focal film distance is 40 inches. Tabletop is usually used unless the width of the bird is greater than 10 cm. Short exposure times are used (< 1/120 sec) and a high mA (> 300 mA) with a low kVp (45-65 kV).[2] Develop a systematic approach to viewing radiographs so as not to miss anything and remember there can be more than one lesion. Learn normal avian anatomy to recognize abnormal findings. Follow the SOAP (subjective, objective, assessment, plan) format to be thorough in the evaluation of any radiographic abnormalities identified by developing a differential diagnosis list. Develop a plan for further diagnostic tests and treatments.

Evaluation by System

Musculoskeletal system

The musculoskeletal system includes the skull, spine, pectoral girdle, wings, pelvis, and pelvic limbs. The skull is mainly composed of sinuses. The eye itself has bones to support the eye called scleral ossicles. Realize that in parrots, but not all birds, there is a true joint called the craniofacial hinge (or nasofrontal joint) where the beak joins the frontal bone. Unlike mammals, birds have a variable number of cervical vertebrae, varying between 8 and 25 cervical vertebrae. This allows their beak to reach many places and to be used for preening the entire body, prehending food, biting, and stabilization. Most bird species have a notarium, a fusion of the first thoracic vertebrae. Not all bird species have a notarium and the size and number (2-6) of fused vertebrae vary with the species.[12] The synsacrum is a fusion of the caudal thoracic, lumbar, sacral, and caudal vertebrae. Together, the notarium and synsacrum stabilize the spine. The pygostyle is a distal fusion of the caudal vertebrae making a structure for tail muscle attachment. The tail muscles are used for steering during flight. The sternum usually has a prominent keel for pectoral muscle attachment, unless it is a flightless bird (such as an ostrich). Ratites such as ostrich have a flat boat-shaped sternum with no keel. The pectoral girdle in birds consists of three bones including the coracoid bone (which acts as a strut enabling flight), the clavicle and the scapula. The bones of the wing include the humerus, radius, ulna, ulnar and radial carpal bones and major and minor metacarpals, phalanges, and alula (remnants of a thumb). The bones of the hind limb include the femur, tibiotarsus, tarsometatarsus, and phalanges. The digits are numbered from medially to laterally and the number of phalanges in each digit is one more than the digit number. The femur, humerus and some vertebrae are pneumatic bones, meaning they are air filled and connect directly to the respiratory tract. The medullary cavities of these bones are covered in respiratory epithelium and respiratory tumors have been documented within the medullary cavity of the humerus.

Musculoskeletal diseases in birds include arthritis, osteomyelitis, fractures (it is common to see air in nearby soft tissue if a pneumatic bone is freshly fractured), polyostotic hyperostosis (from hyperestrogenism in females due to breeding activity, ovarian or oviductal tumors, Sertoli cell tumors).

Respiratory system

Birds possess an extensive infraorbital sinus; most of their head is sinus. Compared to mammals, birds have a very large trachea, and therefore birds can inhale more air (and bring in more oxygen) than mammals which is required for flight. Birds have complete tracheal rings, and that is why the use of uncuffed endotracheal tubes is recommended in birds to avoid pressure necrosis of the trachea. Birds lack a diaphragm; therefore, they must be allowed to move their sternum up and down or they will suffocate. The syrinx is responsible for sound generation in the bird, they do not possess a larynx. The syrinx is at the bifurcation of the trachea just cranial to the heart. Some male ducks, such as Mallards, have a bony syringeal bulla coming off the right side of the trachea just cranial to the heart - this allows them to have a loud quack.[1] Some birds, such as swans, cranes, spoonbills, and birds of paradise have a coiled trachea. Penguins have a trachea that bifurcates in the cranial cervical area. Anatomic structures of the avian respiratory tract include the trachea, primary bronchus, air sacs, secondary bronchus, parabronchi, air capillaries. Radiographically, the lung itself normally looks spongy or has a honeycomb appearance and that is the end on parabronchi being visualized. Birds have air sacs that store and warm air, and act as bellows. Parrots typically have nine air sacs (two cervicocephalic, one clavicular, two cranial thoracic, two caudal thoracic and two abdominal). The air sacs directly communicate with the respiratory system, thus a tube can be placed through the body wall of a bird directly into the caudal thoracic or abdominal air sac to allow the bird to breathe through this tube. They can even be anesthetized through this tube. Remember that some bones (femur, humerus, some vertebrae) are pneumatic and not only communicate with the respiratory system but are also lined with respiratory epithelium.

Respiratory diseases include pneumonia where the fine detail of the lung is lost, air sacculitis where the thickened walls can be seen radiographically, and subcutaneous emphysema, possibly from a ruptured air sac.

Cardiovascular system

Birds, like mammals, have a four-chambered heart, the apex of which is surrounded by the liver. There is no diaphragm in a bird. The avian heart is larger than a mammalian heart by about 1.5 to 2 times. This allows more oxygen flow for the tremendous work performed during flight. The cardiac silhouette usually lies between the second and fifth rib and constitutes 50% of the width of the cranial coelomic cavity on the VD view measured at the level of the fifth thoracic vertebra.[4] Echocardiography is recommended to assess cardiac size and function.[4] The apex of the heart silhouettes with the cranial part of the liver to make an hourglass shape on the VD view called the cardiohepatic silhouette. It is abnormal to see calcification in the greater vessels, so watch for it in older and obese birds as it is often secondary to atherosclerosis.

Gastrointestinal system

Since birds lack a diaphragm, they possess a coelomic cavity, not an abdominal cavity. Birds do not have teeth. Instead, they have a beak that is variable between species. Parrots are sometimes called hookbills because of their strong, hooked beak. The esophagus of birds is divided into two sections, the cervical esophagus and thoracic esophagus. The two sections are divided by an out-pouching of the esophagus called the crop (ingluvies), generally found on the right side cranial to the thoracic inlet. The size and shape of the crop varies with species, and not all species possess a well-defined crop, or it is completely absent (for example owls and ducks). The ingluvies stores food to be digested later when the bird is up in a tree where it is safe. On VD radiographs the crop lies to the right just cranial to the thoracic inlet. Birds possess a proventriculus, also known as the true glandular stomach. On VD it overlies the left liver lobe. On lateral it is dorsal to the liver. They also possess a ventriculus. On VD it is slightly to the left, can have grit pieces in it. On lateral it is caudoventral to the proventriculus. It is also known as the gizzard which grinds food such as grains. Birds such as raptors that eat soft prey items have a softer, more flaccid ventriculus, whereas birds such as pigeons that eat hard grain whole without shelling it first, have a firm, thick-muscled ventriculus for grinding the grain. On the VD radiograph the liver should not extend beyond a line drawn from the coracoid to the acetabulum and on the lateral it should not extend much past the sternum. On the lateral radiographic view there should be a radiolucent triangle with the base being the liver, the cranial edge the lungs, and the caudal edge the intestines representing an area of just air sacs. If this triangle is obliterated from any direction, then suspect organ enlargement in that direction. On the lateral view of the radiographs, the round spleen is usually seen slightly dorsal and caudal to the proventriculus. The cloaca of birds is the endpoint for three systems: the GI, reproductive and urinary systems.

The most common gastrointestinal disease is a metallic foreign body within the ventriculus. Sometimes a piece of paint can have enough lead or zinc in it to be toxic to the bird, but may only show up as a mineral-dense particle on radiographs, or not at all, because of how thin it is. Loss of the cardiohepatic waist can be due to hepatomegaly, cardiomegaly, enlargement of proventriculus-ventriculus, air sac disease, splenomegaly, enlargement of reproductive tract, or ascites.

Genitourinary system

Parrots have three divisions to their kidney: the cranial, middle and caudal divisions. Correct terminology is to call these areas divisions, not lobes.[13,14] The kidneys of birds are located dorsally in a concavity of the synsacrum and surrounded on three sides by bone, therefore any swelling of the kidney can only expand ventrally or increase pressure within the kidney or increase the pressure on structures running through the kidney such as the sciatic nerve. This is why a budgerigar with a common renal tumor will present lame in one leg. Radiographs of kidneys are best viewed on the lateral view, but they are summated. The gonad (left ovary or both testicles) are located just cranial to the cranial pole of the kidneys (testicle usually summated) on the lateral view.

Genitourinary diseases include renomegaly, ureteroliths, egg binding, gout, or renal calcinosis.

Advanced Avian Imaging

Although this talk is focused on radiographic examination of birds, just a word is included on advanced imaging that is available. Normal examples are described in the literature. These include contrast studies either awake or under anesthesia and intubated (e.g., barium GI or iodinated material in excretory urograms), fluoroscopy, ultrasound (difficult due to extensive air sacs in coelomic cavity of birds), echocardiography, computed tomography (CT), or microCT, magnetic resonance imaging (MRI), and positron emission tomography (PET).

Acknowledgments: Thank you for the hard work put forth by all the interns, residents, students, and veterinary technicians throughout my years of working in academia. You had a part in the cases presented.

Disclosure statement: Nothing to disclose.

References

1. Silverman S, Tell L. *Radiology of Birds: An Atlas of Normal Anatomy and Positioning*. St. Louis, MO: Elsevier Mosby; 2010.

2. Smith BJ, Smith SA. *Atlas of Avian Radiographic Anatomy.* Philadelphia, PA: WB Saunders; 1990.

3. Samour JH, Naldo JL. *Anatomical and Clinical Radiology of Birds of Prey*. Philadelphia, PA: WB Saunders; 2007.

4. McMillan MC. Imaging techniques. In: Ritchie BW, Harrison GJ, Harrison LR, eds. *Avian Medicine: Principles and Application*. Lake Worth, FL: Wingers Publishing; 1994; 246-326.

5. Hanna AL, Logsdon ML, Mattoon JS. Radiographic evaluation of normal and common diseases. In: Greenacre CB, Morishita TM. *Backyard Poultry Medicine and Surgery: A Guide for the Veterinary Practitioner.* 2nd Ed. Hoboken, NJ: Wiley-Blackwell; 2001:173-191.

6. Krautwald ME, Tellhelm B, Hummel GH, *et al. Atlas of Radiographic Anatomy and Diagnosis of Cage Birds*. Berlin, GM: Verlag Paul Bracey; 1992.

7. Konig HE, Korbel R, Liebich HG. *Avian Anatomy: Textbook and Color Atlas.* Essex, UK: 5m Books Ltd; 2016.

8. Orosz SE, Ensley PK, Haynes CJ. *Avian Surgical Anatomy Thoracic and Pelvic Limbs*. Philadelphia, PA: WB Saunders; 1992.

9. Orosz S, Echols S, Redig P. *Avian Surgical Anatomy and Orthopedic Management.* 2nd Ed. Florence, OR: Teton NewMedia; 2024.

10. Baumel JJ, King AS, Breazile JE, et al. *Handbook of Avian Anatomy: Nomina Anatomica Avium.* 2nd Ed. Cambridge, MA: Nuttall Ornithological Club; 1993.

11. The Grey Parrot Anatomy Research Project. Avian Studios. Available at: https://www.avianstudios.com/the-grey-parrot-anatomy-project/. Accessed March 1, 2024.

12. Aires AS, Reichert LM, Muller RT, Andrade MB. Review of morphology, development, and evolution of the notarium in birds. *Anat Rec.* 2022;305(9):2079-2098. doi: 10.1002/ar.24852.

13. Lumeij JT. Nephrology. In: Ritchie, B. W., G. J. Harrison, and L. R. Harrison (eds.). *Avian Medicine: Principles and Application*. Lake Worth, FL: Wingers Publishing; 1994:538-555.

14. King AS. Apparatus urogenitalis (systema urogenitale). In: Baumel JJ, King AS, Breazile JE, Evans HE, et al. *Handbook of Avian Anatomy: Nomina Anatomica Avium.* 2nd Ed. Cambridge, MA: Nuttall Ornithological Club; 1993.

Speaking Squawkish and the Animal Umwelt: How and What We Communicate to Companion Birds

Todd Driggers, DVM

Session #119

Affiliation: From Avian and Exotic Animal Clinic of Arizona, 1911 South Lindsay Rd., Mesa, AZ 85204, USA.

Umwelt

Umwelt (pronounced oomvelt) is the sensory bubble (anatomic and physiologic) that animals in a particular habitat with a specific niche have evolved to live in their ideal and natural world. The evolved senses unique to that animal are often secret from other species and evident to the same species. Inquisitive minds may speculate about animals' number of senses and the combined or multifunctional anatomy and physiology to produce those senses, but there is little doubt of the secret world of the sensory bubble or umwelt. Senses have perception expansive purposes and limit boundaries that allow a specific species to take advantage of a food source, evade becoming prey, and reproduce. Through reproduction, inheritable traits (i.e., anatomy, physiology) favoring survival become traits that stimulate evolving counter traits (i.e., changing senses) in competing animals for survival.

The Immense World is a 13-chapter book that details exceptional senses possessed in the human and animal world. The book forces us to rethink our assumptions of what and how a particular unique anatomy and sense function in nature for a given species' survival. Imagine each species as a house and each window as an adaptive sense the animal uses to perceive its habitat. A garden surrounds that house. The animal's perception of the garden would be unique according to the window's size and location (i.e., the individual's evolved senses).

Recognizing what is important to them, which is often difficult, through their unique senses and evolutionary needs is imperative to help create a life in captivity worth living or the wisdom of letting them exist in their naturally evolved habitat and niche. Learning how birds learn and how to teach them is integral to their welfare. Teaching must be done with knowledge of their natural skills and individual talents. After reading the book The Immense World by Ed Young, I realized that the animal umwelt should be considered when designing a captive existence both physically and socially to positively impact our patients' welfare as individuals.

Our Captive Dilemma

Consider the captive parrot the point of reference for the practical purpose of an avian conference. Birds arrive in our human-centric world, where sight and verbal language are our primary ways of connecting and communicating. Their evolutionary-derived instincts are judged based on how they adapt, connect, and speak in the world we physically and socially constructed. A quote attributed to Albert Einstein, "Everybody is a genius. But if you judge a fish by its ability to climb a tree, it will live its whole life believing that it is stupid." If we judge them through the limited lenses of our expectations and treat them as if our created captive world is ideal, their suffering will be variably manifested according to our treatments' impact on their perception of their existence.

What we understand and are taught can be attributed to some dogmatic oversimplification or extrapolation about the needs or abilities of avian species. Scientific ontogenic and species descriptions of evolutionary biology are scattered, with limited scope and topic, and from primarily obscure sources. If something is understudied, it is incompletely understood. Oversimplification and extrapolation are done out of necessity and are placeholders for science. Despite this difficulty, avian veterinarians and their stewards can improve with scholarly searches and available species content and databases such as that available by Cornell Lab of Ornithology.

As parrot stewards and veterinarians, our limits of understanding the species' natural history and parrot sensory experiences also limit creative solutions that enrich the captive environment. Classic training in form and function may give a unique insight into anatomy and physiology to extrapolate how a particular species may use its evolutionary sensory gifts. Still, it does not relieve us of our responsibility to attempt to gather species-specific resources. The conscious and unconscious minds inform and interpret our patient's experience of captive living. Their welfare is enhanced when their experiences in captivity stimulate their senses. To enrich our patients' lives and advance their welfare, we must ask, "How does that animal's form and function explain how sense serves the animal's survival?" How does understanding the importance of an animal's sensory perception shape the captive experience?

The Need to Recognize Disparity of Captive Care and Connection (The State of Affairs)

By the definition of evolution, veterinarians can agree that captively bred large parrots of one generation are still wild animals. How they perceive their captive environment is through the sensory bubble of potentially millions of years of evolution. Captivity, through their umwelt, would, therefore, be an environment in which their unconscious mind lacks genetic tooling to interpret. Their genetic tooling informs their conscious mind to interpret what their senses perceive and derive appropriate responses to those stimuli.

Learning defined → stimulus → senses perceive subconscious and conscious interpretation → "x" action.

While veterinarians may agree that single generationally captive-bred birds are still wild animals, the bird breeding industry, clients, and many parrot stewards add an easily misunderstood label of domestic. Domestically bred or reared babies refer to where the animal is bred, hatched, and nurtured, not its genetic suitability for being a captive pet (i.e., domestication). They are often sold "a talking dog with feathers" at pet stores, with minimal teaching of an agreed-upon standard of captive care or socialization process specific to learning about how to be what they are. Learning their care welfare needs would depend on the basic science of natural history observations. Comprehensive species-specific lifestyle observations have been minimally reported in the literature. As a result, most recommendations are not generally scientifically based. The market, not science, has driven the need for captive compromises regarding primary care, rights, and welfare to fill those scientific gaps. The pet industry has made necessary assumptions and extrapolations that have proposed practically resourced solutions for unexplored complex needs.

Birds domestically bred and raised are taken from a clutch socialized group, feathers trimmed, and placed in a captive cage to keep them safe. They arrive in homes with stewards who do not know what they do not know and also do not understand the need for essential connections and communications through the umwelt of their bird. Without knowing the need for developing avian-appropriate and meaningful connection and communication, well-intentioned pet owners interact in ways that, over time, may be subconscious or conscious triggers for evolutionarily derived behaviors such as reproduction. Years of not speaking squawkish (i.e., avian communication and connection) lead to loneliness and sometimes unrecognized desperation. The bird signals its desperation to connect meaningfully by feather destruction, mutilation, screaming, and biting, among other things. Fortunately, the best, well-intentioned, and undereducated owner presents this bird to the veterinarian, rescue, or sanctuary with the question, "What is wrong with my bird?" They want a diagnosis and treatment for the symptom of "x" problems with the expectation that medicine, avian collars and jackets, and diet changes will cure it. They want to label the symptom as if the problem rests in this bird's intrinsic nature and not with the industry practices. Maladaptive responses in captive birds result in veterinarians treating symptoms of the lack of understanding of the importance of connection and communication.

Connection and Communication

Birds naturally communicate in species-specific evolutionarily derived vocalizations and body language. Signs, signals, and display rituals may also go unrecognized in captive settings. Physical responses are limited by caging, feather trimming, and physical isolation from social environments that decrease the bird's perceived required survival action or desired outcome. Domestically raised parrots may use the exact words we teach them appropriately because they recognize a scenario that they have been previously positively reinforced. Is the appropriate vocalization and meaning of those words reaching into our world for the attention they socially require? The point is that the language they speak and how they may attempt to get into our world is likely far different than what we, the veterinarians, convey and understand, and what their avian stewards want to believe. Unfortunately, avian thoughts are not knowable except as evidenced by behavior over time. Captive environments for birds with unrecognized connection and communication diminish avian welfare. Birds present nearly daily for physical and psychological chief complaints in avian medical practice.

Applied behavioral analysis (ABA) through functional assessment (FA) has been applied to parrot behavior to encourage positive behavior and help extinguish negative behaviors. Many avian practitioners and behaviorists utilize this technique to listen to and shape behaviors. The benefit of ABA ->FA is that, done correctly, it gives consistent feedback to the parrot regarding how it can control what happens to them. It can teach tricks, shape behavior, and help keep birds socially engaged in a domestic setting to displace some natural behaviors. When they utter words we have taught them or perform a behavior they have learned, it is a request to communicate and connect with us.

How Can We Use Some Tools Generated by the Neurodivergent Community to Help Connect and Communicate?

The potentially sensitive topic of human neurodivergence aims to open a dialogue to consider the parallel connections between people and animals. The author apologizes for incomplete understanding of the topic and appreciates any critique that will give further insight. The author has had the privilege to interact with people and families with neurodivergent family members. While it is an imperfect extrapolation, listening to those family members describe techniques, interactions, and breakthroughs with their neurodivergent members (autistic and various personality disorders), it has become evident that perhaps their wisdom can add value to how we interact not only with our fellow neurodivergent humans but also help families connect and communicate with the exotic pet they steward. Understanding and teaching clients' educational tools and considering perspectives may be the most significant steps we can take to enhance captive avian welfare.

Human learning and communication modes are through four general sensory and cognitive categories (Fig 1). Many humans utilize parts of each category. A more thorough, nuanced discussion and understanding are beyond this author's training.

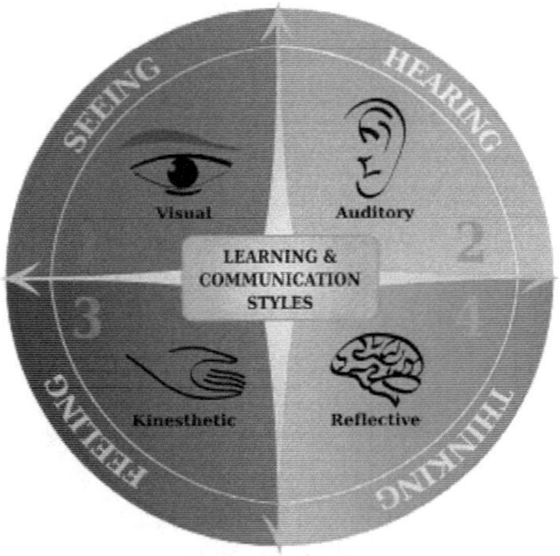

Figure 1. Four sensory and cognitive categories of learning styles. Source: Signing Time. Available at: https://www.signingtime.com/parenting-articles/active-learning-understanding-your-childs-learning-style/. Accessed May 20, 2024.

Many neurodivergent persons tend to be highly dominant in one of the above categories. Teaching to the learner is explicitly tailored to that person's spectral abilities. Child development professionals who study neurodivergent persons have assessment strategies essential to developing learning methods that match teaching and learning styles for successful spectrum goals.

One such neurodivergent condition that we are evolving a better understanding of is autism. While this condition is a spectral disorder, successful treatment varies depending on the spectral degree. Understanding how autistic children sense and perceive the world may allow us to appreciate and benefit from their often incredibly talented umwelt. The goals of many autistic parents for their children are to find a way to connect their neurodivergent children to the highest possible perceived reality to gain self-sufficiency in the social world in which they live. Learning practices for interventional purposes for a neurodivergent person must include knowledge of their sensory-specific aversions and preferences. Words and language are typically less valued and relied upon by autistic persons. Non-verbal language and senses are the more significant methods of connection and communication.

Scripting is one method that some autistic children utilize to communicate. It usually involves using technology with movies, digital tablets, or phones where the learner repetitively watches a given sequence. The watched sequence may result in a particular action that many neurodivergent people utilize to communicate a want or need. They may point to the video sequence or verbalize words that were said during the sequence, resulting in an outcome they want. Pointing toward desired outcomes or objects is a significant learning and communication strategy.

Learning is negatively impacted when most people are emotionally dysregulated or aroused. Interactions during dysregulation are mostly ineffective, so appropriate timing is critically important. Many neurodivergent persons have restricted interests (e.g., social situations and others), making connecting and communication most successful with a plan tailored to that individual. Repetitive behaviors such as vocalizations, pacing, and walking on the toes are adaptive behaviors that serve a function for the neurodivergent person. Educated and connected parents increase the likelihood of successful patient-specific outcomes—consultation with professionals specializing in neurodivergent persons to direct plans, perspectives, assessments, and follow-up. Functional assessment has been utilized to help some neurodivergent persons but is unpopular with much of the neurodivergent community because it fails to consider spectrum-related specifics, as mentioned above.

Ben's Story

This is a story about a talented 24-year-old boy named Ben, whom I recently became inspired by. He is the son of Matt Kredich (my son's university swim coach) and Kim Kredich. Ben has a twin (Miles) and a younger brother, Coleman. He plays many instruments and is a musical savant and professional musician. Ben earned a college degree from the University of Tennessee and lives independently near the campus in Knoxville, Tennessee. Ben also happens to have autism and was diagnosed at a young age. He loves animals, speaks to them like friends, and interprets what he feels they are saying in various accents. He tells his family they can't understand dogs because dogs speak in "barkish." His family and friends describe him as inclusive, kind, wondrous, happy, and perhaps odd in intriguing ways. Ben learns something every day, and that makes him happy. Whenever Ben hears a new word, he says, "That is a new one." His brother Coleman says Ben sees more beauty in a blade of grass than most couldn't see in a sunset. Miles, his twin, and Ben have synesthesia, which is when one sense is activated, and an unrelated sense gets activated. A note on the piano might activate his brain to see orange or blue when he plays. The whole song is like a series of colors through his experiences. When he goes to different cities, he notices the key to a familiar sound in that area. To Ben, Chicago is in the key of D because of the sound of the trains rumbling down the track. Knoxville is in the key of C because of the chirping bugs in the woods. Ben is a bit neophobic (i.e., afraid of new things) but has learned to take risks.

The Kredich family has advocated for Ben as a child and still today. His mother adopted the phrase they coined for his life, the "dignity of risk." This expression has driven opportunities for his life as a toddler, preschooler, elementary, and through college. Ben learned that being Ben was safe with his family and friends. His father, Matt, a skilled human and renowned coach, describes Ben's world as vibrations, sound, and music.

Resonance is defined as a reinforcement of a sound, as a musical tone in a vibrating body added to by another body. Ben's connections with his people create resonance because, through his environmental, sensory, and umwelt world, musical vibrations combine with other beings' vibrations, making the pairing bigger and better for the whole. Resonance was true for animals and people willing to have the Ben experience. Ben vibes with animals in natural ways. He struggles to connect with new people but tries to because Ben asks many questions and remembers obscure answers people give him, even when said in passing. If you are lucky enough to know Ben, he might surprise you by asking how your pet, Coco, the cat, is doing. Ben's questions are his way of coming into your world to fight off his loneliness by being socially affected by autism.

As a young child, Ben was thought to be, well perhaps, a little different developmentally than his twin brother, and the Kredich family sought help from a doctor who made the diagnosis and gave him his label. Ben, the autistic, was not given much hope. To his family, he was just Ben, the little boy who was a gift from God. His observant family was also very musical and sang to him, which usually was a way to calm him down. Ben lost his language for a while and seemed to respond with emotions, not unlike a bird who feels unheard by its stewards. Kim, his mom, connected with Ben (i.e., she called floor time) by giving him playful obstruction. She felt these interactions taught her how he communicates (i.e., circles of communication) by listening to and evaluating his responses to the "playful obstructions." His mother sang a potty song to him and noticed that sometimes it helped relax him, and other times Ben was inconsolable or tortured by the same song. Ben was taught this potty song in the C key, and any different key his mother sang to Ben sounded like scraping metal. His family freed Ben that day as his primary sensory bubble, or his umwelt was vibrational musicality, and their most profound connections were when they resonated (i.e., matched vibrations). They began releasing Ben so he could reach his potential and relieve his loneliness.

Ben's parents recognized his social struggles relating to humans and desired that he would grow into being his best independent self. They didn't focus on an autistic label but implemented actions seeking connection and communication. Ben's talents were unlocked through time, dedication, patience, and playful obstruction. They strived to see Ben by testing his umwelt and discovered that of the musical world of vibrations.

Birds' connection and communication conflicts with humans come from a misunderstanding or misinterpretation of the sensory bubble, style of learning interaction, and body and verbal language. Agreed-upon body language information, play(ful) obstructions, watching individual bird behaviors, positive reinforcement, that is compelling wild freedom nature provides through daily self-maintenance and survival. Through the eyes of an average pet owner, they seem surprised that the parrots aren't grateful for the food, water, and cage they provide, claiming to be a maid, a cook, and a tolerator of noise. Implementing a flexible, functionally equivalent plan of connection and communication can help extinguish symptoms over time.

How Does This Story Apply?

Like Ben, captively bred birds are asked to live and understand two social worlds: the natural world (i.e., the neurodivergent brain or the habitat and niche from which they have evolved) and the human social world. Parrots are very social, connect, and communicate in species-specific squawkish (i.e., behavior and vocally). Parroting (i.e., parenting a parrot) and parenting require a plan and consistent action to discover who they are as an individual. Whether parroting or parenting, we are responsible for connecting with them into their instinctual psychosocial world and adapting our captive social world to meet our functionally relevant, equivalent physical and social equivalents. Few pet stewards achieve preventive enrichment. A failure to plan is a plan to fail.

Like Ben, birds bred in captivity ask questions not in the form of a question but of a physical response. They try to connect with us (e.g., squawkish questions) by vocalizing, physical touch, and flight or movement. We often interpret and reinforce behaviors through our humanist lenses, which are intended to be understood through squawkish language. If these behavioral queries are not recognized, they may become frustrated and exaggerate, so we react and more easily understand that there is a problem. The question they ask with movement may be "Do I matter?" or "Do you see me?" to recognize their need for social connection and release me to move like a bird does. The younger Ben also communicated through his frustration, tears, and fits when what he was trying to communicate was not being understood. Like Ben, connection and communication are potent ways to exert some control in a complicated and challenging-to-understand world.

Human verbal and body language have human meaning and interpretation specific to our cultures. Unfortunately, when birds talk in our language, it is not clear to their stewards whether the words are literal or imitative. It makes us continue to force our language on them instead of trying to speak squawkish. Mitch Hedberg, a deceased comedian, once told this joke relating to bird language: "I had a parrot that could talk, but it didn't say I'm hungry, so it died." Punchlines and dark humor speak exaggerated truths in life. Birds usually use body language and get verbally excited and loud to express themselves. They learn when we match the intensity of our body language, voice energy, or loud volume to their own. For example, if you want to teach your bird to talk, you need to resonate vibrational energy. Smash your finger while hammering, and the energy of pain results in an instant four-letter vocabulary. When your bird says that four-letter word you taught them, that is your bird trying to connect in our world. Please speak to your bird with passion (i.e., with vibrational energy and resonance) to communicate in their world. Get loud and high-pitched if you want to connect to their world. The author's experience is that they link back into our world by matching our loudness or saying our exact words or expressions amazingly in context.

We connect with our birds, which imprint highly on us through our physical touch. Physical touch early and in excess connects to their world where they misunderstand sexual intent. Our hands communicate physical promises in their world that our bodies can't keep and that our minds never consider the impact. We connected in their world of squawkish physical touch. Too much physical touch can be misperceived by parrots as a desire for reproductive activity. Done in excess, birds may become sexually frustrated and stimulated, which can also lead to medical conditions such as cloacal prolapse and have negative psychological impacts.

Flight and movement are avian body language communications regarding what a bird might think, want, or want to escape. Fight or flight are responses to fear or desired circumstances that are evolutionary permitted by freedom. We prohibit them from communicating and expressing their instinctual reactions by feather trimming or cage confinement. Attempted flight and pain from aggressive feather-trimming techniques or stress from unmitigable circumstances can morph into learned helplessness. Domesticated house pets such as dogs or cats may connect with captive parrots, be entertained by them, and engage them by calling them into the room and attempting to feed them, bite them, or talk to them. If you have a dog or cat with predatory or instinctual characteristics, it may also create fear in their squawkish language. Never trust your human interpretation. Never allow direct physical contact, as even with apparently good relationships, situational traumas can instantly end a bird's life. If birds have no control of their movement by either feather trimming or unwanted visitors, they cannot escape (even though the cage protects them), which also restricts their evolutionary instinct to fly.

Our communication could be better based on what industry practices have conveyed through education or lack thereof. The reinforcers of our verbal and physical connections need to be understood. The industry has focused on avian care, not avian welfare. It has focused on keeping them alive for us, not helping them live a meaningful life. It is not our perception of their life and interest that matters; it is their perception that gives them meaning. Ben's life improved when he was genuinely seen through his musicality connection. Playful obstructions that Ben's mom set up for him provided time to observe what made him happy or sad, think, learn, and create.

While many parents of neurodivergent youths may not favor ABA and functional assessment, they are effective processes for creating playfully obstructive interactions with captive parrots. This structured time requires the identification of antecedents, behaviors, and consequences that can be encouraged, reinforced, or ignored based on a desired response. The most common reason people don't connect with their birds is they need to understand their pet in time to connect and consider how their bird individually communicates. Planning and implementing relational strategies that have functional meaning through all the senses of their companion may reveal their characteristics, enabling understanding of their motivators. Connecting with a bird is challenging in its captive environment. In parrots, FA is designed around accomplishing a task or a goal and has the added benefit of not stimulating them reproductively. It is unpopular with neurodivergent youths as it is seen as rigid when not applied with the appropriate amount of finesse and wisdom, as can be seen in autistic spectral learners. In structured interactions, parents and professional neurodivergent team members consider spectral factors and goals to achieve the best possible outcomes. General awareness of senses and the concept of umwelt may help guide us to see veterinary patients as unique individuals who require our perspectives towards their care and welfare. Planned behavior interactions with parrots are essential for their social well-being. Functional assessment and tailored behavior plans will help to enhance parrot welfare. We must work with our clients to improve their welfare needs by whatever means necessary.

Permission Statement: The Kredich family has granted permission for Dr. Todd Driggers to promote positive messages about Ben Kredich, their family, and their challenges with autism at the Association of Avian Veterinarians Annual Conference. Ben was a big animal advocate and connected with them. The family promotes advocacy for those affected by neurodivergent conditions and will continue to show their love through these endavors.

The Ain't Doing Right (ADR) Backyard Hen

Renée Schott, DVM, CWR

Session #161

Affiliation: From The Wildlife Rehabilitation Center of Minnesota, 2530 Dale St. N, Roseville, MN 55113, USA.

Abstract: Backyard hens continue to grow in popularity for their appeal as a local source of fresh eggs; however, they often become a beloved pet and part of the family. When these hens present as sick, families want veterinary care but are often not prepared to spend as much as on other companion species. This masterclass, by a chicken veterinarian enthusiast, reviews the most common presentations for sick pet hens and presents some quick diagnostics to help guide owners' decisions in treatment options.

Introduction

Backyard chickens are increasingly allowed in cities and clients are keeping them for fun and eggs. Urban clients will often develop strong emotional bonds and consider hens as pets on the same level as their companion cats and dogs; however, preventive care is not common so these pets will usually present sick without any prior medical history. Chickens are adept at hiding signs of disease and novice keepers will likely miss subtle clinical signs of disease.[1] As a result, sick chickens often should be considered an emergency. Please note that many traditional infectious diseases are not covered in this masterclass because there is a plethora of literature, treatments and quarantine recommendations from the production poultry literature. This talk focuses on the hen that often presents as an emergency with generalized symptoms.

Veterinary Examination and Diagnostics

History

Collect all history from the owner. This information might provide a clue to the current condition or could potentially prompt a husbandry conversation after the crisis. Be sure to obtain the following information.

- Signalment:
  - Age, breed, standard or bantam
- Acquisition:
  - When was the bird obtained?
  - Was it obtained from a private party or hatchery? Did the previous owner obtain the bird from a hatchery?
  - If from a hatchery, were any vaccinations given?
  - Was the bird quarantined? How long (and where)?

- Husbandry (ideally house visit, or ask for pictures):
  - What are winter housing conditions (e.g., temperature, humidity, substrate, roosts, nest boxes, frequency of cleaning, airflow)?
  - What are summer housing conditions (e.g., temperature, humidity, substrate, roosts, nest boxes, frequency of cleaning, airflow)?
  - Is the environment free range, fenced, netting on top, or other arrangement?
  - How many birds are in the flock and what are their age, sex, size, breed and personalities? Where is this bird in the pecking order?
  - When was the last time a new bird was added to flock?
  - What are food and water characteristics?
    - Is clean water available and changed daily? How many places is it provided? Is the water out of environmental elements (e.g., heat, cold)?
    - Is fresh feed offered daily? How many places is it offered?
    - What kind of feed is offered?
    - Is a calcium supplement offered?
    - Are any other foods offered?
    - How often and how many treats are offered? Are table scraps offered?

- Laying history:
  - For how many years has the bird been laying?
  - What is the frequency of egg laying?
  - When was the last time the hen laid?
  - What are the size of eggs?
  - Are there any abnormalities on the eggs (e.g., blood on shell, calcium deposits, abnormal shape or size, shell-less eggs)
  - Has recent broodiness been observed?

- Are there prior medical issues?

- When did the owner first start to notice something off with hen and what was noticed?

- How is the bird's appetite?

- How are the feces and urates?

- What is the bird's activity level?

Physical examination

Perform as in any other species and pay attention for any of the following abnormalities.

- Hands-off exam:
  - Look for any dyspnea (e.g., tail bobbing, any obvious/exaggerated inhalation/exhalation, open-mouth breathing).

- Beak and mouth:
  - Look for any plaques, stringy saliva, color of mucous membranes, beak discoloration.

- Eyes:
  - Look to see if the eyes are clear, open and bright. Or are they half closed, sunken, and dull?
  - Look for swellings around the eyes.

- Comb or wattle:
  - Look at its color and size. Ask the owner if it is redder and larger or paler and smaller versus when the hen is laying normally.

- Feathers and skin:
  - Look for lice, mites, any areas of feather loss or wounds. Are all feathers preened and out of sheaths? Are they in good condition and fresh or worn and damaged? Are there stress bars?

- Crop:
  - Look to see if the crop is full. Is it distended? Palpate and note what you feel (e.g., feed, hay, mats, liquid, gas).

- Auscultation:
  - Listen to heart for murmurs. Listen to the lungs and airsacs for abnormal respiratory sounds.

- Vent:
  - Is the vent clean? Are feathers around vent full of feces? Is there evidence of diarrhea?

- Caudal and ventral coelom:
  - Palpate to determine if it is enlarged or distended. Is there fluid or soft tissue? Is it painful?
  - Ascites and/or coelomic distention is an incredibly common abnormal physical exam finding in the ADR hen.

2-view radiographs

Ventrodorsal and lateral views are required with any ADR hen. They can quickly be done in a low-stress manner without anesthesia. Sedation (0.5 mg/kg midazolam ± 2 mg/kg butorphanol) are great to use in any hen but especially necessary in stressed individuals or those in respiratory distress.

Evaluate the entire radiograph and be sure to evaluate the opacities of the caudal coelom and relative location of the ventriculus to the rest of the coelomic contents.

Ultrasound

Ultrasound is beneficial for internal examination and for therapeutic coelomic centesis. Airsacs make ultrasound difficult to interpret and perform.

Blood lead levels

Lead is in surprising places in the environment and many chickens may have detectable blood lead levels. Any blood lead level above 10 μg/dl should be chelated and the eggs should be discarded because lead can pass into the egg and humans if eaten.[2] Chickens with lead toxicity may show classic clinical signs (e.g., weakness, GI stasis, abnormal brain neurologic signs), or they may present clinically normal.

Other diagnostics

A CBC and chemistry may be helpful depending on what other clinical exam findings are found. Cytology of the crop (i.e., Gram stain to look for yeast overgrowth) or other cytology of wounds may be helpful.

Common Diagnoses

Secondary co-morbidities

Owners may come in suggesting the following conditions are the primary problem for the hen; however, they are almost always a secondary co-morbidity. Search and find the primary pathology!

- Gastro-intestinal parasites
- Lice, mites
- Crop stasis
- Diarrhea of unknown origin

Primary pathology

- Neoplasia:
  - Neoplasia is common in any area, but often will arise in the reproductive tract.
  - Exploratory celiotomy can sometimes reveal carcinomatosis, which usually grossly appears as many small masses involving many or all organs, especially the GI tract.

- Reproductive problems:
  - Ectopic eggs (i.e., yolks only or partially/completely shell) are when developing egg(s) enter the coelomic cavity through an oviductal rupture or reverse peristalsis from oviductal obstruction (e.g., masses, impactions, salpingitis), cloacal or unknown reasons.[3] Ectopic eggs are often termed internal laying and may result in a sterile or non-sterile coelomitis.
  - Impacted oviduct can be caused by salpingitis (sterile,[3] caused by *E. coli* or *E. coli* is found secondarily[4]), cystic hyperplasia, metritis, neoplasia or unknown etiologies that may cause caseated material to form, often in layers around misshapen/ruptured/soft-shelled/partially formed egg pieces.[3] This material can cause oviduct impactions if it gets big enough. This material can be laid whole or in small pieces and these are often termed lash eggs. Lash eggs are most common in older, prolific laying breeds.

- Gout or urolithiasis (less common):[5]
  - This condition is often due to laying feed fed to hens (or roosters) not in lay.
  - Predisposing factors include infectious bronchitis, prior renal damage, and water deprivation.

- Elevated blood lead levels (less common):
  - Chickens are stoic and often do not have overt clinical signs related to mildly elevated lead levels but likely have subclinical/underlying problems caused by the lead. Lead will also make recovery from any other condition more difficult so it should be chelated whenever it is above 10 μg/dl.

Uncommon Diagnoses

- Egg-binding, dystocia:
  - This diagnosis is uncommon in poultry (occurs in small bantams).[3]
- Primary crop impaction:
  - Crop stasis or GIT stasis is almost always due to an underlying disease process, unless the hen is a meat breed.

Treatments

Legalities

In the United States (and in many other countries[6]), all chickens are considered meat animals even if owned by a vegan on a farm sanctuary. It does not matter if the bird or eggs will ever be used as food. In the United States, simply being this species makes it a food animal in the eyes of the U.S. Food and Drug Administration.[7]

- Below is the list of drugs that are prohibited from extralabel use; these drugs may never be used in chickens.
  - Antivirals
  - Chloramphenicol
  - Clenbuterol
  - Diethylstilbesterol
  - All fluoroquinolones, including enrofloxacin (any -oxacin)
  - All glycopeptides, including vancomycin
  - All nitroimidazoles, including metronidazole
  - All nitrofurans, including nitrofurazone
    - Indexed drugs (e.g., Supralorin-F, Virbac)
- The Food Animal Residue Avoidance Databank (FARAD) has addressed egg residues.[7]
  - Information can be found on the FARAD website: http://www.farad.org/eldu-and-withdrawal-time.html.

Supportive Care

Supportive care is necessary with any diagnosis and will make an incredibly huge difference in the attitude of the hen.

- Subcutaneous fluids:
  - Provide 50 mg/kg/day (for additional benefit add vitamin B complex 1-3 ml/L; if the patient has been inappetant for a few days, add 30 mEq/L KCl). Monitor hydration twice per day and supplement with subcutaneous fluids as needed.

- Analgesia:
  - This is recommended in almost all conditions. The author prefers meloxicam (1 mg/kg q12h), recognizing this can make a gout case worse.

- Nutrition:
  - Often these hens will be anorexic for some time. Switching to a high-quality feed (e.g., Roudybush Maintenance Crumbles, Roudybush Inc, Woodland, CA) or moistening the food may entice hens to eat. Feel free to use insects (e.g., mealworms, earthworms) as well as fresh fruits and vegetables to entice the bird to eat. Monitor weight daily and if the patient is losing weight or is anorexic, gavage feeding is necessary. If prolonged anorexia exists, start with 30 ml/kg/feeding Emeraid Intensive Care Omnivore® (Lafeber LLC, Cornell, IL). If the bird is only off feed for a few days, moisten high-quality feed, and blend and gavage 30 ml/kg/feeding. Feeding twice per day is usually adequate.

Treatment of specific conditions

- Neoplasia:
  - Provide supportive care.
  - Coelomcentesis if ascites; ultrasound can help guide.
  - If legal in one's country, deslorelin (Supralorin-F, Virbac) may help slow the progression of some neoplasias.[8,9]

- Reproductive problems:
  - Provide supportive care.
  - If legal in one's country, deslorelin is implanted as soon as possible. This will put the hen out of lay,[10] stopping the problem from compounding and buying some time.
  - Two weeks after implantation, exploratory celiotomy may be required to clean out any eggs laid internally, lash eggs, or other gross pathology.
  - Owners must be warned that implantation with deslorelin is often needed for the rest of the hen's life, since most disorders will recur after the implant wears off.
  - Salpingectomy may be needed for recurrent lash egg or salpingitis.

- Systemic antibiotics:
  - These are only indicated when a bacterial infection is found and the bird would benefit from administration of systemic antibiotics.
  - Examples of appropriate antibiotic use include presence of pneumonia and worsening infected wounds, among other conditions.
  - Examples of inappropriate antibiotic use include most pododermatitis. Usually these are found as a chronic condition and the abscess is already walled off from the body. Non-infected wounds are another situation that would not require antibiotic use.

- Euthanasia:
  - Euthanasia is common and humane in the sick backyard hen if appropriate treatment is not an option. Many reproductive conditions are chronic and require treatments that may not be available in your country. Ensure your client understands that euthanasia is a treatment option.

Conclusions

Unfortunately, the most common disorders in the aging backyard hen are chronic, terminal conditions that will require life-long treatment, care and possibly surgery. With a dedicated owner, these hens can go on to live years, but may not survive if neoplasia is resistant to deslorelin. Regardless, it is important to have a conversation with the owner to determine the relationship the hen has in the family. It is important to be open and honest about prognosis, cost and treatment options from the first visit.

References

1. Crespo R, Senties-Cue G. Postmortem survey of disease conditions in backyard poultry. *J Exot Pet Med.* 2015;24:156-163.

2. Leibler JH, Basra K, Ireland T. et al. Lead exposure to children from consumption of backyard chicken eggs. *Environ Res.* 2018;167:445-452.

3. Echols MS. Soft tissue surgery. In: Greenacre CB, Morishita T, eds. *Backyard Poultry Medicine and Surgery.* Hoboken, NJ: John Wiley & Sons; 2015:220-262.

4. Morishita T, Porter R. Gastrointestinal and hepatic diseases. In: Greenacre C, Morishita T, eds. *Backyard Poultry Medicine and Surgery: A Guide for Veterinary Practitioners.* Hoboken, NJ: John Wiley & Sons; 2015:181-203.

5. Applegate TJ. Backyard poultry nutrition. In: Greenacre C, Morishita T, eds. *Backyard Poultry Medicine and Surgery: A Guide for Veterinary Practitioners.* Hoboken, NJ: John Wiley & Sons; 2015:72-81.

6. Whitehead ML, Roberts V. Backyard poultry: legislation, zoonoses and disease prevention. *J Small Anim Pract.* 2014;55:487-496.

7. Marmulak T, Tell LA, Gehring R, et al. FARAD digest: egg residue considerations during the treatment of backyard poultry. *J Am Vet Med Assoc.* 2015;247:1388-1395.

8. Keller KA, Beaufrère H, Brandão J, et al. Long-term management of ovarian neoplasia in two cockatiels (*Nymphicus hollandicus*). *J Avian Med Surg.* 2013;27:44-52.

9. Mans C, Pilny A. Use of GnRH-agonists for medical management of reproductive disorders in birds. *Vet Clin North Am Exot Anim Pract.* 2014;17:23-33.

10. Eusemann BK, Sharifi AR, Patt A, et al. Influence of a sustained release deslorelin acetate implant on reproductive physiology and associated traits in laying hens. *Front Physiol.* 2018;9:1-11.

The Unwell Waterfowl: Common Presentations, Diagnostics and Treatments

Renée Schott, DVM, CWR

Session #154

Affiliation: From The Wildlife Rehabilitation Center of Minnesota, 2530 Dale St. N, Roseville, MN 55113, USA.

Abstract: Wild and domestic waterfowl present with a variety of ailments; however, their unique morphology and physiology present challenges not encountered in other families of birds. This case-based, interactive talk is chocked full of knowledge nuggets to help you get through your next waterfowl case.

Introduction

Waterfowl are frequently encountered as patients in wildlife rehabilitation settings, alongside domestic counterparts kept by individuals for companionship or the production of eggs and meat. Although ducks are most commonly kept as companions, geese and occasionally swans are also kept. There are many commonalities in presentations, diagnostics, and treatments. An hour is not enough time to cover everything; a thorough discussion of waterfowl management and other common conditions can be found elsewhere.[1,2]

Veterinary Examination and Diagnostics

History

Collect all history from the owner. This information might provide a clue to the current condition or could potentially prompt a husbandry conversation after the medical issue is addressed. Be sure to obtain the following information.

- Signalment:
  - Age, breed, class (e.g., bantam, light, medium, heavy). What feeding style and perching style does the individual have (e.g., dabbler, diver, grazer)?
- Acquisition:
  - When was the bird obtained?
  - Was it obtained from a private party or hatchery? Did the previous owner obtain the bird from a hatchery?
  - If from a hatchery, were any vaccinations given? Vaccinating waterfowl is not as common as vaccinating chickens.
  - How long (and where) did the owner quarantine after obtaining the bird?

- Husbandry (ideally house visit, or ask for pictures):
  - What are winter housing conditions (e.g., temperature, humidity, substrate, roosts, frequency of cleaning, airflow)? Although waterfowl generally are more tolerant of colder temperature than chickens, they still can get frostbite on their feet and legs.
  - What are summer housing conditions (e.g., temperature, humidity, substrate, roosts, frequency of cleaning, airflow)?
  - Is the environment free range, fenced, netting on top, or other arrangement?
  - How many birds are in the flock and what are their age, sex, size, breed and personalities? Where is this bird in the pecking order? Are there any other species in the flock?
  - When was the last time a new bird was added to flock?
  - What are food and water characteristics?
    - Is clean drinking water available and changed daily? How many places is it provided? Is the water out of environmental elements (e.g., heat, cold)?
    - Is clean swimming always available and changed daily? Water must be deep enough to allow the bird's head and neck to be submerged and sloped to allow easy access.
    - Is fresh feed offered daily? How many places is it offered?
    - What kinds of feed is offered?
    - Is a calcium supplement offered (e.g., for laying breed)?
    - Are any other foods offered?
    - How often and how many treats are offered? Are table scraps offered?

- Laying history (if applicable):
  - For how many years has the bird been laying?
  - What is the frequency of egg laying?
  - When was the last time the hen laid?
  - What are the size of eggs?
  - Are there any abnormalities on the eggs (e.g., blood on shell, calcium deposits, abnormal shape or size, shell-less eggs)
  - Has recent broodiness been observed?

- Are there prior medical issues?

- When did the owner first start to notice something off and what was noticed?

- How is the bird's appetite?

- How are the feces and urates?

- What is the bird's activity level?

- What is the client's complaint?

Physical examination

Perform a physical examination as would be performed in any other species. Pay attention to the following potential abnormalities.

- Hands-off exam:
  - Look for any dyspnea (e.g., tail bobbing, any obvious/exaggerated inhalation/exhalation, open mouth breathing); also watch for abnormal neurologic symptoms as the patient explores the room.
- Bill and mouth:
  - Look for any plaques, stringy saliva, color of mucous membranes, or bill discoloration.
- Eyes:
  - Look to see if eyes are clear, open and bright. Or are they half-closed, sunken, and dull?
  - Look for swellings around the eyes.
- Feathers and skin:
  - Look for lice, mites, areas of feather loss or wounds. Are all feathers preened and out of sheaths? Are they in good condition and fresh or worn and damaged? Are there stress bars?
- Crop:
  - Look to see if the crop is full. Is it distended? Palpate and note what you feel (e.g., feed, hay, mats, liquid, gas).
- Auscultation:
  - Listen to the heart for murmurs. Listen to each air sac individually because each can insulate sounds from the others.
- Vent:
  - Is the vent clean? Are feathers around the vent full of feces?
- Caudal and ventral coelom:
  - Palpate to determine if it is enlarged or distended. Is there fluid or soft tissue? Is it painful?
- Feet:
  - Examine feet closely, including the plantar side of the feet. Look for calluses and pododermatitis.
- Ambulation:
  - Watch the patient ambulate on a non-slip surface and over obstacles. This helps discern if there are problems (e.g., old fractures, arthritis) in the caudal limbs or hips.
- Body conditions score:
  - Note a body condition score, including pectoral muscle mass and fat deposits.

2-view radiographs

Ventrodorsal and lateral views can be obtained in a low-stress manner without full anesthesia, but sedation is recommended (i.e., 2-5 mg/kg midazolam, 2-5 mg/kg butorphanol, + 2-5 mg/kg ketamine) and is especially necessary in stressed individuals or those in respiratory distress.

Evaluate the entire radiograph and be sure to evaluate the opacities of the caudal coelom and relative location of the ventriculus to the rest of the coelomic contents. It is not uncommon for waterfowl to ingest foreign bodies.

Ultrasound

Ultrasound is beneficial for internal examination and for therapeutic coelomic centesis. Airsacs make this a little more challenging.

Blood lead levels

Lead is in surprising places in the environment and many wild waterfowl may have detectable blood lead levels, which likely means domestic ducks in the same environment also have detectable blood lead levels. Any blood lead level above 10 µg/dl should be chelated and the eggs should be discarded because lead can pass into the egg and humans if consumed.[3] Wild waterfowl often have lead toxicity due to accidentally eating lead sinkers, lures, and spent shot.

Common Diagnoses

Secondary co-morbidities

Owners may come in suggesting one of the following conditions as a primary problem but these are almost always a secondary co-morbidity. Search and find the primary pathology!

- Pododermatitis:
  - This is usually caused by inappropriate substrate, not enough or not deep enough swimming water, or other conditions that cause unbalanced weight distribution. The foot lesions must be treated in addition to the underlying issue. Toes that become luxated or have osteomyelitis may need to be amputated. When amputating, amputate the associated web so it's not flopping around. Limits in toe/webbing amputation in wild waterfowl for release are described.[4]
- Difficulty standing and walking
- Coccidiosis:
  - This is common in free-ranging waterfowl that have access to wetlands. Severe overgrowth of coccidia, especially in young birds, often causes lethargy, lack of thriftiness, and depressed mentation.
- Crop stasis:
  - This is often caused by lead toxicity. Hydration and chelation are often enough to move the impacted food through the GI tract.

- Aspergillosis:
    - This is a fungal infection in the airsacs, which is common in waterfowl that are immunocompromised and/or exposed to large amounts of spores.

Primary pathology

- Trauma:
    - This is common in wild waterfowl and not uncommon in domestics. General avian trauma guidelines and treatment should be followed. Noted differences below:
    - Scalping injuries (caused by forced copulation or predator) are common. Administer supportive care. When the patient is stable enough for surgery, perform an advancement flap after aggressive debridement (e.g., including scraping the necrosis off the skull). It is ok to wait even a few weeks to do this. Keep the wound covered with a thin layer of SSD and inside so flies cannot access the wound. Often, waiting for second intention healing on a waterfowl scalping results in the eyelids being pulled dorsally, leaving the patient unable to fully close its eyes.
    - Liver rupture or lacerations are common with whole-body traumas in waterfowl. If any dyspnea is noted, or extremely hard air sacs on auscultations in an animal with a history of whole-body trauma, it may be safe to assume there is liver trauma and associated intracoelomic bleeding. Take care to administer appropriate supportive care in a low stress environment so blood clots can remain in place.
    - Knuckling:
        - Soft tissue trauma to the legs, including but not limited to constriction injuries, can cause a temporary loss of proprioception in the foot or inflammation of the tendons preventing the patient from voluntarily placing the foot in a normal position. To prevent the leg tendons from contracting, and to keep the patient ambulatory, a walking splint is easily applied to the knuckling joint.

- Neoplasia:
    - If the bird is a domestic egg-laying breed or an aged individual, neoplasia is not uncommon.

- Reproductive problems:
    - These are common in domestic duck breeds kept for eggs. See "The ADR Backyard Hen" for more information.

- Elevated blood lead levels (less common):
    - Waterfowl are stoic and often do not have overt clinical signs related to mildly elevated lead levels but likely have subclinical or underlying problems caused by the lead. Lead will also make recovery from any other condition more difficult so it should be chelated whenever the blood lead level is > 10 μg/dl.

Treatments

Legalities

In the United States (and in many other countries[5]), many waterfowl species are considered meat animals even if owned by a vegan on a farm sanctuary. It does not matter if the bird or eggs will ever be used as food. In the United States, often simply being this species makes it a food animal in the eyes of the Food and Drug Administration.[6]

- Below is the list of drugs that are prohibited from extralabel use; these drugs may never be used in waterfowl.
  - Antivirals
  - Chloramphenicol
  - Clenbuterol
  - Diethylstilbesterol
  - All fluoroquinolones, including enrofloxacin (any -oxacin)
  - All glycopeptides, including vancomycin
  - All nitroimidazoles, including metronidazole
  - All nitrofurans, including nitrofurazone
  - Indexed drugs (e.g., Supralorin-F, Virbac)
- The Food Animal Residue Avoidance Databank has addressed egg residues.[6]

Supportive Care

Supportive care is necessary with any diagnosis and will make an incredibly huge difference in the attitude of the bird.

- Subcutaneous fluids:
  - Provide 50 mg/kg/day (for additional benefit add vitamin B complex 1-3 ml/L; if the patient has been inappetant for a few days, add 30 mEq/L KCl). Monitor hydration twice per day and supplement with subcutaneous fluids as needed.
- Analgesia:
  - This is recommended in all conditions. See various available formularies for the most up-to-date dosing recommendations.
- Nutrition:
  - Often birds will be anorexic for some time. Switching to a high-quality feed (e.g., Mazuri Waterfowl Maintenance, Mazuri Exotic Animal Nutrition, St. Louis, MO) or floating the food in water may encourage the bird to eat. Feel free to use insects (e.g., mealworms, earthworms) as well as fresh greens to entice the bird to eat. Monitor weight daily and if the patient is losing weight or is anorexic, gavage feeding is necessary. If prolonged anorexia exists, start with 30 ml/kg/feeding an elemental diet such as Emeraid Intensive Care Omnivore® (Lafeber LLC, Cornell, IL). If the bird is only off feed for a few days, moisten high-quality feed, and blend and gavage 30 ml/kg/feeding. Feeding twice per day is usually adequate.

- Physical therapy:
    - For any condition in which ambulation or hindlimbs are affected, swim therapy is essential to regain leg strength, take pressure off the feet, prevent keel lesions, and promote range of motion. Swim therapy should be performed every day to every other day. The length of each session is dependent on the patient waterproofing and endurance. At minimum, each patient should be observed for the first 10 min and then checked up on every 10 min after that until they are proficient and swimming safely. Enrichment such as floatable pellets, greens, or live minnows are great additions to encourage movement. Water should be changed and the pool cleaned after every session to preserve waterproofing. Depth of the water should be deeper than the feet unless the patient is having difficulties navigating, in which case the water level can be lowered.

References

1. Greenacre CB, Morishita TY. *Backyard Poultry Medicine and Surgery: A Guide for Veterinary Practitioners.* 2nd Ed. Hoboken, NJ: Wiley-Blackwell; 2021.

2. Hernandez SM, Barron HW, Miller EA, et al. *Medical Management of Wildlife Species A Guide for Veterinary Practitioners*. Hoboken, NJ: Wiley-Blackwell; 2019.

3. Leibler JH, Basra K, Ireland T. et al. Lead exposure to children from consumption of backyard chicken eggs. *Environ Res.* 2018;167:445-452.

4. Duerr RS, Purdin GJ. *Topics in Wildlife Medicine Volume 4: Orthopedics*. St. Cloud, MN: National Wildlife Rehabilitators Association; 2017.

5. Whitehead ML, Roberts V. Backyard poultry: legislation, zoonoses and disease prevention. *J Small Anim Pract.* 2014;55:487-496.

6. Marmulak T, Tell LA, Gehring R, et al. Egg residue considerations during the treatment of backyard poultry. *J Am Vet Med Assoc.* 2015;247(12):1388-1395.

Deslorelin Implants: How, Why and Why Not to Implant Them

Anna Osofsky, DVM, DABVP (Avian Practice)

Session #201

Affiliation: From Carrollton West Pet Hospital, Carrollton, TX 75007, USA.

Abstract: Deslorelin implants are GnRH super-agonists commonly used in avian practice to manage reproductive-related disorders. While in mammals these implants tend to suppress gonadal hormone production for a reliable 6-12 mo, in avian species they are much less predictable in duration of effect and, in some individuals, they appear to have no effect at all. This paper reviews the current literature regarding deslorelin implant use, the legal issues surrounding the use of this drug in birds in the United States, diseases for which it can be considered as a treatment option, and the technique the author uses for implant placement. A video of the procedure is provided and case examples are discussed in the lecture.

Introduction

Clinical signs secondary to reproductive up-regulation or disease are a common cause for patient presentation to the avian veterinarian. Some of these diseases are obviously reproductive in origin, such as chronic egg laying, hypersexuality, ovostasis, yolk coelomitis, oviductal or cloacal prolapse, ovarian cysts, and ovarian and testicular neoplasms.[1-3] Other clinical signs that may or may not be related to reproductive hormones include hyperlipidemia, feather-damaging behavior, self-mutilation, and aggression.[4] There is even one case in the literature in which a white-crowned pionus parrot (*Pionus senilis*) with a history of traumatic brain injury had seizures that occurred associated with reproductive behaviors.[5]

While treating mammals with reproductive disease is often as simple as surgical gonadectomy, treating reproductive diseases in birds can be quite complex. The ovary and testicle in avian patients are very closely attached to major vasculature making surgical removal extremely risky. Removal of the oviduct and shell gland (i.e., salpingohysterectomy) can be used to help control many reproductive diseases in hens but even this procedure carries significant expense as well as surgical and anesthetic risk, especially in smaller species. In addition, salpingohysterectomy will often lead to down-regulation of the ovary but, since the ovary is still present, there is the potential for persistence of hormonal behaviors as well as internal laying or yolk coelomitis.[1]

The first step in addressing non-life-threatening reproductive disorders in avian patients, such as chronic egg laying, hypersexuality, and aggression is instituting appropriate changes in the environment that might be stimulating reproductive behavior. The Association of Avian Veterinarians has a client education handout called "Managing Chronic Egg-Laying in Your Pet Bird" that can be a very helpful adjunct to client education. Potential environmental factors contributing to reproductive issues include the presence of a conspecific mate, immediate removal of eggs when birds are interested in sitting on them, availability of nesting sites (sometimes these are unexpected places such as under the couch or snuggled on an owner's chest under a sweatshirt), allowing shredding of paper or wood, feeding warm, soft foods (especially if the owner is feeding these directly by hand), over-abundance of high-calorie food items (e.g., seeds, nuts, other carbohydrates), excessively long photoperiod, allowing regurgitation behaviors (common in macaws that like to pump on their owner's hand or budgerigars regurgitating to a mirror), allowing masturbatory behavior (on particular toys or the owner—some birds will masturbate on anything making this hard to control), inadequate mental stimulation (e.g., lack of foraging opportunities), and inappropriate petting (e.g., stroking of the back or ventral abdominal regions).[1] Even in cases that require immediate medical management, part of the treatment plan should address environmental factors that are contributing to the medical problem.[1]

Fortunately for the avian veterinarian and our patients/clients, hormone therapies in the form of gonadotropin-releasing hormone (GnRH) super-agonists are available, often allowing for successful medical management of reproductive diseases of avian patients. As with most vertebrates, the hypothalamo-pituitary-gonadal axis regulates avian reproduction. GnRH released from the hypothalamus stimulates the production of luteinizing hormone (LH) and follicle-stimulating hormone (FSH) from the anterior pituitary. LH stimulates the gonads to produce estrogens, progestins, and androgens. The role of FSH in avian species is not completely clear although it does seem to be a part of steroidogenesis and ovarian folliculogenesis.[2] Environmental cues such as photoperiod, food availability, and presence of a mate stimulate the central nervous system to release GnRH in a pulsatile fashion from the hypothalamus; GnRH is short-lived with a half-life of 3-4 min.[1] Birds have been found to have three different forms of GnRH (aGnRH-1, 2, and 3). While all mammalian GnRH is a conserved/identical decapeptide across all species, aGnRH-1 differs from its mammalian counterpart by one amino acid and aGnRH-2 differs by three amino acids. There are also three different GnRH receptor subtypes in mammals while two have been found in birds.[1]

Synthetic GnRH super-agonists used in avian medicine include leuprolide acetate and deslorelin acetate. Because the endocrine system functions in a negative feedback manner, the prolonged presence of GnRH will initially upregulate gonadal hormone synthesis and release but eventually down-regulate it by reducing the number of GnRH receptors in the brain. The initial increase in gonadal hormone production is sometimes called the flare-up effect (discussed in greater detail below).[6] One-month depot leuprolide was the first GnRH super-agonist used in avian medicine but it is not commercially available in a single-use form, it has been shown to work for only approximately 2-3 wk in Hispaniolan Amazon parrots (*Amazona ventralis*) and cockatiels (*Nymphicus hollandicus*),[1] and can become quite costly and inconvenient to the owner due to the need for frequent injections to manage chronic disease.[3] Deslorelin acetate was the subsequent GnRH super-agonist used routinely in avian medicine and its use is discussed in more detail below.

Slow-release deslorelin implants such as Suprelorin (Virbac, Carros, France) were developed as a subcutaneous injection for reversible chemical castration of dogs. There are two sizes: 4.7 mg and 9.4 mg, which are meant to last for 6 and 12 mo, respectively; both sizes are available and legal to use in birds, including poultry, in much of the world. In the United States, only the 4.7-mg implants are available as Suprelorin F (Virbac AH, Inc. Fort Worth, TX, USA); they are legally marketed as a Food and Drug Administration (FDA) Indexed Minor Use/Minor Species product under MIF 900-013 for the management of adrenal gland disease in domestic ferrets. While Suprelorin F has been used in many species besides ferrets in published case reports and research projects in the United States, it is still not FDA-approved for use in these species. In addition, this product has not been evaluated for withdrawal times for food-producing birds such as chickens and ducks so its use in these species is of significant concern. Thus, not all veterinarians in the United States are comfortable using Suprelorein F in any animals except ferrets. Some clinicians are comfortable using it in non-poultry avian species but not food-producing species. Ultimately, the decision to use this product in birds in the United States rests with the individual clinician.

Summary of the Deslorelin Literature

Japanese quail (*Coturnix coturnix japonica*)

There are three published journal articles describing studies that investigate the use of subcutaneous deslorelin implants in Japanese quail. The 4.7-mg implants were found to reversibly decrease egg laying in six out of 10 birds for 70 days.[2] In a follow-up study, two 4.7-mg implants or one 9.4-mg implant were tested; seven out of 10 birds treated with the two 4.7-mg implants stopped laying 1 wk after implant placement and did not start laying again for 100 days. In the 9.4-mg implant group, seven out of 10 birds also ceased laying eggs but onset of action was prolonged at 5-12 wk and lasted to the end of the study at 182 days.[7] Thus, although the 9.4-mg implant lasted longer, its effect on laying was unpredictable. A 2022 study investigated the use of 4.7-mg implants in conjunction with cabergoline, a prolactin inhibitor, to see whether the flare-up effect could be prevented; in the quail studies, during the flare-up, birds laid more eggs than normal and the eggs were often abnormal in appearance due to their shortened time in the reproductive tract. This is a therapeutic concern for birds being treated for active reproductive tract disease, such as ovostasis or yolk coelomitis. Nine out of nine quail stopped laying in the deslorelin/cabergoline group and seven out of nine birds stopped laying in the deslorelin/placebo group; both groups ceased egg laying within 5 days of implant placement and did not resume laying for the 44-day study duration. The cabergoline did not have a significant effect on egg laying/flare-up effect.[6]

Chickens (*Gallus gallus*)

Two studies have been performed on chickens; all implants were placed subcutaneously. One study, by Noonan, was presented only as a conference abstract using actively laying 2-y-old birds. There were 10 birds in each study group: 4.7-mg implant, 9.4-mg implant, and placebo. By the second week, all the implant birds had ceased laying; the 4.7-mg group stopped laying for 125-237 days (mean = 180 days), the 9.4-mg group had a range of 229-357 days (mean 319 days); the placebo group continued to lay as normal.[8] The second study, by Eusemann, used 10 14-wk-old and 10 21-wk-old hens to receive 4.7-mg deslorelin implants; the implants stopped follicle production for as short as 7 to as long as 16 wk (end of study).[9] The cause for the difference in effectiveness between these two studies is unclear but Eusemann speculated that the cause could have been the older age of the birds in the Noonan study or perhaps different layer lines were used. Comparison is somewhat difficult since the Noonan study is an abstract only.

Pigeons (*Columba livia*)

There is one study with pigeons in which 4.7-mg deslorelin implants were placed intramuscularly in 10 male and 10 female birds after which LH was measured in all birds and egg production was measured in the females. The implants significantly suppressed serum LH concentrations for 84 days in females and 56 days in males. Egg production was stopped for at least 49 days in the hens.[10]

Cockatiels

If there were ever a species that needed deslorelin implants, it would be the cockatiel. It is not surprising that there are several papers describing its use in this species. In one study, 13 females were given 4.7-mg implants subcutaneously. None of the implanted birds laid eggs within 180 days of implantation and only five of the 13 birds laid any eggs for the duration of the 365-day study. The first egg was laid between 192 and 230 days post implant. Interestingly, and somewhat astoundingly, the clutch size described in the placebo birds in this study was 3 to 29 (median = 12).[3] Two other case reports on cockatiels by Nemetz and Keller describe the use of 4.7-mg deslorelin implants to treat ovarian neoplasia in three birds; the implant aided in the medical management of two of the three birds.[11,12] The two birds in the Keller case report were also treated with periodic therapeutic coelomocentesis to remove the ascitic fluid that is common in avian patients with ovarian neoplasia. In the Nemetz paper, the author gives a useful description of what to monitor, aside from lack of egg laying, to test the success of the implant: normal plasma triglycerides, lack of broody behavior at home, down-regulated ovarian silhouette on radiographs, lack of osteomyelosclerosis (polyostotic hyperostosis) on radiographs, down-regulated weight, and lack of odiferous droppings.

Domestic turkeys (*Meleagris gallopavo*)

Two turkeys from a zoological institution were treated with two 4.7-mg or two 9.4-mg deslorelin implants placed intramuscularly to reduce aggressive interactions between the birds and towards the staff of the facility. The two 4.7-mg implants reduced aggression starting 2 wk post implantation and lasted 3.5-5 mo. The two 9.4-mg implants started working 2 wk post implantation and lasted 7 mo.[13]

Retrospective clinical studies

There are two retrospective studies that describe the use of deslorelin clinically. In a largely parrot patient population, 4.7-mg implants were used subcutaneously in 96 different birds for problems ranging from dystocia, prolapse, and pathologic fractures to feather-destructive behavior, aggression, and dermatitis.[4] The most common interval between implants was 3 mo; some birds received seven implants over a 2-y period. In some cases, if the clinical problem was seasonal, only one implant was needed annually. Most birds showed at least some clinical improvement. A retrospective study looking at the clinical use of deslorelin implants in zoos reported that some avian species appeared to become desensitized to deslorelin implants with repeated use including tufted puffins (*Fratercula cirrhata*), Charadriiformes, yellow-knobbed curassow (*Crax daubentoni*), and Galliformes. Other birds did not appear to become desensitized including Psittaciformes and Accipitriformes (e.g., bald eagle, *Haliaeetus leucocephalus*).[14]

Other literature

As mentioned above, 4.7-mg deslorelin implants were used to manage seizures secondary to a traumatic brain injury in a white-crowned pionus parrot. Seizure onset coincided with reproductive behaviors at home; the bird was treated with eight subcutaneous deslorelin implants total over a 2-y period; they were placed every 2-6 mo as needed to help manage the onset of hormonal behaviors, and thus, the seizures.[5]

There is an excellent, although slightly outdated, review article with a clear and concise summary of the neuroendocrine control of avian reproduction and the use of GnRH-agonists to manage reproductive disease in birds.[1] There are also several conference presentations dealing with deslorelin implant usage but this author did not have access to those abstracts and therefore they are not reviewed here. There is one paper summarized in the Mans review that is important to mention. Deslorelin implants (4.7 mg) were used in nine budgerigars (*Melopsittacus undulatus*) tentatively diagnosed with hormone-producing testicular tumors; diagnosis was based on brown cere hypertrophy, polyostotic hyperostosis, and a soft tissue mass in the coelom. Seven of the nine birds improved with the treatment which had to be repeated approximately every 20 wk.[1]

While some authors describe intramuscular use of the implant, subcutaneous application is the most commonly used technique since the implant needle is an 11-gauge needle. No case reports described significant complications or adverse effects associated with the deslorelin implants. Consistent with the Van Sant review of 96 patients, this author has only seen occasional minor complications associated with the implant such as one patient (i.e., umbrella cockatoo, *Cacatua alba*) removing the implant within 2 wk of placement, one implant causing focal skin necrosis and irritating the patient (i.e., Quaker parrot, *Myiopsitta monachus*) so it had to be removed, and one patient (i.e., cockatiel) bleeding moderate amounts at the time of implant placements on three different occasions. In the author's experience, a single 4.7-mg implant often causes cessation of egg laying for a year in seasonal layers but may last only 4-6 mo in birds that lay year-round or have chronic, ongoing reproductive tract disease. The author has also had several patients that were experiencing ovostasis at the time of implant placement and then went on to lay the eggs within a week. It is important to warn owners that the implant may cause a flare-up effect leading to an increase in hormonal behaviors for 5-7 days after implant placement.

Implant Placement Technique Used by the Author

Once it has been determined that implant placement is appropriate, the patient is sedated with 1 mg/kg midazolam and 1 mg/kg butorphanol delivered IM. In the author's experience, intranasal sedation does not lead to a predictable level of sedation so IM administration is preferred for this procedure. A small area (to prevent excessive cooling of the patient) on the dorsum between the spine and the scapula is wiped down with alcohol to part the feathers and allow a thin layer of lidocaine gel application; this is allowed to sit while the implant is prepared. The author uses two restrainers for this quick procedure: one to hold the patient upright and one to hold the wings; the patient's head must be flexed towards the ventrum to get it out of the way of the implant applicator. The skin is then aseptically prepared, usually with one pass of chlorhexidine scrub and one pass of alcohol. Exam gloves rather than surgical gloves are used for implant placement since the implant applicator is not sterile. The implant is placed subcutaneously between the spine and the scapula from a cranial to caudal direction, so that the implant is less likely to fall out of the resultant skin defect should the wound dehisce. The bevel of the implant needle should be visible under the skin as it is being introduced; the needle is introduced subcutaneously until the bevel is completely covered by skin plus an additional 2 mm or so before pushing down the plunger. This is to prevent the implant from getting dragged out by the needle when it is removed. Remove the needle slowly. If the implant does start to come out of the skin defect, having sterile atraumatic forceps handy to gently manipulate it back into the subcutaneous pocket is useful. Bleeding is controlled with gentle pressure using gauze, though usually it is minimal. The skin defect is closed with a small drop of tissue glue. The author has found that using 3.5x magnification loupes has reduced the incidence of bleeding during application, especially in smaller species. The author has placed implants in patients as small as parrotlets (*Forpus* spp.) without complications. Occasionally, two 4.7-mg implants are used at once, in which case both can be placed side by side through the same skin defect. When previous implants have not been absorbed, the author will place the new implant adjacent to the previous implants(s). In the author's experience, it can take a year or longer for an implant to cease to be visible. The author routinely performs this procedure on an outpatient basis in the exam room while the clients watch because it is usually quick and well tolerated.

The author has used this same technique to place microchips in small avian patients (< 100 g) that must have a microchip for international travel. There is an increased risk that the microchip might come out versus the traditional IM placement but it is certainly safer to place it subcutaneously.

Some additional "in my experience" information from colleagues board-certified in Avian Practice by the American Board of Veterinary Practitioners is provided here. The reviewer of this paper commented that she has placed 4.7-mg deslorelin implants in a zebra finch (*Taeniopygia* spp.) and several canaries (*Serinus canaria*); these birds are even smaller than a parrotlet. Two international colleagues commented that they generally use the 4.7-mg implant even though they have access to the 9.4-mg implant; they feel that extending the time of efficacy is generally not worth the expense. They have not noted that the 9.4-mg implant takes longer to have an onset of action, contrary to the Japanese quail study[7] reported above.

Conclusions

Deslorelin implants are considered safe with no known medical contraindications in avian patients. Their efficacy can be quite variable among species and even individuals within a species. The authors of the research reviewed above speculate as to why avian patients have such variable responses to deslorelin while in mammals it has been consistently efficacious. Theories include the differences in the structures of avian GnRH molecules whereas all mammals have the same molecule or perhaps that there are differences in the expression of GnRH receptors in avian species vs mammalian.[14] In addition, there may be stronger pituitary resistance to desensitization in avian species or the higher metabolic rates of avian species may shorten the duration of efficacy.[10] Regardless, given that deslorelin implants have proven quite safe and relatively easy to implant, having them available as one tool in managing complex and life-threatening reproductive-associated diseases in avian patients has proven invaluable.

References

1. Mans C, Pilny A. Use of GnRH-agonists for medical management of reproductive disorders in birds. *Vet Clin North Am Exot Anim Pract.* 2014;17:23-33.

2. Petritz OA, Sanchez-Migallon Guzman D, Paul-Murphy J, et al. Evaluation of the efficacy and safety of single administration of 4.7-mg deslorelin acetate implants on egg production and plasma sex hormones in Japanese quail (*Coturnix coturinix japonica*). *Am J Vet Res.* 2013;74(2):316-323

3. Summa NM, Sanchez-Migallon Guzman D, Wils-Plotz EL, et al. Evaluation of the effects of a 4.7-mg deslorelin acetate implant on egg laying in cockatiels (*Nymphicus hollandicus*). *Am J Vet Res.* 2017;78(6): 745-751.

4. Van Sant F, Sundaram A. Retrospective study of deslorelin acetate implants in clinical practice. *Proc Annu Conf Assoc Avian Vet.* 2013;211-220.

5. Kabakchiev C, Laniesse D, Jame F, et al. Diagnosis and long-term management of post-traumatic seizures in a white-crowned pionus (*Pionus senilis*). *J Am Vet Med Assoc.* 2020;250(10):1145-1152.

6. Ruckl A, Thompson DL, Hatt JM. Effect of the prolactin inhibitor cabergoline and the gonadotropin releasing-hormone agonist deslorelin in the suppression of plasma prolactin concentrations and egg laying in quail (*Coturnix japonica*). *J Avian Med Surg.* 2022;36(1):39-52.

7. Petritz OA, Sanchez-Migallon Guzman D, Hawkins MG, et al. Comparison of two 4.7-milligram to one 9.4-milligram deslorelin acetate implants on egg production and plasma progesterone concentrations in Japanese quail (*Coturnix coturnix japonica*). *J Zoo Wildl Med.* 2015;46(4):789-797.

8. Noonan B, Johnson P, de Matos R. Evaluation of egg-laying suppression effects of the GnRH agonist deslorelin in domestic chickens. *Proc Annu Conf Assoc Avian Vet.* 2012;321.

9. Eusemann BK, Sharifi AR, Patt A, et al. Influence of sustained release deslorelin acetate implant on reproductive physiology and associated traits in laying hens. *Front Physiol.* 2018;9:1846.

10. Cowan ML, Martin GB, Monks DJ, et al. Inhibition of the reproductive system by deslorelin in male and female pigeons (*Columba livia*). *J Avian Med Surg.* 2014;28(2):102-108.

11. Nemetz L. Deslorelin acetate long-term suppression of ovarian carcinoma in a cockatiel (*Nymphicus hollandicus*). *Proc Annu Conf Assoc Avian Vet.* 2012;37-42.

12. Keller KA, Beaufrere H, Brandão J, et al. Long-term management of ovarian neoplasia in two cockatiels (*Nymphicus hollandicus*). *J Avian Med Surg.* 2013;27(1):44-52.

13. Molter CM, Fontenot DK, Terrell SP. Use of deslorelin acetate implants to mitigate aggression in two adult male domestic turkeys (*Meleagris gallopavo*) and correlating plasma testosterone concentrations. *J Avian Med Surg.* 2015;29(3):224-230.

14. Agnew MK, Asa CS, Franklin AD, et al. Deslorelin (Suprelorin®) use in North American and European zoos and aquariums: taxonomic scope, dosing, and efficacy. *J Zoo Wildl Med.* 2021;(52)2:427-436.

Section 2

Plenary Session

Stephanie Lamb, DVM, DABVP (Avian Practice), ***AAV President***
Session Chair

Avian Cervical Intervertebral Disease and Implications Learned from Anatomy Studies

M. Scott Echols, DVM, DABVP (Avian Practice), and
Shawn Miller, PhD

Session #107

Affiliation: From Medical Center for Birds, Oakley, CA 94561, USA (Echols), Parrish Creek Veterinary Hospital and Diagnostic Center, Centerville, UT 84014, USA (Echols), Scarlet Imaging, Murray, UT 84157, USA (Echols), School of Biological Sciences, University of Utah, Salt Lake City, UT, USA (Miller).

Abstract: Cervical intervertebral diseases are poorly documented and understood in birds and may include spondylosis and more (e.g., sclerosis, degeneration, bursitis, and infection). With newer advanced computed tomography (CT) imaging, visualizing intervertebral diseases has become more commonplace with companion and managed collection birds. One author (MSE) has recognized a trend towards intervertebral diseases localized to the mid-cervical region, which begs the question regarding why at this location. Observations of grey parrot (*Psittacus erithacus*) cervical skeletal structures, from the Grey Parrot Anatomy Project and other avian species may offer possible explanations as to why the mid-cervical region is different and may be at more risk of intervertebral disease. Birds donated to the project were processed using either BriteVu® (to evaluate vasculature) or scarlet tissue stain (i.e., to evaluate soft tissue structures) and then micro CT scanned for detailed evaluation. Detailed dissections of the grey parrot neck are conducted by an anatomist (SM). The presentation covers the clinical case CT findings, associated cervical intervertebral disease and anatomic information uncovered from the ongoing anatomy study.

Pharmacokinetics of Oral Acetaminophen with Concurrent Silymarin Administration in Orange-Winged Amazon Parrots (*Amazona amazonica*)

Lucyanne Megan, DVM,
Hugues Beaufrère, Dr med vet, PhD, DABVP (Avian Practice), DipECZM (Avian), Dipl ACZM,
Heather Knych, DVM, PhD, DACVCP,
David Sanchez-Migallon Guzman, LV, MS, DipECZM (Avian), DipECZM (Small Mammal), Dipl ACZM

Session #146

Affiliation: From the William R. Pritchard Veterinary Medical Teaching Hospital, University California-Davis, Davis, CA 95616, USA (Megan), Department of Medicine and Epidemiology, School of Veterinary Medicine, University of California-Davis, Davis, CA 95616, USA (Beaufrère, Guzman), K. L. Maddy Equine Analytical Chemistry Laboratory, University of California-Davis, Davis, CA 95616, USA (Knych).

Abstract: Acetaminophen is an antipyretic and analgesic drug commonly used in human and veterinary medicine. In humans, it is used to manage acute and chronic pain.[1] Acetaminophen undergoes complex metabolism that depends on species, age, and dose administered. At supratherapeutic doses, accumulation of the toxic metabolite *N*-acetyl-*p*-benzoquinone imine can result in hepatotoxicity. Silymarin, an antioxidant flavonoid complex, has been shown to ameliorate these toxic effects in several species.[2,3] Acetaminophen pharmacokinetics have been evaluated in turkeys,[4] chickens,[4] and geese.[5] Currently, there are no reports on the pharmacokinetics or safety of acetaminophen administration in any psittacine species. This study evaluates the pharmacokinetics of acetaminophen (Tylenol) administered via oral gavage to orange-winged Amazon parrots (*Amazona amazonica*). A pilot study was performed on eight birds, with two birds each receiving doses of 5, 10 and 20 mg/kg of acetaminophen and 10 mg/kg of acetaminophen with silymarin (35 mg/kg). A single-dose study was then performed at 100 mg/kg with concurrent silymarin administration (50 mg/kg) via oral gavage. Plasma concentrations of acetaminophen and its two primary metabolites (APA-glucuronide and APAP-sulfate) were measured by liquid chromatography-tandem mass spectrometry at eight time points over 24 hours. Pharmacokinetic analysis was performed using noncompartmental analysis. Blood chemistry panels were performed before and 24 hours after acetaminophen administration to monitor for hepatotoxic effects. This study will serve as a basis for future research evaluating pharmacodynamics of acetaminophen alone and in combination with other analgesic medications.

References

1. Mazaleuskaya LL, Sangkuhl K, Thorn CF, et al. PharmGKB summary: pathways of acetaminophen metabolism at the therapeutic versus toxic doses. *Pharmacogenet Genomics*. 2015;25(8):416-426.

2. Kim YC, Na JD, Kwon DY, et al. Silymarin prevents acetaminophen-induced hepatotoxicity via up-regulation of the glutathione conjugation capacity in mice. *PLOS ONE*. 2018;49:235-240.

3. Ihedioha JI, Anyogu DC, Ogbonna ME. The effects of silymarin on acetaminophen-induced acute hepatic and renal toxicities in domestic pigeons (*Columba livia*). *J Avian Med Surg*. 2020;34(4):348-357.

4. Neirinckx E, Vervaet C, De Boever S, et al. Species comparison of oral bioavailability, first-pass metabolism and pharmacokinetics of acetaminophen. *Res Vet Sci*. 2010;89(1):113-119.

5. Sartini I, Łebkowska-Wieruszewska B, Gbylik-Sikorska M, et al. Acetaminophen pharmacokinetics in geese. *J Am Vet Med Assoc*. 2022;260(12):1-8.

Establishing the Pharmacokinetics of Trazodone in Hispaniolan Parrots (*Amazona ventralis*)

Haley Straub, DVM,
Thomas N. Tully Jr., DVM, MS, DABVP (Avian Practice), DipECZM (Avian)
Levent Dirikolu, DVM, MVSc, PhD,
Andreas F. Lehner, MS, PhD,
Justin Zyskowski,
John Buchweitz, MS, PhD, DABT

Session #115

2024 House Officer Manuscript Competition Winner

Affiliation: From the Louisiana State University College of Veterinary Medicine, Department of Veterinary Clinical Sciences (Straub, Tully, Jr.), Department of Comparative Biological Sciences (Dirikolu) Skip Bertman Dr, Baton Rouge, LA 70803, USA, Michigan State University Veterinary Diagnostic Laboratory, Section of Toxicology, Lansing, MI 48910-8104, USA (Lehner, Buchweitz, Zyskowski).

Abstract: The objective of this study was to establish the pharmacokinetics of a single oral dose of trazodone in the Hispaniolan Amazon parrot (*Amazona ventralis*). Trazodone is a selective serotonin antagonist and reuptake inhibitor used commonly in human and veterinary medicine as an antidepressant behavioral modification medicine.[1-3] A single oral dose of compounded trazodone hydrochloride solution (20 mg/ml) at 50 mg/kg was administered to seven healthy adult Hispaniolan Amazon parrots. The subjects ranged in age from 10-15 y and weighed 228.0- 323.0 g. Blood was collected at baseline (i.e., 2 wk before study), 1, 2, 4, 6, 10, and 14-hour post-drug administration. Plasma concentrations of both trazodone and its active metabolite m-chlorophenylpiperazine (mCPP) were determined via liquid chromatography tandem mass spectrometry (LC-MS/MS). Non-compartmental pharmacokinetic analysis was completed. The half-life ($t_{1/2}$) ± SD of trazodone was 1.89 ± 0.49 hours, and the $t_{1/2}$ ± SD of mCPP metabolite was 1.9 ± 0.55 hours. Maximum serum drug concentrations, or Cmax (ng/ml), were 738.3 ± 285.3 for trazodone. Times to achieve Cmax (hour) were 1 hour post-dosing for trazodone, and 2 hours post-dosing for the m-CPP metabolite. While this study did not establish the behavioral effects of trazodone, no adverse effects were observed throughout the 48-hour period following drug administration. Our results indicate a 50 mg/kg single oral dose of trazodone to Hispaniolan parrots may be considered a safe dose. Plasma concentrations are comparable to previously published values in humans, dogs, horses, and pigeons for up to 14 hours following dosing.[4,5] This study indicates that further studies are needed to establish the pharmacodynamics and efficacy of trazodone in the medical management of behavioral problems in psittacine species.

References

1. Siebert L. Pharmacotherapy for behavioral disorders in pet birds. *J Exotic Pet Med.* 2007;16(1):30-37.

2. Visser M, Ragsdale M, Boothe D. Pharmacokinetics of amitriptyline HCL and its metabolites in healthy African grey parrots (*Psittacus erithacus*) and cockatoos (*Cacatua species*). *J Avian Med Surg.* 2015;29(4):275-281.

3. Mertens P. Pharmacological treatment of feather picking in pet birds. *Inst Anim Hygiene Ethol Anim Welf.* 1997:209-210.

4. Desmarchelier M, Beaudry F, Ferrell S, Frank D. Determination of the pharmacokinetics of a single oral dose of trazodone and its effect on the activity level of domestic pigeons (*Columba livia*). *Am J Vet Res.* 2019;80(1):102-109.

5. Vatassery GT, Holden LA, Hazel DK, et al. Determination of trazodone and its metabolite, 1-m-chlorophenyl-piperazine, in human plasma and red blood cell samples by HPLC. *Clin Biochem* 1997;30:149-153.

Use of Social Media in Building Your Brand and Connecting with Pet Parents

Rachel Siu, DVM

Session #135

Affiliation: From Carrollton West Pet Hospital, 3729 Old Denton Rd, Carrollton, TX 75007, USA.

Abstract: In this lecture, Dr. Rachel Siu, creator of the social media page, Exotic.Pet.Vet, discusses how she grew her pages to have over one million followers and how she connects with pet parents and veterinary professionals worldwide. She discusses methods to create your personal brand, and how you can use it to educate owners and reach new clientele. She also reviews challenges associated with social media, and how to overcome these challenges to make your social media work for you.

Elevating Veterinary Businesses on Social Media

Establishing a distinct and relatable personal brand contributes to the humanization of veterinary services, fosters stronger connections with pet parents, and attracts prospective clients. Once successful, a social media account can also become a second source of income to your veterinary practice.

Personal branding begins with showcasing the expertise and personality of the veterinary professionals behind the practice. Incorporating elements like professional profiles, team introductions, and behind-the-scenes glimpses into daily operations humanizes the clinic and creates a connection between the veterinary team and pet parents. Often, pet owners are left wondering what happens to their loved ones when they're dropped off and are intrigued when learning about the behind-the-scenes of veterinary medicine, as evidenced by the success of the many veterinary medicine TV shows.[1]

Content creation that highlights the individual passions, interests, and unique stories of the clinic's veterinary professionals not only adds a personal touch but also helps to differentiate the clinic from competitors. This personal narrative can include educational posts, success stories, and anecdotes that resonate with pet owners, reinforcing the trust and credibility of the veterinary practice.

Many veterinarians and veterinary hospitals have their own social media, but very few become successful to the point of receiving worldwide recognition. So, what sets these accounts apart?

Building a Personal Brand

As the first step in building a social media presence, you must find your brand and aesthetic. Your social media should be cohesive and easily identifiable. For your handle, you can simply use your practice's name or your name, or you can think of something catchy within your niche. Although choosing themed colors may look nice, it's not everything. The most important part of branding is to find your niche in the industry. Do you have a primary focus or something in your practice that sets you apart, such as an interest in a particular field? For myself and many of us here, our focus is exotic animals, but many other successful accounts share information primarily on ophthalmology, dentistry, surgery, etc.; these accounts often grow quickly because the presenters are credible and pet owners hope to gain specific information from them. There are so many dog-and-cat veterinary accounts out there, so if you can find something that sets you apart, you are more likely to attract followers.

When thinking of your brand, it's also important to consider its personality. Some veterinary accounts are popular due to their bluntly honest educational cases, others for showcasing an attractive figure dancing to different beats, while some perform creative skits that are meant to elicit a laugh. A simple way to determine this for your brand is to ask yourself: what is your goal? Personally, I'm not much of a dancer, nor am I creative or funny enough for skits, but one thing I certainly believe in is education.

The Differences Between Social Media Platforms

It is important to familiarize oneself with the different social media channels. I primarily focus on Instagram, TikTok, YouTube, and Facebook, but there are additional channels out there such as Twitter, Snapchat, and LinkedIn that others have success with. Each channel tends to have different strengths and weaknesses. Facebook is a versatile platform that allows for various types of content, ranging from photos, videos, Facebook Live, stories, and user-generated-content such as pet owners posting their own reviews and pictures to your page and their own posts. Facebook has also started to pay out for ads on longer videos, and owns Instagram, so you can sync your accounts and post on both at the same time.

In the past, Instagram primarily focused on photos, but they are shifting more towards reels, or videos. A strength of Instagram is the ability to connect with other accounts. Recently they introduced a collaborative post option where you can share a post with another page, which will share it with the followers of both pages, increasing your reach. TikTok is a highly popular social media platform famous for its short-form video content. Short and snappy videos do very well; be sure to cut out any lulls in a video because viewers often have short attention spans and get bored easily. Since it is so easy to scroll, your videos must have a good hook at the beginning, or viewers will lose interest. For more long-form content, how-to videos and storytelling or narratives do very well. A significant strength of TikTok is the ability to duet or stitch another user's video to add your personal commentary. Instagram also has a similar feature called remix and sequence. Some of my best-performing content is me simply commenting with my veterinary interpretation of another user's video of their pet.

YouTube is all about videos, and although it used to be focused on horizontal, long-form content, recently they've created YouTube Shorts, where you can repost your vertical, shorter videos as well. One thing I love about YouTube is that your videos rarely grow old. People are always searching for content on YouTube, and a video that I posted months ago may suddenly become viral. YouTube also has one of the highest payouts because of this, so my income continues to steadily rise month-to-month, whereas on TikTok and Facebook, income seems to be more sporadic.

How to Grow Your Following

One of the best aspects of social media is connecting with other people. Some of my closest friends I gained because of social media. On your personal brand's social media, it's important to follow the right accounts. Try to follow other accounts within your niche, such as other veterinarians, or let's say, bird accounts, if you happen to specialize in birds. If they make an interesting post, be sure to share it onto your stories (Instagram) or repost it (TikTok). Your followers will be interested in that content and those pages are more likely to share your content in the future. Just be sure to always tag them and give credit when it's due. Tagging them will also notify them that you shared their content, so they'll see that you're supporting them.

Don't be afraid to add interesting commentary on their posts. Not only will you increase their engagement, but their followers will see your comment, and may even follow you. In the past, I've done joint giveaways to grow both of our followings, and lately, I've even been able to meet up with some of my long-time social media friends in person. One rule I recommend is to keep your personal account and veterinary account separate; try to follow your family and friends on your personal account only. One rule of thumb from many top social media accounts is to always have more followers than the number of people you follow; this creates credibility as your followers know you're not just following whoever and whatever.

Once you create your account, try to post often and on a consistent basis. The most successful accounts post a few times a week. Some social media managers say to post every day; however, to some followers, this may become overwhelming. Alternatively, if you only post once a month, then people are less likely to follow you to see what you're up to.

Collaborations

Things you should be wary of are bots and scams. Often, pages will offer to do collaborations with you, but they may not be legitimate. These pages can appear to have thousands of followers, but they may have bought these followers to appear well-established. You can tell if an account is a bot account because they will have thousands of followers but only a few likes on each of their posts. This is why engagement is so much more important than the number of followers you have, and why we should be striving for interactions instead.

Brands interested in collaboration will usually ask to reach out to you via email, but I've also messaged brands that I had a personal interest in to see if they'd like to collaborate. Some brands that I've worked with in the past include Pretty Litter, Litter Robot, Vetigel, Moichor, FIGS scrubs, and Netflix. When I had a smaller account, I would negotiate my own contracts with brands, but now, I have a managing company that does this for me. Once you have an established account, brand collaborations are a great way to make money to fund your social media. Because of my social media-based second income, I was able to cut my hours at the clinic and instead put that time towards content creation.

Creating Content

You may be asking yourself: where do I begin? Social media used to be as simple as posting a picture and caption, but now many platforms have shifted towards video. There are countless video-editing apps out there; I use an app on my phone called CapCut. With this app, I'm able to quickly add multiple videos and pictures together, cut out the boring parts, mute them, and add voiceovers and background music if needed. Videos with captions tend to perform better because many people watch videos on silent at work, on the toilet, or even while their partners are sleeping! Recently I've also been watermarking every video as there are many accounts out there that will reshare your content as their own.

Now, what about captions and hashtags? Hashtags are an effective way to categorize and discover content. There's a lot of debate on how to put hashtags to the most effective use; do you use the common hashtags or something more niched? I like to do a few of each but try not to overdo it. Most people put hashtags at the end of their captions, but you can also intersperse them throughout to highlight a thought. As for captions, they can be as simple or in-depth as you like. Long captions are great for diving more in-depth into a short video; however, if you've included most of your information in your video, then a short caption may be all you need.

Direct Messages

Most social media accounts have messenger features, and there are different ways to use them depending on your goals. Some business accounts actively monitor their messages and use them as another way to interact with their clients and customers. Some prefer to have an automated reply saying that they do not monitor their messages, and it is best to email or call if you need to get in contact with them.

Every single day I get message requests asking me for veterinary advice. I used to kindly respond to them saying that it is illegal to give medical advice without a valid client-patient relationship; however, I had some rude replies and people not respecting my boundaries, so now I simply do not reply. In addition, it would take hours to respond to everyone now that I have so many followers. On Instagram, TikTok, and Facebook, if you do not reply to the chat, then it appears that you never opened it. If you use an automated reply, then users can keep messaging you over and over, which can lead to harassment. I know when you are first starting out, it is tempting to reply and engage with your followers, but the line of giving personal veterinary advice is blurry and not something I want to cross.

Protecting Your Identity

One question you need to answer is how public do you want your name to be? Some veterinarians on social media want privacy, so they simply go by their first or last name, and do not include the place they practice. You can take this to an extreme, and go by a pseudonym, like my friend "@tacticalveterinarian" who has thousands of followers but has not once shown their face. Many of you probably walked by them at ExoticsCon and didn't even know it. On the other hand, if you do include your name and the practice you work at, you can use your social media to attract clientele. I've had pet owners drive hours from out-of-state to come see me. The downside of this is that your life is slightly on display. Fortunately, I have not had any stalkers yet because I don't post much about my personal life, but I've heard stalking is possible.

Asking for Owners' Consent

If you are planning to post pictures of clients' pets, it is extremely important to get consent. There are a couple of ways to do this. I've worked at practices that simply have a social media release as part of their intake paperwork. This is great for sharing cute pictures of your patients that you see throughout the day. If you're sharing something more medically complicated, I personally ask the owners and have them sign a specific social media form.

When choosing whether to ask a client, I always consider the individual case. I usually only ask pet parents that I 'vibe' with, as the last thing you want to do is aggravate an already stressed-out client. Sometimes I don't ask about posting risky procedures until after they are over, and I know there was a positive outcome. There are different ways to phrase this, but I usually tell the owners, "I have a page where I educate pet parents, vet students, and veterinarians. Rocky has been so good for blood draws in the past. Is it ok if I take a video and use it for education?" or "I got a good picture of before and after the mass removal. Is it ok if I share this on our social media since Fluffy did so well, and it's such a good example?" Most owners are excited for their pet to be a star, and you may even gain a follower! Especially with exotic animals, owners are happy you're trying to educate other veterinarians because they understand exotic pets are often overlooked in veterinary school. I have never had anyone say no in my years of managing my social media, but I do believe patient and client selection is key.

Summary

In summary, building a personal brand on social media is a powerful tool for veterinary professionals to connect with pet parents on a deeper level. By humanizing the practice, showcasing individual expertise, and actively engaging with the community, veterinary businesses can establish a memorable and relatable presence that resonates with pet owners, ultimately fostering stronger client relationships and long-term loyalty.

References

1. Johnson S. The Top 8 Vet Shows and 5 Things I Learned by Watching Them. SJ Insights. 2020. Available at: sjinsights.net/2020/04/05/the-top-8-vet-shows-and-5-things-i-learned-by-watching-them/. Accessed March 28, 2024.

Pharmacokinetics of Single-Dose Oral Atorvastatin and its Metabolites in Cockatiels (*Nymphicus hollandicus*)

Hugues Beaufrère, Dr med vet, PhD, DABVP (Avian Practice), DipECZM (Avian), Dipl ACZM,
David Sanchez-Migallon Guzman, LV, MS, DipECZM (Avian) DipECZM (Small Mammal), Dipl ACZM,
Heather Knych, DVM, PhD, DACVCP

Session #141

Affiliation: From the Department of Medicine and Epidemiology, School of Veterinary Medicine, University of California-Davis, Davis, CA 95616, USA (Beaufrère, Guzman), the K. L. Maddy Equine Analytical Pharmacology Laboratory, School of Veterinary Medicine, University of California-Davis, Davis, CA 95616, USA (Knych).

Abstract: The objectives of this study were to evaluate the plasma concentrations and determine pharmacokinetic parameters of atorvastatin and its primary active metabolites (para- and ortho-hydroxyatorvastatin) following administration of a single oral dose in cockatiels (*Nymphicus hollandicus*). Fourteen adult cockatiels (seven males, seven females) of about 2 y of age were used in this study. A compounded oral suspension of atorvastatin 10 mg/ml was administered via oral gavage at 20 mg/kg to each bird. Blood samples were collected at eight different time points from 0 to 24 hours post-administration in a balanced incomplete block design with three blood sampling per bird and six replicates per time point. Plasma concentrations of atorvastatin, para-hydroxyatorvastatin, and ortho-hydroxyatorvastatin were determined by liquid chromatography-tandem mass spectrometry. Pharmacokinetic analysis was performed using non-compartmental and compartmental analysis. Estimated T_{max} for atorvastatin, para-hydroxyatorvastatin and ortho-hydroxyatorvastatin were all 3 hours. Estimated C_{max} were 152.6 ng/ml, 172.4 ng/ml, and 68.8 ng/ml, respectively. Elimination half-lives were 4, 6.8, and 4.6 hours, respectively. Compared to orange-winged Amazon parrots, atorvastatin resulted in higher plasma concentrations in cockatiels likely related to better absorption, but half-life was shorter. Plasma concentrations and pharmacokinetic profile support the therapeutic use of atorvastatin at the dose evaluated in this species based on human pharmacokinetic data. A starting dose of 20 mg/kg PO q12-24h could be used to treat lipid disorders pending more data on multi-dose use and hypolipidemic efficacy.

Generation of Reference Intervals and Evaluation of Seasonal Variation in Clinical Pathology Parameters of Backyard Laying Hens

Alison Cummings, DVM, DABVP (Avian Practice),
Carolyn Cray, PhD,
Fabiano Montiani-Ferreira, DVM, PhD, Dipl BCVO,
Laurie Hess, DVM, DABVP (Avian Practice)

Session #171

Affiliation: From Avian and Exotic Animal Medical Center 5989 Meijer Drive Suite 11, Milford, OH 45150, USA (Cummings), Department of Pathology & Laboratory Medicine, University of Miami Miller School of Medicine, Miami, FL 33136, USA (Cray, Montiani-Ferreira), Department of Veterinary Medicine, Universidade Federal do Paraná, Curitiba, Brazil 80035-050 (Montiani-Ferreira), Division of Comparative Pathology (Montiani-Ferreira), Veterinary Center for Birds and Exotics, Bedford Hills, NY 10507, USA (Hess).

Abstract: In recent years, more backyard poultry are being kept as pets in the United States. Popular backyard chicken breeds, bred for high egg production and friendly temperament, differ from the birds in commercial egg and meat production typically used in veterinary research studies. Increased interest in chickens as pets has created a need for more breed-specific veterinary care and antemortem laboratory evaluations, such as complete blood count (CBC), biochemical panel, and protein electrophoresis (EPH) assessments, for these animals.[1-11] For this purpose, 48 clinically healthy young laying hens from four different flocks, representing seven popular backyard breeds, were evaluated. Reference intervals for CBC, biochemistry panel, and EPH were established per the guidelines of the American Society for Veterinary Clinical Pathology. Paired samples (i.e., summer versus winter) were utilized to assess for the effects of seasonal reproductive fluctuations on these results, with the assumption that the inflammation and physiological stress occurring during increased reproductive activity would significantly impact their laboratory results. Notably, significant seasonal differences were seen in all measured values of the CBC, as well as concentrations of glucose, calcium, aspartate transaminase, creatine kinase, and phosphorus in the biochemistry panel; and both the albumin/globulin ratio and concentrations of alpha-1 and gamma globulins in the EPH. These findings suggest the need for season-specific reference intervals when evaluating hematologic results from backyard chickens.

References

1. Board MM, Crespo R, Shah DH, et al. Biochemical reference intervals for backyard hens. *J Avian Med Surg.* 2018;32:301-306.

2. Elkhoraibi C, Blatchford RA, Pitesky ME, et al. Backyard chickens in the United States: a survey of flock owners. *Poult Sci.* 2014;93:2920-2931.

3. Crespo R, Shivaparasad HL. Interpretation of laboratory results and values In: Greenacre CB, ed. *Backyard Poultry Medicine and Surgery: A Guide for Veterinary Practitioners*. Hoboken, NJ: John Wiley & Sons, Inc.; 2021:515-532.

4. Cray C. Protein electrophoresis of non-traditional species: a review. *Vet Clin Pathol.* 2021;50:478-494.

5. Facon C, Roman Y, Guerin J-L. Assessment of inflammatory status in poultry using plasma protein electrophoresis as a diagnostic tool. *Revue de Médecine Vétérinaire.* 2014;165:305-312.

6. Tothova C, Sesztakova E, Bielik B, et al. Changes of total protein and protein fractions in broiler chickens during the fattening period. *Vet World.* 2019;12:598-604.

7. Arzour-Lakehal N, Boudebza A. Interpretation of protein electrophoresis on avian samples. *Int J Poult Sci.* 2021;20:179-187.

8. Vergneau-Grosset C, Beaufrere H, Ammersbach M. Clinical biochemistry In: Speer BL, ed. *Current Therapy in Avian Medicine and Surgery.* 1st Ed. St. Louis, MO: Elsevier; 2016;486-501.

9. Rezende MS, Mundim AV, Fonseca BB, et al. Profile of serum metabolites and proteins of broiler breeders in rearing age. *Braz J Poult Sci.* 2017;19:583-586.

10. Friedrichs KR, Harr KE, Freeman KP, et al. ASVCP reference interval guidelines: determination of de novo reference intervals in veterinary species and other related topics. *Vet Clin Pathol.* 2012;41:441-453.

11. Scanes CG. Blood. In: Scanes CG, ed. *Sturkie's Avian Physiology*. 6th Ed. Waltham, MA: Elsevier Academic Press; 2015:167-191.

Do Birds Enjoy Flying? An Analysis of Affect Following Flight in Galahs (*Eolophus roseicapilla*)

Jorg Massen, PhD,
Kate Malone, MSc,
Roos de Vries, MSc,
Mandy Beekmans, MSc,
Yvonne van Zeeland, DVM, MVR, PhD, DipECZM (Avian),
DipECZM (Small Mammal), CPBC, Christine Oei

Session #175

Affiliation: From the Division of Animal Behaviour and Cognition, Department of Biology, Faculty of Sciences (Massen, Malone, de Vries), Division of Zoological Medicine, Department of Clinical Sciences (Beekmans, van Zeeland), Farm Animal Health, Department Population Health Sciences, Faculty of Veterinary Medicine, Utrecht University, Utrecht, The Netherlands.

Abstract: Different bird species have been observed performing seemingly useless aerial acrobatics, which may in fact be considered play, indicating that birds enjoy themselves during flight; however, birds cannot report (verbally) on their emotions, which makes the assignment of such emotions notoriously difficult.[1] Recently, researchers started to adopt a componential view on emotions, which contains behavioral, psychologic and physiologic elements that can be measured in animals.[2-4] Integrating these three elements, we examined whether flying generates positive affect in 17 captive-bred galahs (*Eolophus roseicapila*) that lived in an indoor-outdoor aviary (L × W × D: 8 × 6 × 4 m) and actively participated in a flight demonstration together with other members of their flock. Specifically, we measured behavior, cognitive bias, and fecal glucocorticoid metabolites, and compared measurements of days on which they did or did not fly. Results showed a high preference for flight in all individuals, if given the opportunity as well as increased appetitive behaviors after flying in comparison to non-flight days. In addition, birds were more likely to respond optimistically in a cognitive bias task after having had the ability to fly, and conversely, more likely to respond pessimistically when flight opportunity was withheld. Although no differences were found between fecal glucocorticoid concentrations on flight and non-flight days, a decrease in glucocorticoid concentrations could be noticed upon increasing precipitation, as also found in other bird species.[5] These findings suggest that galahs prefer to fly and that flight positively affects their emotional state, therefore emphasizing the need for further studies into flight as a potential ethologic need for birds.

References

1. Tinbergen N. On aims and methods of ethology. *Zeitschr Tierpsychol.* 1963;20(4):410-433.

2. Paul ES, Harding EJ, Mendl M. Measuring emotional processes in animals: the utility of a cognitive approach. *Neurosci Biobehav Rev.* 2005;29(3):469-491.

3. Mendl M, Burman OH, Paul ES. An integrative and functional framework for the study of animal emotion and mood. *Proc Royal Soc B Biol Sci.* 2010;277(1696):2895-2904.

4. Massen JJM, Behrens F, Martin JS, et al. A comparative approach to affect and cooperative decision-making. *Neurosci Biobehav Rev.* 2019;107:370-387.

5. Frigerio D, Dittami J, Möstl E, et al. Excreted corticosterone metabolites co-vary with ambient temperature and air pressure in male Greylag geese (*Anser anser*). *Gen Comp Endocrinol.* 2004;137(1):29-36.

Section 3

Hot Topics Session

Christal Pollock, DVM, DABVP (Avian Practice);
Brenna Fitzgerald, DVM, DABVP (Avian Practice)
Session Chairs

Reference Interval Creation for a Novel Renal Biomarker in Healthy Amazon Parrots (*Amazona ventralis*) and Quaker Parrots (*Myiopsitta monachus*)

Adam A. Moreno, DVM,
Thomas N. Tully, Jr., DVM, MS, DABVP (Avian Practice), DipECZM (Avian),
Chin-Chi Liu, MS, MApStat, PhD,
J. Jill Heatley, DVM, MS, DABVP (Avian Practice), DABVP (Reptile and Amphibian Practice), Dipl ACZM

Session #139

Affiliation: From the Louisiana State University School of Veterinary Medicine, Department of Veterinary Clinical Sciences, Skip Bertman Dr., Baton Rouge, LA 70803, USA (Moreno, Tully, Jr, Liu), Texas A&M School of Veterinary Medicine, Department of Small Animal Clinical Sciences, College Station, TX 77843, USA (Heatley).

Abstract: Renal disease is often identified as a cause of morbidity and mortality in avian patients; however, antemortem early detection of renal disease in avian patients is difficult. Anatomic and physiologic differences between mammals and birds mean the use of commonly employed diagnostic testing such as measurement of blood urea nitrogen (BUN) and serum creatinine, urinalysis, and ultrasonography are either nondiagnostic or difficult to achieve. Symmetric dimethylarginine (SDMA) is considered a more sensitive marker for renal disease in humans, dogs, and cats; however, SDMA has not been assessed for diagnostic use in any psittacine species. In this study, we establish reference ranges for SDMA in both Hispaniolan Amazon parrots (*Amazona ventralis*, HAP) and Quaker parrots (*Myiopsitta monachus*, QP). Blood was collected from 23 HAP and 33 QP maintained in research facilities. Analysis of SDMA through a commercially available immunoassay (IA), as well as creatinine, BUN, uric acid, phosphorus, calcium, sodium, potassium, and chloride, were determined through IDEXX Laboratories. Plasma SDMA concentrations ranged from 6-15 µg/dl and 3-15 µg/dl for the HAP and QP, respectively. Sex was a confounding factor for the QP population, but sex did not have a significant effect on SDMA for the HAP population. No significant correlations occurred for SDMA concentrations and other parameters in either of the psittacine species. Our results show proof-of-concept for the IA SDMA and provide reference intervals for SDMA in HAP and QP. Further investigation is required to determine the validity of this methodology and the predictive power of SDMA in the detection of renal impairment for parrots and other companion birds.

Introduction

Renal disease is often diagnosed in companion birds and avian species maintained in zoological collections.[1-4] Unfortunately, clinical signs of renal disease are generally non-specific and include dehydration, weakness, regurgitation, and muscle wasting.[5,6] Polyuria and polydipsia are not always evident in birds with renal compromise.[5,6] Hind limb paresis, particularly unilateral paresis with no history of trauma suggests but is not pathognomonic for inflammatory or expansile kidney disease.[4,7] In mammalian species, identification of azotemia (i.e., elevations in blood urea nitrogen and creatinine beyond a species-specific threshold) and changes in urinalysis values, particularly decreased urine-specific gravity and proteinuria are considered sensitive testing parameters for identification and assessment of disease and severity.[8] Due to anatomic and physiologic differences between the avian and mammalian urinary systems, these diagnostics lack the same utility in avian patients.[5,9-15]

Several fundamental anatomic and physiologic differences between the renal system of mammals and birds make the use of mammalian markers for renal health limited in their clinical utility. Differences in anatomy and physiology, a comprehensive exploration that is beyond the scope of this paper, include nephron structure, nitrogenous waste composition, renal portal blood flow, osmoregulation strategies, post-renal handling of urine, and urinary excretion.[5,14]

Specialized imaging has been used to diagnose various renal disorders; however, these diagnostic modalities can be cost-prohibitive and may require sedation and/or anesthesia. Renal biopsy, which is still considered the reference standard for assessing kidney disease in birds, is much more invasive than phlebotomy, is costly, and necessitates general anesthesia and specialized equipment.[11] In addition, serial monitoring of renal function via biopsy could further exacerbate kidney damage.

Symmetric dimethylarginine (SDMA) is a byproduct of proteolysis excreted exclusively by the kidneys.[16] Importantly, SDMA is thought to be biologically inert, produced in a stable manner, and cleared through renal excretion with little to no further metabolism in the kidney.[17] In humans, dogs, and cats, increased blood SDMA concentrations correlate closely with increased concentrations of BUN and creatinine.[18-23] SDMA accurately reflects glomerular filtration rate (GFR) in dogs and cats, and increased blood SDMA concentrations occur with as little as a 25% decrease in kidney function.[24-26] Experimental induction of tubular damage in rats showed that serum SDMA, measured with the SDMA immunoassay (IA), was an earlier biomarker for the detection of gentamycin-induced toxicity than were serum concentrations of cystatin C, BUN, and creatinine.[27] Therefore, SDMA is a promising analyte for avian species that offers both high sensitivity and high specificity for renal disease. While the physiology of SDMA excretion has not been described in birds, protein arginine methyltransferases (PRMT) are well conserved in vertebrates, suggesting that mammalian methods for quantifying plasma SDMA concentrations should be possible in birds. In greater flamingos (*Phoenicopterus roseus*) a reference interval demonstrated that increased SDMA concentrations correlated with renal disease on necropsy.[28] In chickens (*Gallus gallus*), SDMA was measurable in plasma, and a reference range for the species was established.[29] SDMA shows promise as a renal marker for avian antemortem diagnosis of kidney disease; however, this assay has yet to be validated or assayed in parrots. The goal of our study was to establish reference intervals for plasma SDMA in healthy Hispaniolan Amazon parrots (*Amazona ventralis*, HAP) and Quaker parrots (*Myiopsitta monachus*, QP).

Methods

Animals

This study was approved by Institutional Animal Care and Use Committees (IACUC) at both Louisiana State University for HAP and Texas A&M University for QP under the project numbers: IACUC AM-22-145 and TAMU IACUC AUP 2020-0223, respectively. A total of 23 adult HAP (7 male 13 female, and 3 unsexed individuals) were included in this study. The birds were part of a research flock at Louisiana State University School of Veterinary Medicine, Baton Rouge, LA, USA. The birds were singly housed, non-breeding, had no recent history of disease, and showed no external signs of disease when physically examined prior to the study. The birds were maintained in individual stainless-steel cages (1 × 0.75 × 1 m) in a temperature- and humidity-controlled environment. The HAP were provided a 12-hour photoperiod and fed a commercially available pelleted diet (Kaytee Exact Natural Parrot and Conure food, Kaytee Products, Chilton, WI, USA) and chlorinated tap water was provided ad libitum.

Thirty-five QP (15 female, 17 male) were included in the study and maintained at the Schubot Center for Avian Health Aviaries at Texas A&M University. Hatch dates were known for each QP. Sex was determined by PCR karyotype of DNA from blood samples. Parrots were housed in groups of one to six birds within cages measuring 2-8 × 4 × 4 ft composed of 1 × 1 inch wire mesh, at 73 ± 5 °F with 12-hour light cycles. Each parrot's diet consisted of approximately 1/6 cup ZuPreem FruitBlend® Flavor once daily (Premium Nutritional Products, Mission, KS, USA), 1 Classic Nutri-Berries twice weekly (Lafeber Company, Cornell, IL, USA), and daily water ad libitum. Leg bands allowed for the identification of individual birds.

Sample collection

Blood samples were collected on 2 separate days for the HAP population. The birds were not fasted prior to blood collection. The blood samples were obtained on both days in the morning. Prior to blood collection for each bird, a physical examination was performed. Any abnormal nasal or ocular discharge, abnormal respiratory sounds or sneezing, abnormal urates, urine, or feces, birds receiving any medications in the past 14 days, or birds with a history of urinary tract disease were used as exclusion criteria. The birds were manually restrained for blood collection using a towel. A total of 1 ml of blood was collected from the right or left jugular vein using a 3-ml syringe and 26-gauge needle. The needle was removed from the syringe and the blood was inserted into two 1-ml lithium heparin tubes without a serum separator (BD Microtainer, Becton, Dickinson and Company, Franklin Lakes, NJ, USA). Tubes were labeled with the patient ID and placed in a cooler until centrifugation at 6,000 rpm for 5 min. Plasma was then decanted using a bulb pipette into a 1-ml cryovial. The samples were then stored in a -20 °C freezer for 4 days. The samples were then shipped overnight to IDEXX laboratories (IDEXX Reference Laboratories, 401 Industry Road, Suite 500, Louisville, KY, USA) with cold packs for analysis of SDMA (the high throughput immunoassay [ELISA]), uric acid (UA), BUN, creatinine, phosphorus (P), potassium (K), sodium (Na), chloride (Cl), and calcium (Ca). Results were returned electronically within 24 hours. Insufficient samples were collected for SDMA analysis from two individuals and SDMA and UA for one individual. Blood was collected again from these birds the following day and the methodology for extraction and submission was replicated.

The methodology was similar for the QP population. Based on physical examination, 34 apparently, non-fasted healthy QPs had ~0.5 ml blood obtained from the right jugular vein in the morning of 3 consecutive days. Samples were immediately stored in a lithium heparin microtainer and then centrifuged at 6000 rpm for 10 min within 2 hours of collection. Plasma was then aspirated and placed via bulb pipette into a 1-ml cryovial and stored in a -20 °C frost-free freezer for up to 15 days prior to shipment for analysis. The samples were then shipped overnight to IDEXX laboratories for the same analysis as described above for the HAP population.

Statistical analysis

Data analyses were performed using JMP Pro 17.0.0 (JMP Statistical Discovery, Cary, NC). A generalized linear model was used with gender and age as the fixed effects for the evaluation of SDMA within HAP and QP, respectively. Assumptions of the parametric models, normality of residuals, and homoscedasticity, were accessed by examining standardized residual and quantile plots. The normality of the residuals was also confirmed via Shapiro-Wilk test. The relationship of SDMA and other parameters was also evaluated via Pearson correlation coefficients within each species. In all hypothesis testing conducted in this study, significance was set at $P < 0.05$, and 95% confidence intervals were employed. Descriptive 90% reference intervals were established in accordance with American Society of Veterinary Clinical Pathology (ASVCP) guidelines as recommended.[30] The 90% reference interval and their limits of SDMA were established based on normal distribution via MedCalc® Statistical Software version 22.007 (MedCalc Software Ltd, Ostend, Belgium).

Results

Study population

A total of 23 HAP and 34 QP were sampled in this study. Two of the QP samples yielded an error in the reporting of SDMA, reducing the total number of QP included in the results to 32.

Reference intervals

The 90% reference interval (RI) established for SDMA in this population was 6.21-13.36 µg/dl for HAP and 3.59-11.97 µg/dl for QP. The SDMA concentration in this HAP population followed a normal distribution with a mean value of 9.78 µg/dl (Fig 1). The data for the QP population was separated by sex and followed a normal distribution both as an entire population and when separated by sex. The details of the descriptive statistics are listed in Tables 1 (HAP) and 2 (QP). There was one outlier identified in the QP population (SDMA = 15 µg/dl). This individual was not excluded from the data analysis.

Sex and age

Sex had a significant effect on SDMA ($P = 0.0412$, Female: 8.75 ± 2.81 µg/dl, Male: 6.93 ± 2.00 µg/dl) levels with QP. The male QP has on average lower SDMA concentrations than females (Table 2). Age did not have a significant effect on SDMA ($P > 0.05$) in QP. Sex was not found to have an impact on SDMA in HAP ($P > 0.05$), and as the specific age of the birds in this population is unknown, age as a variable could not be tested.

Correlation to other variables

No significant correlations occurred between SDMA concentrations and other analytes, including UA, BUN, P, K, Na, Cl, Ca, and Ca:P in either species ($P > 0.05$). Of the other analytes tested, occasional slight deviations in electrolyte concentrations from established reference ranges for other psittacine species were seen; however, none were deemed clinically relevant.[34,35] Two individuals in the HAP population and one in the QP population showed mild elevations in UA above established, comparable reference ranges.[34,35] One of the two HAP that had elevated UA had the highest SDMA values in the population, suggesting that this individual may have underlying, subclinical urinary disease. At the time of submission, this individual remains alive with no clinical concerns. The other had a mid-range value for SDMA (9 µg/dl). The QP with hyperuricemia (12.8 mg/dl) had an SDMA concentration (10 µg/dl) within the reference range established here.

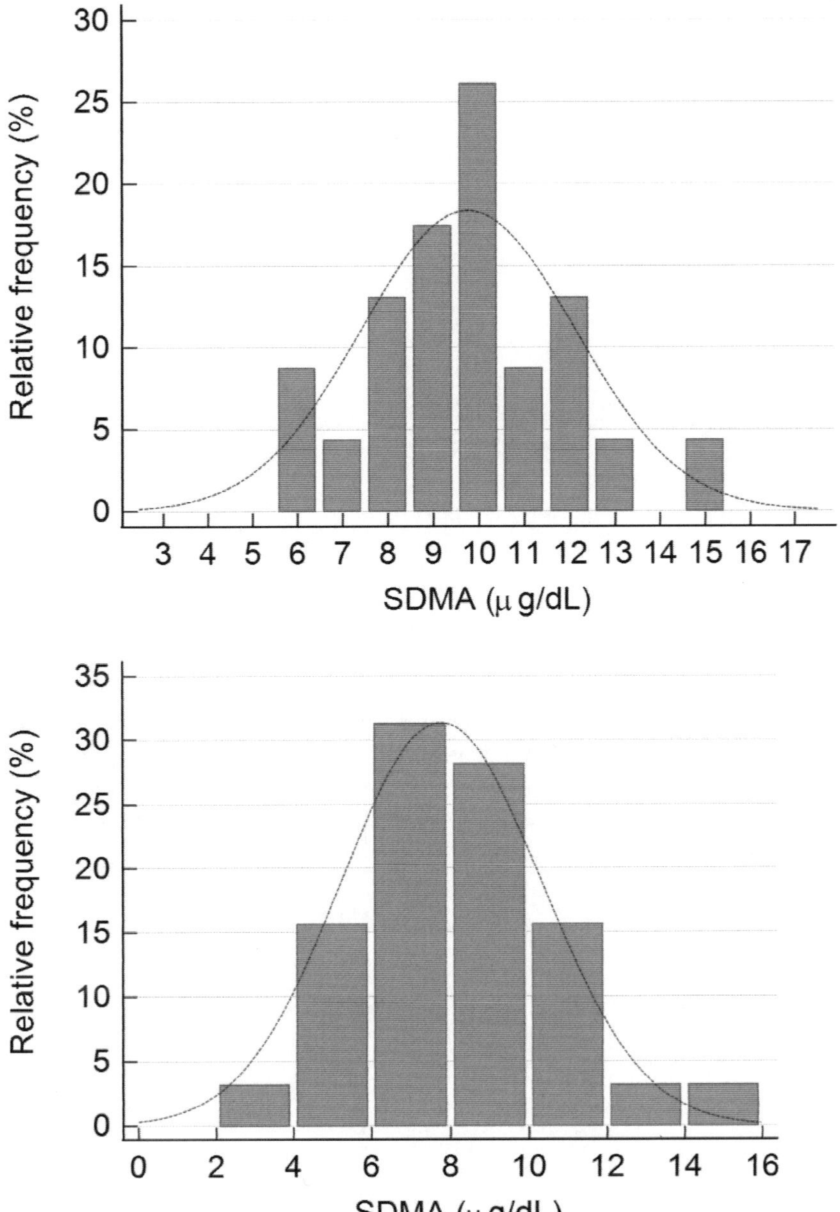

Figure 1. Histogram showing plasma symmetric dimethylarginine (SDMA) concentrations (μg/dl) of the Hispaniolan Amazon parrot (*Amazona ventralis*) population (top) and the Quaker parrot (*Myiopsitta monachus*) population (bottom). Data follow a normal distribution.

Table 1. Statistical summary of plasma symmetric dimethylarginine (SDMA) concentrations of 23 Hispaniolan Amazon parrots. Data were normally distributed (Shapiro Wilk *P* value = 0.6148) and no sex effect was identified. A 90% reference interval was created based on normal distribution.

Descriptor	SDMA (µg/dl)
Sample size	23
Minimum	6
Maximum	15
Median	10
Mean	9.78
Standard deviation	2.17
Normality	Yes
Reference interval	
Lower limit	6.21
90% confidence interval of lower limit	4.90-7.52
Upper limit	13.36
90% confidence interval of upper limit	12.05-14.67

Table 2. Statistical summary of plasma symmetric dimethylarginine (SDMA) of 32 Quaker parrots (*Myiopsitta monachus*). The data were normally distributed (Shapiro Wilk *P* = 0.1096, female and male; 0.2520, female; 0.3446, male). Males had lower SDMA concentrations (1.82 ± 0.84 µg/dl in difference, *P* = 0.0412). Normality was assessed with a Shapiro-Wilk test. A 90% reference interval was created based on normal distribution independence. One suspected outlier was included in this analysis (15 µg/dl).

	SDMA (µg/dl)		
	All	Female	Male
Sample size	32	15	17
Minimum	3	5	3
Maximum	15	15	10
Median	7.5	9	7
Mean	7.78	8.75	6.93
Standard deviation	2.55	2.81	2.00
Normality	Yes	Yes	Yes
Reference interval			
Lower limit	3.59		
90% confidence interval of lower limit	2.30-4.89		
Upper limit	11.97		
90% confidence interval of upper limit	10.67-13.27		

Discussion

The principal goal of this study was to evaluate if the commercially available IA method of measuring SDMA would work for psittacine blood, and to establish species-specific reference ranges for two psittacine species. The reference intervals established here show similar upper limits, suggesting possibly that the upper limit of SDMA in other psittacine species may also be in this range. The effect of sex on SDMA identified in the QP population is not well understood from this data. The QPs are housed in small flocks and engage in social behaviors. It is possible that egg-laying, follicular development, and hormonal influences may influence SDMA that was not recorded in the singly housed HAP. SDMA is not known to have a strong sex discrepancy in other species.[31]

Assessment of renal disease in birds

Renal disease is often diagnosed in avian species. Unfortunately, the diagnosis is frequently determined during postmortem examination.[3,4] Acute renal injury, nephritis, renal neoplasia, and chronic renal disease are examples of disease conditions that occur in psittacine birds.[3,4,7,13] Currently, there are no reliable antemortem blood tests that assess renal function, especially early in the disease process.[5] Glomerular filtration rate is the most reflective of renal function; however, direct measurement of GFR is challenging and not clinically practical, so clinicians and researchers often rely on proxy biomarkers of GFR.[31]

Biomarkers used routinely in mammalian patients are BUN and creatinine. Creatinine, while filtered and actively secreted by the avian kidney, fails as a marker for renal health in birds because birds excrete creatine primarily, before the conversion to creatinine. This leads to plasma creatinine levels below the limit of detection by commonly used assays.[5] While BUN is excreted fully through glomerular filtration in euhydrated birds, reabsorption by the renal tubules is nearly complete during a state of dehydration. BUN is not a sensitive marker for renal disease; however, has been suggested as an indicator of hydration status.[5,10] In addition, while the kidney regulates electrolyte concentrations, and abnormalities or changes in electrolyte concentration can be identified in mammalian renal disease, these changes do not appear predictable across avian species, and the effect on plasma electrolytes by kidney disease is not well-studied in birds.[5,32-34]

UA levels are used to evaluate kidney function in birds. However, this biomarker is unreliable due to a low sensitivity and specificity in avian species. Birds are uricotelic with UA representing 70-80% of nitrogenous waste in birds and is most frequently measured and used as an indicator of renal function in birds.[5,35] The main challenges with interpreting UA levels in birds include the mechanisms of UA excretion and dietary influence. UA is not eliminated by glomerular filtration and therefore not predictive of glomerular filtration rate (GFR). Instead, UA is actively secreted into the proximal tubules and is independent of urine flow. Therefore, blood concentrations of UA reflect the functional capacity of the renal tubules, rather than filtration capability.[14] Because of this, the diagnostic utility of UA is best when screening for or tracking trends in tubular damage as is seen with tubular nephrosis, dysfunctional proximal convoluted tubules, and interstitial nephritis.[1,5] Blood UA concentrations are also influenced by diet. High protein meals common among carnivorous and insectivorous birds are associated with a post-prandial increase in blood UA concentrations.[36] Thus, blood samples destined for UA determination are best obtained from fasted carnivorous or insectivorous avian species for proper interpretation. An experimental trial in which 26 cockatiels (*Nymphicus hollandicus*) were fed isocaloric diets with variable levels of crude protein (CP), found that those fed a high protein diet (70% CP) had a significant increase in plasma UA concentrations without evidence of dietary-induced kidney dysfunction.[37] Another study assessing dietary protein content and plasma UA concentrations found that parakeets fed a moderate protein diet (13.2-25%) did not affect the plasma UA concentrations.[38] The effects of diet and hydration of the UA are influenced heavily by the species and their evolutionary history. Moreover, hyperuricemia only occurs when renal function has dropped to below 30% of the original function, at which point the capacity for clinicians and caretakers to identify the cause of renal insult and intervene is typically unsuccessful.[5]

Additional diagnostic testing commonly used to identify renal disease includes urinalysis, diagnostic imaging, and renal biopsy.[5,11,39-41] Because birds lack a bladder and urine mixes with fecal material in the cloaca, urinalysis via cystocentesis is not possible and urinalysis of voided urine is often non-diagnostic due to fecal contamination. Catheterization of the ureter has been described but is technically difficult and requires anesthesia.[36-38] Urinalysis, when performed with methodologies used in mammalian species, is prone to inaccuracy in birds due to fecal contamination, and the separation of UA from the liquid urine in the form of a colloid suspension.[5] Additionally, in most species, urine, through reverse peristalsis, renters the caudal gastrointestinal tract for water and electrolyte resorption as an adaptation to conserve water.[42] In cases of polyuria, collection of excreted urine is possible; however, collection of a sample uncontaminated by feces is difficult in non-polyuric patients.[5] USG is also highly variable among avian species and has a narrow clinical relevance with the exception of cases in which the USG is repeatedly low or in which the USG fails to increase with water deprivation.[3]

Assessment of urine is also made challenging in birds because the composition of ureteral urine and voided urine vary significantly, unlike that in mammals, due to variation in the structure of avian kidneys, osmoregulation and handling of water. Avian kidneys contain nephrons with and nephrons without a loop of Henle. The loopless nephrons consist of a glomerulus, proximal, and distal tubule and are radially around a central efferent vein.[14] The nephrons are arranged to empty into collecting ducts in a manner the does not allow for the nephrons to utilize the countercurrent multiplier system.[14,43] More complex nephrons with a developed loop of Henle and convoluted proximal tubule exist primarily in the deeper cortex of the kidney.[43] While the ratio of looped nephrons to loopless nephrons varies between species, all studied species, the loopless nephrons outnumber the looped, with approximately 85% being loopless and 15% being looped.[44] There is also great variation in the length of the loop of Henle of looped nephrons, essentially forming a continuum from loopless to large, looped nephrons.[14]

Even with a limited number of looped nephrons, avian kidneys do possess the ability to concentrate urine. Medullary cones containing tubules, loops of Henle, collecting ducts, and vasa recta arranged in parallel, allowing for countercurrent flow of fluid.[43] With UA, the primary nitrogenous waste product in birds, being insoluble in water, it plays no role in the solute composition of the medullary cones. Instead, the concentration of urine in the kidney is controlled primarily by sodium chloride. The avian countercurrent system can only concentrate urine osmolality to 1.5-2.5 times that of plasma, compared to 25-30 times for mammals.[14] Birds conserve water not through concentration of urine in the kidney, but rather through retrograde peristalsis of urine from the cloaca into the colon, where water is reabsorbed. In addition, because urates precipitate out of solution and do not contribute to the osmolarity of urine as urea does, the urine-to-plasma osmolar ratios between birds and mammals cannot be accurately compared because their end-products of nitrogenous waste differ in their contribution to urine osmolality.[14,45]

The kidneys of birds are supplied by three branches of the inferior aorta as well as a venous blood supply through a renal portal system. Blood flow from the renal portal system can be regulated by the autonomous nervous system. A smooth muscle valve in the common iliac vein, when open, allows blood to flow directly into the inferior vena cava. Adrenergic stimulation closes this valve and directs blood flow to the renal parenchyma.[46]

Birds have glomeruli that when compared individually to those in mammals, have a lower single nephron glomerular filtration rate (SNGFR); however, total GFR between mammals and birds of similar size is nearly comparable, with the disparity offset by birds having a larger number of nephrons. To complicate this, whole-kidney GFR in birds is highly variable between species and influenced greatly by dehydration with a decrease in GFR as high as 65% compared to GFR in a euhydrated bird.[14]

The current reference standard for antemortem diagnosis of renal disease in birds is a renal biopsy; however, this procedure necessitates the use of general anesthesia, training, and specialized equipment.[24] Moreover, performing an endoscopic renal biopsy is not practical or possible for all clinicians or small patients and is unsafe for serial monitoring or critical patients. In addition, collecting renal biopsy samples is only possible in areas visualized by the endoscope and the amount of renal tissue biopsied is small resulting in a concern of missing the lesion and, or poor sensitivity of this method. Identifying a sensitive, specific, and non-invasive tool for assessing kidney function in birds is of paramount importance to the veterinary community for antemortem assessment of renal health.

SDMA measurement methodology

Liquid chromatography-mass spectrometry (LC-MS) is considered the reference standard for measurement of SDMA due to its accuracy in evaluated species; however, this methodology is more expensive, more time-consuming, and less readily accessible compared to the IA methodology.[47] A study investigating plasma levels of SDMA in chickens using both the reference standard liquid chromatography-tandem mass spectrometry (LC-MS/MS) and a commercially available IA to determine SDMA concentrations found a low correlation between IA and the LC-MS/MS.[29] The IA used in this study serves as a potential source of bias as SDMA levels using LC-MS/MS were not measured; however, the commercially utilized IA is the most used methodology in canine and feline patients due to increased accessibility and decreased cost compared to the LC-MS/MS. Deviations in plasma SDMA concentrations between the LC-MS/MS and the IA methodology exist in dogs and cats; however, the analyte is still routinely measured using the IA in clinical settings. In addition, published studies measuring SDMA as proof-of-concept in other species such as large felids (*Acinonyx jubatus*, *Panthera tigris*),[48,49] the Hermann's tortoise (*Testudo hermanii*),[50] and the greater flamingo (*Phoenicopterus* roseus)[28] have utilized the IA methodology. While the reference intervals established in chickens showed disagreement between the two methodologies, further work is needed to investigate the source and clinical significance of bias because the IA methodology remains the most accessible to clinicians today. Unless the IA methodology fails to be validated when measured in birds with known or induced renal disease, or the LC-MS/MS methodology is investigated in psittacine patients, validated, and made commercially available, the IA methodology remains the most practical for establishing a species-specific reference range.

Study limitations

A limitation of this study is the small sample size used for each population. Age and sex were not known for all members of the HAP population, limiting sex or age-related relationships within the population. Comparison of birds with known kidney disease to those without kidney disease, either naturally acquired or clinically induced, was also beyond the scope of this study. In addition, the IDEXX IA was not validated for either species in this study.

Clinical relevance

An annual complete blood count and plasma biochemistry panel is often a core component of wellness examinations for both companion birds and birds in zoological collections. SDMA, or other sensitive plasma markers of kidney dysfunction in birds, if validated, could serve as a key component of screening diagnostics, and a powerful indicator for tracking the response to treatment and or progression of kidney disease in birds. Establishing proof-of-concept and assessing the validity of SDMA in non-mammalian species is an understudied, and necessary area of research for veterinarians. Establishing a reference range for SDMA in more avian species, as well as validating the analyte's predictive power in varying stages of disease, will improve the diagnostic capabilities of clinicians with avian patients and improve the quality of life of these patients by offering an early detection option for renal diseases while providing the diagnostic information necessary to develop viable treatment options. SDMA reference ranges have successfully been established in greater flamingos (*Phoenicopterus roseus*) and domestic chickens (*Gallus gallus*).[32,33] Increased SDMA concentrations using the IA methodology in the flamingo study correlated with elevations in UA and renal lesions identified during postmortem examination. Validation work to show how plasma SDMA levels change in patients with known renal disease, or comparison of plasma SDMA to histopathology findings of the kidneys is a necessary next step in assessing the clinical utility of this biomarker in predicting kidney function in our avian patients.

For both HAP and QP, SDMA concentrations may provide a clinically useful reference range for the two psittacine species, and if this analyte is validated in these species, these reference ranges will pose as tool for clinicians and researchers working with these species. The upper end of the SDMA reference interval of the HAP and QP is similar to that of other animal species, which suggests clinical usefulness of this analyte.[27,28,32,33] Interestingly, no correlation was found between UA and SDMA in either population. Patients were deemed healthy based on physical examination; however, no confirmatory testing (e.g., histopathology, glomerular filtration rate, or imaging) was performed to screen for kidney disease within either population. The single outlier in the QP population may represent a true outlier or a patient with subclinical kidney disease. Comparison of SDMA and UA in patients with documented kidney disease would better serve to compare the predictive power of SDMA in screening for and monitoring kidney disease. Further investigation into the predictive power of SDMA in avian patients with varying degrees of kidney dysfunction is required to further identify the clinical use of this analyte.

References

1. Siller WG. Renal pathology of the fowl -- a review. *Avian Pathol.* 1981;10:187-262.

2. Schneider RR, Hunter DB, Waltner-Toews D, Barker IK. A descriptive study of mortality at the Kortright waterfowl park: 1982-1986. *Can Vet J.* 1988;29:911-914.

3. Phalen D. Avian renal disorders. In: Fudge AM, ed. *Laboratory Medicine Avian and Exotic Pets.* Philadelphia, PA: WB Saunders; 2000:61-68.

4. Drury RR, Dorrestein G. Psittacines, Coliiformes, Musophagiformes, Cuculiformes. In: Terio KA, McAloose D, St. Leger J, eds. *Pathology of Wildlife and Zoo Animals.* Cambridge, MA: Academic Press; 2018: 775-798.

5. Scope A, Schwendenwein I. Laboratory evaluation of renal function in birds. *Vet Clin North Am Exot Anim Pract.* 2020;1:47-58.

6. Echols MS. Evaluating and treating the kidneys. *Clin Avian Med.* 2006;2:451-492.

7. Van Toor AJ, Zwart P, Kaal GT. Adenocarcinoma of the kidney in two budgerigars. *Avian Pathol.* 1984;13:145-150.

8. Hokamp JA, Nabity MB. Renal biomarkers in domestic species. *Vet Clin Pathol.* 2016;45(1):28-56. doi: 10.1111/vcp.12333.

9. Cojean O, Larrat S, Vergneau-Grosset C. Clinical management of avian renal disease. *Vet Clin North Am Exot Anim Pract.* 2020;23(1):75-101.

10. Lumeij JT. Plasma urea, creatinine, and uric acid concentrations in response to dehydration in racing pigeons. *Avian Pathol.* 1987;16:377-382.

11. Lierz M. Avian renal disease: pathogenesis, diagnosis, and therapy. *Vet Clin North Am Exot Anim Pract.* 2003;6(1):29-55.

12. Fudge AM. Avian clinical pathology-hematology and chemistry: In: Altman RB, Clubb SL, Dorrestein GM, eds. *Avian Medicine and Surgery.* Philadelphia, PA: WB Saunders; 1997:142-157.

13. Orosz S, Dorrestein GM, Speer BL. Urogenital disorders. In: Altman RB, Clubb SL, Dorrestein GM, eds. Avian Medicine and Surgery. Philadelphia: WB Saunders; 1997:614-644.

14. Braun EJ. Osmoregulatory systems of birds. In: Scanes C, ed. *Sturkie's Avian Physiology.* London, UK: Academic Press; 2015: 285-300.

15. Scanes CG, Witt J, Ebeling M, Heckmann D. Quantitative comparison of avian and mammalian physiologies for parameterization of physiologically based kinetic models. *Front Physiol.* 2022;13:858386.

16. Bedford MT, Richard S. Arginine methylation an emerging regulator of protein function. *Mol Cell.* 2005;18:263-272.

17. Schwedhelm E, Böger RH. The role of asymmetric and symmetric dimethylarginines in renal disease. *Nat Rev Nephrol.* 2011;7:275-285.

18. Bode-Boger SM, Scalera F, Kielstein JT, et al. Symmetrical dimethylarginine: A new combined parameter for renal function and extent of coronary artery disease. *J Am Soc Nephrol.* 2006;17:1128-1134.

19. Dixon JJ, Lane K, Dalton RN, et. al. Symmetrical dimethylarginine is a more sensitive biomarker of renal dysfunction than creatinine. *J Crit Care.* 2013;17:423.

20. Nabity MB, Lees GE, Boggess M, et al. Symmetric dimethylarginine assay validation, stability, and evaluation as a marker for early detection of chronic kidney disease in dogs. *J Vet Intern Med.* 2015;29(4): 1036-1044.

21. Hall JA, Yerramilli M, Obare E, et al. Comparison of serum concentrations of symmetric dimethylarginine and creatinine as kidney function biomarkers in cats with chronic kidney disease. *J Vet Intern Med.* 2014;28(6):1676-1683.

22. Hall JA, Yerramilli M, Obare E, et al. Serum concentrations of symmetric dimethylarginine and creatinine in dogs with naturally occurring chronic kidney disease. *J Vet Intern Med.* 2016;30(3):794-802.

23. Hall JA, Yerramilli M, Obare E, et al. Comparison of serum concentrations of symmetric dimethylarginine and creatinine as kidney function biomarkers in healthy geriatric cats fed reduced protein foods enriched with fish oil, L-carnitine, and medium-chain triglycerides. *Vet J.* 2014;202(3):588-596.

24. Nabity MB, Lees GE, Boggess M, et al. Symmetric dimethylarginine assay validation, stability, and evaluation as a marker for early detection of chronic kidney disease in dogs. *J Vet Intern Med.* 2015;29(4): 1036-1044.

25. Hall JA, Yerramilli M, Obare E, et al. Comparison of serum concentrations of symmetric dimethylarginine and creatinine as kidney function biomarkers in cats with chronic kidney disease. *J Vet Intern Med.* 2014;28(6):1676-1683.

26. Hall JA, Yerramilli M, Obare E, et al. Serum concentrations of symmetric dimethylarginine and creatinine in dogs with naturally occurring chronic kidney disease. *J Vet Intern Med.* 2016;30(3):794-802.

27. Hamlin DM, Schultze AE, Coyne MJ, et al. Evaluation of renal biomarkers, including symmetric dimethylarginine, following gentamicin-induced proximal tubular injury in the rat. *Kidney360.* 2021;3(2):341-356.

28. van Zanten TC, Xie S. Evaluation of symmetric dimethylarginine as an endogenous marker of renal disease in greater flamingos (*Phoenicopterus roseus*). *J Zoo Wildl Med.* 2023;53(4):644-653.

29. Kane LP, Keller KA, Murphy R, et al. Adult chicken hens (*Gallus gallus*) have measurable circulating plasma symmetric dimethylarginine via liquid chromatography-tandem mass spectrometry. *Am J Vet Res.* 2023;84(9):ajvr.22.12.0222.

30. American Society of Veterinary Clinical Pathologists. ASVCP Quality Assurance and Laboratory Standards Committee (QALS) Guidelines for the Determination of Reference Intervals in Veterinary Species and other related topics: SCOPE. Available at: https://cdn.ymaws.com/www.asvcp.org/resource/resmgr/QALS/Other_Publications/RI_Guidelines_For_ASVCP_webs.pdf. Accessed February 1, 2024.

31. Sargent HJ, Elliott J, Jepson RE. The new age of renal biomarkers: does SDMA solve all of our problems? *J Small Anim Pract.* 2021;62:71-81.

32. Blaxland JD, Borland ED, Siller WG, Martindale L. An investigation of urolithiasis in two laying flocks of fowls. *Avian Pathol.* 1980;9:5-19.

33. Wideman RF, Mallinson ET, Rothenbacher H. Kidney function of pullets and laying hens during outbreaks of urolithiasis. *Poult Sci.* 1983;62:1954-1970.

34. Pegram RA, Wyatt RD, Smith TL. Oosporein-toxicosis in the turkey poult. *Avian Dis.* 1982;26:47-59.

35. Radin MJ, Swayne DE, Gigliotti A, Hoepf T. Renal function and organic anion and cation transport during dehydration and/or food restriction in chickens. *J Comp Physiol B.* 1996;166:138-143.

36. Lumeij JT, Remple JD. Plasma urea, creatinine, and uric acid concentrations in relation to feeding in peregrine falcons (*Falco peregrinus*). *Avian Pathol.* 1991;20:79-83.

37. Koutsos EA, Klasing KC, Smith J, Woods LW. Adult cockatiels (*Nymphicus hollandicus*) metabolically adapt to high protein diets. *J Nutr.* 2001;131:2014-2020.

38. Angel R, Ballam G. Dietary protein effect on parakeet reproduction, growth and uric acid. *Proc Annu Conf Nutr Adv Grp Am Zoo Aquar Assoc.* 1995.

39. Lumeij JT. Pathophysiology, diagnosis and treatment of renal disorders in birds of prey. In: Lumeij JT, Remple D, Redig PT, et al., eds. *Raptor Biomedicine III.* Lake Worth, FL: Zoological Education Network; 2000:169-178.

40. Hofbauer H, Krautwald-Junghanns M-E. Transcutaneous ultrasonography of the avian urogenital tract. *Vet Radiol Ultrasound.* 1999;40(1):58-64.

41. Radin MJ, Hoepf TM, Swayne DE. Use of a single injection solute-clearance method for determination of glomerular filtration rate and effective renal plasma flow in chickens. *Lab Anim Sci.* 1993;43(6):594-596.

42. Williams JB, Braun EJ. Renal compensation for cecal loss in Gambel's quail (*Callipepla gambelii*). *Comp Biochem Physiol.* 1996;113A(4):333-341.

43. Castotti G, Lindberg KK, Braun EJ. Functional morphology of the avian medullary cone. *Am J Physiol Regul Integr Comp Physiol.* 2000;279:R1722-R1730.

44. Braun EJ, Dantzler WH. Function of mammalian-type and reptilian-type nephrons in kidney of desert quail. *Am J Physiol.* 1972;222(3):617-629.

45. Taylor MW. Clinical significance of the avian cloaca: interrelationships with the kidneys and the hindgut. In: Speer BL, ed. *Current Therapy in Avian Medicine and Surgery.* St. Louis, MO: Elsevier; 2016:329-344.

46. Burrows ME, Braun EJ, Duckles SP. Avian renal portal valve: a reexamination of its innervation. *Am J Physiol.* 1983;245(4):H628-H634.

47. Relford RL, Robertson JE, Clements CA. Symmetric dimethylarginine: improving the diagnosis and staging of chronic kidney disease in small animals. *Vet Clin North Am Small Anim Pract.* 2016;46(6):941-60.

48. Waugh L, Lyon S, Cole GA, et al. Retrospective analysis and validation of serum symmetric dimethylarginine (SDMA) concentrations in cheetahs (*Acinonyx jubatus*). *J Zoo Wildl Med.* 2018;49(3):623-631. doi:10.1638/ 2017-0056.1.

49. Mota SM, Brandao J, Guthrie A. Comparison of blood symmetric dimethylarginine and creatinine as endogenous markers of kidney function in captive tigers (*Panthera tigris*). *J Zoo Wildl Med.* 2021;52(3):628-637.

50. Lehmann V, Altherr B, Pantchev N, et al. Symmetrical dimethylarginine as a diagnostic parameter in Hermann's tortoises (*Testudo hermanii*). *Front Vet Sci.* 2022;22:9:824748. doi:10.3389/fvets.2022.824748.

What We (Don't) Know About Parrot Welfare: A Systematic Literature Review

Andrea Piseddu, MSc,
Yvonne van Zeeland, DVM, MVR, PhD, DipECZM (Avian),
DipECZM (Small Mammal), CPBC,
Jean-Loup Rault, PhD

Session #177

Affiliation: From the Institute of Animal Welfare Science, University of Veterinary Medicine Vienna, Veterinaerplatz 1, 1210 Vienna, Austria (Piseddu, Rault), Division of Zoological Medicine, Department of Clinical Sciences of Companion Animals, Faculty of Veterinary Medicine, Utrecht University, Utrecht, The Netherlands (van Zeeland).

Abstract: Parrots are popular companion animals but show prevalent, and at times severe, welfare issues. Nonetheless, there are no scientific tools available to assess parrot welfare. The aim of this systematic review was to identify valid and feasible outcome measures that could be used by caregivers as welfare indicators for companion parrots. From 1848 peer-reviewed studies retrieved, 98 studies met our inclusion and exclusion criteria (i.e., experimental welfare studies, captive parrots). For each outcome collected, validity was assessed based on the statistical significance reported by the authors, as other validity parameters[1] were rarely available for evaluation. Feasibility was determined by considering the need for specific instruments, veterinary-level expertise or handling the parrot. A total of 1512 outcomes were evaluated, of which 572 had a significant p-value and were considered feasible. These included changes in behavior (e.g., social interactions, facial and body displays, activity level, flying/climbing, exploration, foraging, maintenance behaviors), body measurements (e.g., body weight, plumage condition) and abnormal behaviors (e.g., phobic and fear-related behaviors, incessant screaming, oral, whole body or locomotor sterotypies), amongst others. A high risk of bias due to underreporting of intra- and inter-rater reliabilities and assessor blindness undermined the internal validity of these outcomes. Moreover, a strong taxonomic bias, a predominance of studies on parrots in laboratories, and an underrepresentation of companion parrots jeopardized their external validity. Nevertheless, this systematic review helped to summarize the current state of scientific knowledge on aspects and provides a promising starting point for validating a set of welfare indicators in parrots.

References

1. Hooijmans CR, Rovers MM, De Vries RB, et al. SYRCLE's risk of bias tool for animal studies. *BMC Med Res Methodol.* 2014;14(43):1-9.

Hepatopathy and Dystrophic Calcification in a 20-Year-Old Eclectus Parrot (*Eclectus roratus*)

Rachael Hill, DVM, and
Bianca Murphy, DVM, DABVP (Avian Practice)

Session #183

Affiliation: From Aloha Pet and Bird Hospital, 968 East Eau Gallie, Satellite Beach, FL 32937, USA.

Abstract: A 20-y-old male eclectus parrot presented due to progressive hyporexia and lethargy. Initial diagnostics included hematology, biochemistry, and sedated orthogonal radiographs. Hematologic assays identified a relative heterophilia and moderately elevated bile acids. Radiographic findings included diffuse, multifocal regions of mineralization, caudoventral displacement of the soft tissues of the intestinal peritoneal cavity and decreased air sac visibility secondary to a markedly distended proventriculus. Contrast fluoroscopy identified severe hypomotility, intestinal dilation, and confirmed that all regions of mineralization were not associated with the GI tract. Further antemortem screening included avian bornavirus PCR, *Mycobacterium* sp. PCR, and AGAA serology, all of which were negative. Despite parenteral supportive care, this bird continued to decline and was euthanized 11 days after presentation. Gross necropsy findings were impressive linking the regions of dystrophic calcification to multifocal regions of granulomas, severe hepatomegaly, and diffuse gas dilation of the entirety of the GI tract. Histopathology identified pyogranulomatous coelomitis with acid-fast organisms and chronic cholangiohepatitis. Radiographically visible dystrophic calcification is currently under reported in avian literature. Therefore, these lesions require the consideration of wide-ranging differential categories notably neoplastic and infectious etiologies. This case provides a specific pathogen to consider in parrots with this finding.

References

1. Donovan TA, Garner MM, Phalen D, et al. Disseminated coelomic xanthogranulomatosis in eclectus parrots (*Eclectus roratus*) and budgerigars (*Melopsittacus undulatus*). *Vet Pathol.* 2021;59(1):143-151. doi: 10.1177/03009858211045931.

2. McRee AE, Higbie CT, Nevarez JG, et. al. Mycobacteriosis in captive psittacines: a brief review and case series in common companion species (*Eclectus roratus*, *Amazona oratrix*, and *Pionites melanocephala*). *J Zoo Wildl Med.* 2016;48(3):851-858. doi: 10.1638/2016-0176.1.

3. Speer B. *Current Therapy in Avian Medicine and Surgery.* St. Louis, MO: Elsevier; 2016.

Male and Female Hispaniolan Amazon Parrots (*Amazona ventralis*) Differ in Their Corticosterone Response to Sedatives

Meridith L. Helms-Pack, BS,
Thomas N. Tully, Jr., DVM, MS, DABVP (Avian Practice), DipECZM (Avian),
Breonna S. Freeman, MBS,
Keegan R. Stansberry, BS,
Chin-Chi Liu, MS, MApStat, PhD,
John A. Tuminello, BS,
Christine R. Lattin, PhD

Session #106

Affiliation: From the Louisiana State University School of Veterinary Medicine, Skip Bertman Dr., Baton Rouge, LA 70803, USA (Helms-Pack, Tully, Freeman, Liu, Tuminello), Louisiana State University Biological Sciences, 202 Life Sciences Bldg., Baton Rouge, LA 70803, USA (Stansberry, Lattin).

Abstract: A previous study showed that plasma corticosterone levels significantly increased when Hispaniolan Amazon parrots (*Amazona ventralis*) were restrained.[1] The sedatives midazolam and butorphanol are being used more often in veterinary medicine to improve the experience of patient handling for both the patient and caregiver.[2] It was hypothesized that a clinically relevant dose of 2 mg/kg of both midazolam and butorphanol administered intranasally would significantly decrease the plasma corticosterone response in restrained Hispaniolan Amazon parrots (n = 11) when compared to controls (n = 11). Ten male and 12 females weighing 250-350 g, of unknown age were used in this study. Blood samples of 0.5 ml from the right jugular vein were taken < 3 min upon entering the room and 15 and 30 min after treatment administration. Corticosterone was extracted from the blood plasma using ELISA plates (Arbor Assays [K014-H1]). Data were logarithmically transformed to meet normality criteria. The normality of residuals from the models was confirmed using Shapiro-Wilk tests. When a fixed effect was detected, Tukey's post hoc comparisons were performed with least square means. Significance was set at $P < 0.05$. When we analyzed results by sex, we found that males that received sedatives (n = 3) had a significantly decreased plasma corticosterone response to restraint than male controls (n = 7), female controls (n = 4), and females that received sedatives (n = 8). These results suggest a sex-specific effect in the metabolism of midazolam and butorphanol comparable to results that have been seen in American kestrels (*Falco sparverius*).[3] Pharmacokinetic studies on these sedatives with respect to sex is recommended.

References

1. Parks SN, Tully Jr. TN, Settle AL, Lattin CR. Handling and restraint induce a significant increase in plasma corticosterone in Hispaniolan Amazon parrots (*Amazona ventralis*). *Am J Vet Res. 84(5):ajvr.22.12.0223.*

2. Doss G, Mans C. Avian sedation. *J Avian Med Surg*. 2021;35(3):253-268.

3. Guzman DSM, Drazenovich TL, KuKanich B, et al. Evaluation of thermal antinociceptive effects and pharmacokinetics after intramuscular administration of butorphanol tartrate to American kestrels (*Falco sparverius*). *Am J Vet Res*. 2014;75(1):11-18.

Diagnostic Value of Serum Amyloid A in Differentiating the Inflammatory Disorders in Birds

Haerin Rhim, DVM, MS, PhD,
Myungsoo Kim, DVM, MS,
Seulgi Gim,
Jae-Ik Han, DVM, MS, PhD, DACCM

Session #138

Affiliation: From the Louisiana State University School of Veterinary Medicine, Skip Bertman Dr., Baton Rouge, LA 70803, USA (Rhim), Jeonbuk Wildlife Center, Jeonbuk National University, Gobong-ro 79, Iksan, 54596, Republic of Korea (Kim, Gim, Han).

Note: Previously published in Rhim H, Kim M, Gim S, Han J-I. Diagnostic value of serum amyloid A in differentiating the inflammatory disorders in wild birds. *Front Vet Sci.* 2024;11:1284113.

Abstract: It is very common for rescued birds to have inflammation caused by trauma or infection. Nonetheless, sometimes it is not easy to determine the presence of potential diseases through physical examination due to the nature of hiding clinical symptoms. The possibility of serum amyloid A (SAA) for diagnosing inflammation in birds is known,[1-4] but there is no information on the extent of the increase in disease in a large number of wild birds. In this study, the value of SAA as a diagnostic marker was evaluated based on the severity of diseases. A total of 291 plasma samples from six species were included. All samples were divided into four groups: severe, moderate, and mild injuries, and clinically healthy based on clinical examinations and evaluated by groups. SAA was measured by a chicken SAA ELISA kit (Eagle Bioscences, Amherst, NH, USA), and compared with other indices including leukocyte count, albumin, and albumin/globulin ratio. The difference between groups was evaluated by Kruskal-Wallis test followed by post-hoc tests using Mann-Whitney test with Bonferroni correction. The SAA concentration in each group was significantly different between groups in all species ($P < 0.05$). The more severe the injuries, the higher SAA level and WBC counts, and the lower albumin and A/G ratio. In subjects for whom follow-up tests were performed, SAA decreased rapidly following treatment while some birds showed a rapid increase before other traditional indicators altered or clinical symptoms appeared. This indicates that SAA can be used as a useful diagnostic biomarker for more detailed evaluation of patients.

References

1. Chamanza R, Toussaint MJ, van Ederen AM, et al. Serum amyloid A and transferrin in chicken. A preliminary investigation of using acute-phase variables to assess diseases in chickens. *Vet Quart.* 1999;21:158-162.

2. Pal Singh P, Singh S, Sunil Kumar BV, et al. Serum amyloid A (SAA) mRNA expression in chicken and quails in response to bacterial stress. *Comp Immunol Microbiol Infect Dis.* 2020;69:101421.

3. Caliendo V, McKinney P, Bailey T, et al. Serum amyloid A as an indicator of health status in falcons. *J Avian Med Surg.* 2013;27:83-89.

4. Kovács BM, Toussaint MJ, Gruys E, et al. Evaluation of goose serum amyloid A acute phase response by enzyme-linked immunosorbent assay. *Acta Veterinaria Hungarica.* 2007;55:349-357.

Chemotherapeutic Management of a Testicular Tumor in a Cockatiel

Natalie Antinoff, DVM, DABVP (Avian Practice)

Session #144

Affiliation: From Texas Avian & Exotic Hospital, 2700 W. State Hwy 114, Grapevine, TX 76051, USA.

Abstract: A 17-y-old 89-g male cockatiel was presented for respiratory distress. A physical exam revealed coelomic distension and respiratory crackles. A focused assessment with sonography in trauma (FAST) scan revealed a soft tissue density and absence of free fluid. A barium GI series was performed and identified a 4.0 cm × 2.3 cm ovoid soft tissue opaque mass in the caudodorsal coelom, with ventral displacement of the gastrointestinal tract. Fine needle aspiration was performed but cytology was non-diagnostic. There was no response to Lupron and meloxicam therapy and a course of furosemide and enrofloxacin was initiated, again with no change in condition. After 7 wk of no response to treatment, the patient's weight had increased to 95 g and the mass was palpably nodular. As a trial, a course of chlorambucil chemotherapy (10 mg/kg PO twice/wk) was initiated and continued for 2 y 7 mo, until the time of death. One month after initiation of the chemotherapy, the owner reported less distress, the patient tolerated handling well, and the coelomic distension had decreased although the mass was still palpable. Six months after the initial presentation, repeat radiographs revealed no evidence of the previously identified mass, and bloodwork was within normal limits. Rechecks were performed quarterly and the patient lived for 2 y and was ultimately euthanized at 2 y 11 mo after initial presentation due to recurrence of the mass. Necropsy identified the mass as a seminoma with metastasis.

References

1. Gosbell M, Luk K, Macwhirter P. Chronic lymphocytic leukemia in a cockatiel (*Nymphicus hollandicus*). *J Avian Med Surg.* 2021;35(3):341-349.

2. Saled A, Beaufrere H, Tully T, et al. Bilateral seminoma with hepatic metastasis in a cockatiel (*Nymphicus hollandicus*). *J Avian Med Surg.* 2011;25(2):126-131.

Pharmacokinetics of a Single-Dose Oral Pravastatin in Cockatiels (*Nymphicus hollandicus*)

Marina Liles, DVM,
David Sanchez-Migallon Guzman, LV, MS, DipECZM (Avian), DipECZM (Small Mammal), Dipl ACZM,
Hugues Beaufrère, DVM, PhD, Dipl ACZM, DABVP (Avian Practice), DipECZM (Avian)

Session #133

Affiliation: From Department of Veterinary Medicine and Epidemiology, School of Veterinary Medicine, University of California-Davis, Davis, CA 95616, USA.

Abstract: Dyslipidemias are prevalent in companion psittacine species. Lipid accumulation can also lead to hepatic lipidosis, adipocytic tumors, and atherosclerotic diseases, and are among the most common conditions in geriatric parrots. Parrots from the genera *Amazona*, *Nymphicus*, and *Psittacus* are particularly affected by atherosclerosis. Statin drugs are the mainstay medical treatment in human therapy, and pravastatin administered to hypercholesterolemic mice and humans decreased cholesterol levels by up to 37%. Pravastatin is a short half-life statin and is considered less effective than long half-life statins such as atorvastatin and rosuvastatin. Unlike other statin drugs, pravastatin has the advantage of being water-soluble and is the most water-soluble of all statins. This study aims to report plasma concentrations of pravastatin in cockatiels at varying doses and subsequently use that data to evaluate if similar efficacy can be reached by administering pravastatin in the drinking water in a separate investigation. A pilot study of 12 cockatiels was performed to determine the lowest pravastatin dose necessary to reach target plasma concentrations. Data were being analyzed at the time of writing and were subsequently used to perform a complete pharmacokinetics study with 14 cockatiels.

References

1. Beaufrere H, Ammersbach M, Reavill DR, et al. Prevalence of and risk factors associated with atherosclerosis in psittacine birds. *J Am Vet Med Assoc.* 2013;242(12):1696-1704.

2. Lorza-Gil E, Salerno AG, Wanschel AC, et al. Chronic use of pravastatin reduces insulin exocytosis and increases beta-cell death in hypercholesterolemic mice. *Toxicology.* 2016;344-346:42-52.

3. Hadjiisky P, Hermier D, Truffert J, et al. Effect of pravastatin, a HMG CoA reductase inhibitor, on blood lipids and aortic lipidosis in cholesterol-fed white Carneau pigeons. *Biochim Biophys Acta.* 1993;1181(3): 279-286.

4. Hague WE, Simes J, Kirby A, et al. Long-term effectiveness and safety of pravastatin in patients with coronary heart disease: sixteen years of follow-up of the LIPID study. *Circulation.* 2016;133(19):1851-1860.

Surgical Arthrodesis of the Intertarsal Joint in a Green-Cheeked Conure (*Pyrrhura molinae*)

Marcie Logsdon, DVM

Session #165

Affiliation: From Washington State University Veterinary Teaching Hospital, 205 Ott Rd., Pullman, WA 99164, USA.

Abstract: A 3-mo-old green-cheeked conure (*Pyrrhura molinae*) presented for evaluation of a congenital malformation of the right tarsus characterized by lateral luxation of the gastrocnemius tendon and associated contracture of the tarsus and pronation of the foot and tarsometatarsus. There was focal erythema and loss of texture on the dorsomedial aspect of the foot and hock from abnormal wear. The defect was reportedly present at the time of hatch. Arthrodesis of the tarsal joint was performed with a modified cross-pinning technique. A 0.09-mm (0.035-inch) intramedullary pin was used to disrupt the articular cartilage. The distal leg was manually realigned. A 0.071-mm (0.028-inch) K-wire pin was placed from the proximal-lateral aspect of the tarsometatarsus across the tarsal joint exiting the distal-medial aspect of the tibiotarsus. A 27-gauge hypodermic needle was used as a second cross pin, inserted from the distal/posterior aspect of the tibiotarsus into the proximal tarsometatarsus. Dental radiographs were used to confirm pin placement. A small porous tape-skin glue splint was used to stabilize the joint. Pins were removed at 2 wk and radiographs showed progressive remodeling; however, there was still a small amount of palpable motion at the tarsal joint. The splint was replaced for another 2 wk. At 4 wk post procedure, the joint was stable on palpation. There was minor lateral deviation at the tarsus and decreased digit function; however, the bird ambulated appropriately on the plantar aspect of the foot. Three months post procedure the patient was doing well and using the foot to walk and climb in its enclosure.

References

1. Meij BP, Hazewinkel HAW, Westerhof I. Treatment of fractures and angular limb deformities of the tibiotarsus in birds by type II external skeletal fixation. *J Avian Med Surg.* 1996;10(3):153-162. Available at: http://www.jstor.org/stable/30133089. Accessed March 1, 2024.

Baseline Thyroxine Values and Thyroid-Stimulating Hormone (TSH)-Stimulation Test with Commercially Available Recombinant Human TSH (rhTSH) in Quaker Parrots (*Myiopsitta monachus*)

Lauren Thielen, DVM, DABVP (Avian Practice),
Sharman Hoppes, DVM, DABVP (Avian Practice),
Nicola Di Girolamo, DVM, PhD, DipECZM (Herpetology), Dipl ACZM, Dipl ACVPM

Session #179

Affiliation: From Cornell University Hospital for Animals, Ithaca NY 14853, USA (Thielen, Di Girolamo), Texas Avian & Exotic Hospital, Grapevine, TX 76051, USA (Hoppes).

Abstract: We evaluated the effect of injections of commercially available recombinant human thyroid-stimulating hormone (rhTSH, Thyrogen) on plasma thyroxine (T4) concentration in healthy Quaker parrots (*Myiopsitta monachus*) to develop a test for diagnosis of hypothyroidism.[1-5] A randomized controlled trial was conducted with changes in T4 as the primary outcome measure. Thirty-two healthy quaker parrots were randomized to receive fresh human rhTSH (8), high-dose frozen human rhTSH (8), low-dose frozen human rhTSH (8), and 0.9% NaCl (8). Venipuncture was performed before and 4 hours post injection for measurement of total T4 by radioimmunoassay analysis and TSH. A TSH stimulation test was completed in 31 of the 32 birds. Median T4 at baseline in the population was 0.14 μg/dl (0.03-0.33 μg/dl). All birds that received rhTSH had an increase in total T4 from baseline. On average, parrots receiving high-dose frozen rhTSH had an increase of 905%. Parrots receiving low-dose frozen rhTSH had an increase of 1185.23%. Parrots receiving high-dose fresh rhTSH had an increase of 944.75%. Parrots receiving control saline had, on average, a 41.7% decrease (0.04-0.14 μg/dl) in T4. All parrots that received rhTSH and had TSH measured after injection (n = 21) had plasma concentrations of TSH > 12.0 ng/ml, while parrots that received saline had TSH ranging from < 0.03 to 0.033 ng/ml. The TSH stimulation test performed with fresh and frozen rhTSH at the doses used in this study resulted in significant increases in T4 in Quaker parrots.

References

1. Schmidt RE, Reavill DR. The avian thyroid gland. *Vet Clin North Am Exot Anim Pract.* 2008;11:15-23.

2. Cole RK. Hereditary hypothyroidism in the domestic fowl. *Genetics*. 1966;53(6):1021.

3. Oglesbee BL. Hypothyroidism in a scarlet macaw. *J Am Vet Med Assoc.* 1992;201:1599-1601.

4. Greenacre CB. Thyrogen for use in TSH testing of healthy and suspected hypothyroid parrots. *Proc Annu Conf Eur Assoc Avian Vet.* 2009:125-126.

5. Greenacre CB, Olsen J, Wilson GH, et al. Thyroid gland function testing with human synthetic TSH. *Proc Annu Conf Assoc Avian Vet.* 2002.

Measurement of Free and Total Thyroxine in Quaker Parrots (*Myiopsitta monachus*)

Lauren Thielen, DVM, DABVP (Avian Practice),
Sharman Hoppes, DVM, DABVP (Avian Practice),
Nicola Di Girolamo, DVM, PhD, DipECZM (Herpetology),
Dipl ACZM, Dipl ACVPM

Session #189

Affiliation: From Cornell University Hospital for Animals, Ithaca, NY 14853, USA (Thielen, Di Girolamo), Texas Avian & Exotic Hospital, Grapevine, TX 76051, USA (Hoppes).

Abstract: Equilibrium dialysis is widely acknowledged as the preferred method for measuring serum-free thyroxine (fT4) levels in humans and animals.[1-5] Radioimmunoassay (RIA) and chemiluminescent techniques are commonly used to measure total thyroxine (TT4) levels in dogs. The purpose of the present study was to establish agreement between (1) free thyroxine measured by chemiluminescence and equilibrium dialysis and (2) total thyroxine measured by chemiluminescence and RIA in Quaker parrots (*Myiopsitta monachus*). As a part of a randomized trial on TSH stimulation testing in Quaker parrots, plasma levels of fT4 were measured by chemiluminescence and radioimmunoassay equilibrium dialysis. Standard agreement analyses were conducted. Paired TT4 measurements were obtained on 63 samples, while paired fT4 measurements were obtained on 52 samples. Median TT4 as measured by RIA was 0.15 µg/dl (0.03-1.98 µg/dl) and by chemiluminescence was 0.05 µg/dl (0.05-1.66 µg/dl). Bland-Altman plots showed a mean difference of -0.18 µg/dl (95% limits of agreement: 0.23 to -0.59 µg/dl) between the two methods. Median fT4 as measured by equilibrium dialysis was 0.85 ng/dl (0.15-5.05 ng/dl) and by chemiluminescence was 0.43 ng/dl (0.13-3.09 ng/dl). Bland-Altman plots showed a mean difference of -0.59 ng/dl (95% limits of agreement: 0.75 to -1.93 ng/dl) between the two methods. Both free and total thyroxine are significantly underestimated when measured by chemiluminescence compared to equilibrium dialysis and RIA.

References

1. Schmidt RE, Reavill DR. The avian thyroid gland. *Vet Clin North Am Exot Anim Pract.* 2008;11:15-23.

2. Cole RK. Hereditary hypothyroidism in the domestic fowl. *Genetics*. 1966;53(6):1021.

3. Oglesbee BL. Hypothyroidism in a scarlet macaw. *J Am Vet Med Assoc.* 1992;201:1599-1601.

4. Greenacre CB. Thyrogen for use in TSH testing of healthy and suspected hypothyroid parrots. *Proc European Assoc Avian Vet.* 2009:125-126.

5. Greenacre CB, Olsen J, Wilson GH, et al. Thyroid gland function testing with human synthetic TSH. *Proc Annu Conf Assoc Avian Vet.* 2002.

A Novel Non-Anesthetic Means to Remove Gastrointestinal Foreign Bodies in Birds

M. Scott Echols, DVM, DABVP (Avian Practice)

Session #198

Affiliation: From Medical Center for Birds, Oakley, CA 94561, USA, Parrish Creek Veterinary Hospital and Diagnostic Center, Centerville, UT 84014, USA, Scarlet Imaging, Murray, UT 84157, USA.

Abstract: Gastrointestinal foreign bodies are frequently identified in both wild and managed avian species. Most often, the foreign bodies accumulate in the ventriculus and are found on radiographs, computed tomography, or other imaging studies. While many small foreign bodies are benign and are often referred to as grit, some contain toxic metals or chemicals and may be associated with local irritation within the gastrointestinal system up to systemic disease. Various means have been suggested to remove these foreign bodies when they are considered a threat to the bird's health. These approaches include feeding non-toxic grit; tube feeding peanut butter, mineral oil and other substances; endoscopic or blind tube placement via an ingluviotomy and flushing under anesthesia; and proventriculotomy and/or ventriculotomy procedures. The author has successfully used a non-invasive protocol for over 2 decades to remove small gastrointestinal foreign bodies in a large variety of avian species. In the protocol, the bird is hospitalized for 2 days or otherwise closely monitored throughout the procedure. The bird is given 50-100 ml/kg SC isotonic fluids q12h during the 2-day tube feeding step. Plain psyllium (Metamucil®) is mixed with any handfeeding or critical care formula 50:50 (dry volume) and combined with water so the solution is thin enough to pass through an appropriately sized feeding needle or tube. The bird is tube fed 2-3% volume/body weight of the Metamucil/feeding formula/water mix q12h for 2 days. This produces a thick gelatinous mixture that traps gastrointestinal particles and creates rubbery tube-shaped fecal portions of the droppings. If fluids are not given during this process, there is a possibility of dehydration and death because body fluid will be drawn into the gastrointestinal system during the tube feeding step. Once the bird completes the 2-day SC fluids and tube feeding cycle, the bird can return to its regular environment. The author will often add 1 tablespoon of Metamucil® per 1 cup of slightly moistened food (such that the Metamucil® sticks to the food) for an additional 3-4 wk and then recheck radiographs to monitor progress. With persistent small foreign bodies, the process can be repeated.

Case Review of Waterfowl that Underwent a Novel Surgical Method for Treatment of Septic Tibiotarsal-Tarsometatarsal Arthritis from 2018 to 2023

Dawn Daugherty, DVM, MPH, and
M. Scott Echols, DVM, DABVP

Session #202

Affiliation: From Parrish Creek Veterinary Hospital and Diagnostic Center, Centerville, UT 84014, USA (Daugherty, Echols), Echols Veterinary Services, Cottonwood Heights, UT 84093, USA (Echols), Scarlet Imaging, Murray, UT 84157, USA (Echols).

Abstract: Lower-limb arthritis is a common disease affecting captive waterfowl. Trauma, lack of exercise, improper substrate, obesity, and genetics are just a few of the many causes. Septic arthritis often results in painful and progressive disease that can diminish quality of life if not properly addressed. Current treatments include systemic and regional antimicrobials including local infusions, regional vascular perfusions and implanted antibiotic-impregnated (e.g., PMMA, dissolvable) beads. These methods may slow progression or stop early cases of septic arthritis; however, chronic septic arthritis cases, especially those involving granulomas or caseous intra and extra-articular material, can be difficult to resolve. A previous novel surgical method was described in eight domestic ducks.[1] This procedure used custom-made drains to resolve septic arthritis of the tibiotarsal-tarsometatarsal joint (TT-TMT). These drain tubes were regularly flushed with dilute chlorhexidine followed by equal parts sterile isotonic fluids, and the procedures were performed with sedation and local anesthetic blocks only or with systemic anesthesia. This presentation details a retrospective case review and follow-up of eight ducks, along with additional patients that underwent this procedure from 2018-2023. The goal of this review is to determine if this novel surgical method is a viable alternative or addition to the current methods being used to treat advanced or chronic cases of septic arthritis in captive waterfowl.

References

1. Echols, MS, Kirk N. Surgical management of septic intertarsal arthritis in waterfowl. *Proc ExoticsCon.* 2020;49.

Surgical Correction of Incomplete Beak Avulsion in Two Parrot Species

Shiwani Tandel, BVSc & AH, MVSc, MVetSc (Cons Med)

Session #173

Affiliation: From Phoenix Veterinary Specialty, 2,3 Vipin Residency, Gokhale Road, Mumbai- 400028, Maharashtra, India.

Abstract: Historically, avulsion fractures of the rhinotheca have guarded prognosis unless repair is performed within 24 hours from the time of injury.[1] Statistical reporting of beak fractures has not been performed and therefore literature on its surgical correction is sparse. Fractures of the beak, if not managed judiciously, may lead to infection, debilitation due to starvation and death of the bird.[2,4] Beakistry[2] is a new specialty in avian medicine and very few papers report specific techniques to correct avulsion fractures of the rhinotheca. Two parrot species, an Alexandrine parakeet and African grey parrot, were presented for partial avulsion at the cranio-facial hinge. In each case, the bird was unable to use its beak to eat. It was painful and was actively bleeding whenever fractured segments were manipulated. A surgical technique, which used a combination of cerclage wires and Kirshner's pins, was used to stabilize the fracture.[3] The wires pass through the skull and the rhinotheca and should be carefully placed, after thoroughly understanding the anatomy of the skull bones, sinuses, nerves and location of the eye.[4] Both cases resulted in a successful union of the fractured maxillary rostrum. The fracture site was assessed every 15 days until the bird could shell seeds. The richer the vascular bed at the site of fracture, the better the chance of the avulsed fragments uniting. The union may take longer than 30-45 days and, in some cases, there may be only a partial union.

References

1. Lothamer C, Snyder CJ, Soukup CMJ, et al. Treatment and stabilization of beak symphyseal separation using interfragmentary wiring and provisional bis-acryl composite. *J Vet Dent.* 2014;31(4):255-262.

2. Fecchio RS. Avian fractured and maloccluded beaks. *Zoo Wild Anim Dent.* 2021;1:87-117.

3. Bennett RA. Surgery of the avian beak. *Proc Annu Conf Am Assoc Zoo Vet.* 2010.

4. Speer B, Powers LV. Anatomy and disorders of the beak and oral cavity. *Vet Clin North Am Exot Anim Pract.* 2016;19:707-736.

Section 4

Wildlife and Conservation Session

Stephany Lewis, DVM, CWR, DABVP (Avian Practice)
Session Chair

New Sedation Protocols in Great Horned Owls (*Bubo virginianus*)

Rachel M. Hirota,
Hugues Beaufrère, DVM, PhD, Dipl ACZM, DABVP (Avian Practice),
Stephanie Chen,
Venus Casagrande,
Michelle G. Hawkins VMD, DABVP (Avian Practice)

Session #101

Affiliation: From the Department of Medicine and Epidemiology (Hawkins, Beaufrère), School of Veterinary Medicine (Hirota), University of California-Davis (Chen, Casagrande), Davis, CA 95616, USA.

Abstract: Free-ranging raptors become stressed with restraint for examination and treatment. Chemical restraint allows for performing procedures while reducing handling stress, adverse physiologic effects and hazards to the handler.[1-3] The objective of this study was to evaluate three sedation protocols for minimizing adverse physiologic responses secondary to manual restraint in great horned owls (*Bubo virginianus*; GHOW). Six adult healthy GHOWs were used in this study. A within-subjects, complete crossover study design was used. Three treatments were administered separately to each GHOW with a minimum 7-day washout period between each treatment. Treatments were administered in the pectoral muscles of each bird. The three treatment protocols were: (1) midazolam (M) 2 mg/kg and dexmedetomidine (D) 0.05 mg/kg IM; (2) M 2 mg/kg and hydromorphone (H) 0.6 mg/kg IM; and (3) M 2 mg/kg, H 0.3 mg/kg, and D 0.02 mg/kg IM. Baseline and sedated values for blood glucose and lactate concentrations, venous blood gas values, heart rate (HR), and respiratory rate (RR) and quality were monitored periodically throughout the sedation period. Sedation protocols also were evaluated for time to sedation, time to initial movement, time to standing with and without ataxia, time to complete recovery, tolerance of sham VD and lateral whole-body radiographic positioning, and sedation quality throughout the sedation period. M + H had significantly lower pO_2 (mean difference =79 mmHg, $P = 0.026$), higher pCO_2 (mean difference = 16 mmHg, $P < 0.001$), and higher HCO_3 (mean difference = 5.9 mmol/L, $P < 0.001$) than baseline. M + D + H has significantly higher pCO_2 (mean difference = 10 mmHg, $P = 0.001$), lower lactate (mean difference = 4.5 mmol/L, $P = 0.001$), and higher HCO_3 (mean difference = 2.7 mmol/L, $P = 0.027$) than baseline. M + H had significantly higher HCO_3 (mean difference = 3.6 mmol/L, $P = 0.003$) than M + D. M + H also had significantly higher potassium (mean difference = 0.9 mmol/L, $P = 0.010$) and lower HCO_3 (mean difference = 3.2 mmol/L, $P = 0.008$) than M + D + H. Time to recumbency differed significantly between protocols ($P = 0.015$), with M + H the longest (8.6 ± 2.9 min). Tolerance to radiographic positioning and all other timings did not differ significantly between sedation protocols. HR and RR did not differ between sedation protocols but significantly decreased over time ($P < 0.001$ and $P = 0.002$, respectively). All sedation protocols provided effective sedation for manual restraint and radiographic positioning with similar cardiopulmonary effects.

References

1. Sosa-Higareda M, Guzman DSM, Knych HK, Hawkins MG. High bioavailability, short half-life, and metabolism into hydromorphone-3-glucuronide following single intramuscular and intravenous administration of hydromorphone hydrochloride to great horned owls (*Bubo virginianus*). *Am J Vet Res.* 2023;84(5):ajvr.22.12.0218.

2. Monopoli MR, Guzman DSM, Paul-Murphy J, Beaufrère H, Hawkins MG. Evaluation of thermal antinociceptive effects of intramuscular hydromorphone hydrochloride in great horned owls (*Bubo virginianus*). *J Avian Med Surg.* 2023;37:209-216.

3. Hawkins MG, Griffenhagen GM. Raptor sedation and anesthesia. *Vet Clin North Am Exot Anim Pract.* 2022;25:135-161.

Survey of Antibody Reactivity to *Aspergillus* in the Common Loon (*Gavia immer*)

Carolyn Cray, PhD,
Tristin Burgess, BVSc, PhD,
John Cooley, Jr., BS,
Emily Fellows, BS,
Thomas Hilling, MS,
Maria Passarelli, VMD,
Shelley Spanswick, BS,
Mark Pokras, DVM

Session #168

Affiliation: From the Division of Comparative Pathology, Department of Pathology & Laboratory Medicine, University of Miami, Miami, P.O. Box 016960 R-46, Miami, FL 33101, USA (Cray), Acadia Wildlife Services, P.O. Box 56, South Freeport, ME 04078, USA (Burgess), Loon Preservation Committee, 183 Lees Mill Rd., Moultonborough, NH 03254, USA (Cooley), Biodiversity Research Institute, 276 Canco Rd., Portland, ME 04103, USA (Fellows), Western Connecticut State University, 181 White Street, Danbury, CT 06810, USA (Hilling), Save Our Seabirds, 1708 Ken Thompson Parkway, Sarasota, FL 34236, USA (Passarelli), Center for Wildlife, P.O. Box 620, Cape Neddick, ME 03902, USA (Spanswick), Cummings School of Veterinary Medicine at Tufts University, 200 Westboro Road, North Grafton, MA 01536, USA (Pokras).

Abstract: The common loon (*Gavia immer*) is a charismatic waterbird considered to be a sentinel of aquatic ecosystem health.[1] In some areas, loons are classified as threatened due to challenges such as climate change, habitat loss, oil spills, and lead poisoning.[1] Rarely kept in captivity, rehabilitation success is poor and is related to an apparent stress-associated susceptibility to aspergillosis. The present study was conducted over 2 y in collaboration with multiple centers in Maine, New Hampshire, and Florida. The goal was to assess the presence of antibody reactivity to *Aspergillus* with secondary testing including *Aspergillus* antigen and gliotoxin detection as well as plasma protein electrophoresis as sample volume permitted.[2-4] Samples were collected during summer months in the northeast from healthy wild adult and juvenile birds as well as clinically abnormal birds presented for possible rehabilitation. In addition, a cohort of samples was obtained from overwintering birds. Gross necropsy results were available for those found dead or euthanized. Of 96 birds, five birds had gross necropsy results consistent with aspergillosis. Only two of five birds displayed antibody reactivity to *Aspergillus* using a recombinant antigen-based ELISA and three tested positive for the presence of gliotoxin. An abnormal electrophoretogram was present in all five of the birds. The presence of antibody reactivity was observed in less than 5% of the clinically normal loons. Overall, aspergillosis is likely uncommon in free ranging birds but can be an opportunistic infection after a stress inducing event. Serologic testing and protein electrophoresis may provide an opportunity to monitor the health of this species.

References

1. Paruk JD, Mager JN, Evers DC. An overview of loon research and conservation in North America. *Waterbirds*. 2014;37(S1):1-5.

2. Cray C, Watson T, Arheart K. Serosurvey and diagnostic application of antibody titers to *Aspergillus* in avian species. *Avian Dis*. 2009;53:491-494.

3. Cray C, Reavill D, Romagnano A, et al. Galactomannan assay and protein electrophoresis findings in psittacine birds with aspergillosis. *J Avian Med Surg.* 2009;23(2):125-135.

4. Reidy L, Desoubeaux G, Cardenas J, et al. Detection of gliotoxin but not bis(methyl)gliotoxin in plasma from birds with confirmed and probable aspergillosis. *J Zoo Wildl Med.* 2022;53(1):60-69.

Repair of Major Metacarpal Bone Using a Modified Tie-In Fixator in a Bald Eagle (*Haliaeetus leucocephalus*)

Len Donato, VMD, DABVP (Avian Practice),
James Cortell, BVMS,
Beatrice DeMarco, VMD, MPH

Session #113

Affiliation: From Department of Clinical Sciences and Advanced Medicine, The University of Pennsylvania School of Veterinary Medicine, 3900 Spruce Street, Philadelphia, PA 19104, USA (Donato), Radnor Veterinary Hospital, 107 North Aberdeen Avenue, Wayne, PA 19087, USA (Donato, Cortell, DeMarco).

Abstract: A 5-y-old bald eagle (*Haliaeetus leucocephalus*) presented with an open wound and palpable crepitus over the right metacarpal bones. Radiographs confirmed mid-diaphyseal transverse fractures of the major and minor metacarpal bones. Surgery was elected to repair the fracture.[1-9] A tie-in fixator (TIF) was used to provide a lightweight and removable appliance, and so the wound could be managed during the healing process. A standard dorsal approach was used to identify and isolate the proximal and distal portions of the fracture of the major metacarpal bone. An intramedullary pin was introduced dorsally on the major metacarpal bone at a 30-degree angle, 1 cm distal to the carpal metacarpal joint. A bend was placed on the pin at the site of entry resulting in the pin having an approximate internal angle of 150 degrees. The pin was passed into the distal fragment until it was seated into the distal, dorsal cortex. The surgical incision was closed in routine fashion. Positive profile pins were placed dorsally most proximal and distal on the major metacarpal bone. The excess intramedullary pin and positive profile pins were bent toward the fracture site over the dorsal aspect of the distal wing. A thermoplastic connecting bar was fashioned to hold the bent portion of the pins together. Radiographs were taken weekly to monitor progress. The fracture healed 6 wk post surgery. The bald eagle was released after proper conditioning. This report demonstrates the successful modification of a TIF for major metacarpal fractures in a large bird of prey.

References

1. Orosz SE, Echols S, Redig PT. *Avian Surgical Anatomy and Orthopedic Management.* 2nd Ed. Florence, OR: Teton NewMedia; 2023.

2. Redig PT, Ponder J. Orthopedic surgery. In: Samour J, ed. *Avian Medicine.* 3rd Ed. St. Louis, MO: Elsevier; 2016:312-351.

3. Burdeaux RR, Wade L. Successful management of open, contaminated metacarpal fractures in an adult snowy owl (*Bubo scandiacus*) with a minimal type II external skeletal fixator. *J Avian Med Surg.* 2018;32(3): 210-216.

4. Dar KH, Dar M, Adil S, et al. Surgical management of compound metacarpal fracture in black kite (Milvus migrans): a case report. *Int J Vet Sci.* 2015;4(2):101-103.

5. Grosso FV. Orthopedic diagnostic imaging in exotic pets. *Vet Clin North Am Exot Anim Pract.* 2019;22: 149-173.

6. Yaseen G, Qureshi K, Hussain A. Results of mini external fixator in metacarpal and phalangeal fractures. *J Pakistan Army Med Corp.* 2016;10(5):715.

7. Lam G, Nelson J, Wei F, et al. Evaluation of threaded external skeletal fixation pin insertion with and without predrilling avian bone. *Vet Comp Orthop Traumatol.* 2020;33:28-35.

8. Slunsky P, WeiB J, Haake A, et al. Repair of a tibiotarsal fracture in a Pomeranian goose (*Anser anser*) with a locking plate. *J Avian Med Surg.* 2018;32(1):50-56.

9. Darrow BG, Weigel JP, Greenacre CB, et al. Ex vivo biomechanical comparison of titanium locking plate, stainless steel nonlocking plate, and tie-in external fixator applied by a dorsal approach on ostectomized humeri of pigeons (*Columba livia*). *J Avian Med Surg.* 2019;33(1):29-37.

Determining the Pharmacokinetic Properties of Two Different Doses of Hydromorphone in Barred Owls (*Strix varia*)

Maryella Cohn,
Laken Russell, DVM,
Haerin Rhim, DVM, PhD,
Kimberly Boykin, DVM-MS,
Mark A. Mitchell, DVM, MS, PhD, DipECZM (Herpetology)

Session #126

Affiliation: From Wildlife Hospital of Louisiana at Louisiana State University School of Veterinary Medicine, Skip Bertman Dr, Baton Rouge, LA 70803, USA.

Abstract: Injured avian wildlife are routinely presented to veterinarians with traumatic injuries, and success with these cases often requires the incorporation of analgesics as a component of the treatment plan.[1,2] Unfortunately, there is a dearth of evidence-based analgesic protocols in avian medicine. To date, those protocols that are published suggest species respond differently to these drugs.[3-6] The barred owl (*Strix varia*) is the most common species of raptor presented to the Wildlife Hospital of Louisiana, and there are no published pharmacokinetic studies for this species. The purpose of this study was to determine the pharmacokinetic properties (i.e., harmonic means, times to maximum concentration, and elimination half-lives) for single intramuscular doses of 0.5 mg/kg and 1.0 mg/kg hydromorphone in 12 adult barred owls. We hypothesized that both doses would provide appropriate concentrations for 12 hours and that there would be significant differences in the pharmacokinetic properties between the doses. Each owl recruited for the study was > 500 g and found to be healthy on examination. Owls were randomly assigned to Group 1 (0.5 mg/kg) or Group 2 (1.0 mg/kg) using a random number generator (random.org). Baseline bloodwork was collected before hydromorphone was injected intramuscularly into the left pectoral muscle. After the injection, subsequent blood samples were collected 5 min, 30 min, 1.5 hours, 3 hours, 6 hours, 9 hours, and 12 hours later.[7] Plasma samples are being processed using liquid chromatography-mass spectrometry. The evidence gained by this research provides us the opportunity to better manage these animals while in our care.

References

1. Hanson M, Hollingshead N, Schuler K, et. al. Species, causes, and outcomes of wildlife rehabilitation in New York State. *PLOS ONE.* 2021;16(9):e0257675.

2. Fix AS, Barrows SZ. Raptors rehabilitated in Iowa during 1986 and 1987: a retrospective study. *J Wildl Dis.* 1990;26(1):18-21.

3. Barrowclough GF, Cracraft J, Klicka J, et al. How many birds are there and why does it matter? *PLOS ONE.* 2016; doi: 10.1371/journal.pone.0166307.

4. Lacasse C, Gamble KC, Boothe DM. Pharmacokinetics of a single dose of intravenous and oral meloxicam in red-tailed hawks (*Buteo jamaicensis*) and great horned owls (*Bubo virginianus*). *J Avian Med Surg.* 2013;27(3):204-210.

5. Riggs SM, Hawkins MG, Craigmill AL, et al. Pharmacokinetics of butorphanol tartrate in red-tailed hawks (*Buteo jamaicensis*) and great horned owls (*Bubo virginianus*). *Am J Vet Res.* 2008;69(5):596-603.

6. Yaw TJ, Zaffarano BA, Gall A, et al. Pharmacokinetic properties of a single administration of oral gabapentin in the great horned owl (*Bubo virginianus*). *J Zoo Wildl Med.* 2015;46(3):547-552.

7. Sosa-Higareda M, Sanchez-Migallon Guzman D, Knych HK, Hawkins MG. High bioavailability, short half-life, and metabolism into hydromorphone-3-glucuronide following single intramuscular and intravenous administration of hydromorphone hydrochloride to great horned owls (*Bubo virginianus*). *Am J Vet Res.* 2023 Mar 11;84(5). doi: 10.2460/ajvr.22.12.0218.

Survey of Serum Copper Levels in Wild and Captive California Condors (*Gymnogyps californianus*)

Hannah Attarian, DVM,
Stephanie Lamb, DVM, DAVBP (Avian Practice),
Jeff Zuba, DVM

Session #147

Affiliation: From Liberty Wildlife, 2600 E. Elwood St., Phoenix, AZ 85040, USA.

Abstract: No reference values are available for serum copper levels in California condors (*Gymnogyps californianus*).[1,2] The objective of this study was to perform an initial survey of serum copper levels in captive condors with no known exposure to copper and wild condors with an unknown level of exposure to copper. The use of this information will help to evaluate the role of copper in the body of California condors and allow better evaluation of potential abnormal elevations of copper in the body and the potential impact of copper toxicity. Serum samples were obtained from zoos and wildlife facilities. Samples were collected into red or royal blue top laboratory tubes and evaluated by the California Animal Health and Food Safety Laboratory for trace element screening. A total of 235 serum samples were collected with 14 results from captive condors and 221 results from wild condors. Male and female birds ranging from 1-42 y were evaluated from October 2014 through May 2023. Serum copper levels ranged from 0.069-1.7 ppm, with captive birds ranging from 0.15-0.27 ppm and wild birds ranging from 0.069-1.7 ppm. Further evaluation is needed to establish normal copper serum reference ranges, to determine the potential for or lack of copper toxicosis, and to determine the role of copper in the body of this species.

References

1. Franson JC, Lahner LL, Meteyer CU, Rattner BA. Copper pellets simulating oral exposure to copper ammunition: absence of toxicity in American kestrels (*Falco sparverius*). *Arch Environ Contam Toxicol.* 2021;62(1):145-153.

2. Finkelstein ME, Glucs ZE, Bakker VJ, et al. California condor poisoned by lead, not copper, when both are ingested: a case study. *Wildl Soc Bull.* 2023;47(3):e1485.

Surgical and Orthotic Management and Complications Encountered Treating an Emu (*Dromaius novaehollandiae*) Rotational Leg Deformity

Lauren Thielen, DVM, DABVP (Avian Practice), and
Selena Tinga, DVM, PhD, Dipl ACVS (Small Animal)

Session #181

Affiliation: From Cornell University Hospital for Animals, Ithaca NY 14853, USA.

Abstract: A 3-mo-old emu presented for supination of its left foot and lameness. At the initial assessment, radiographs diagnosed bilateral valgus deformity of the tarsal joints centered at the proximal physes/metaphysis of the tarsometatarsal bones. Shoes were constructed from casting and bandaging material to keep the toes from pronation and hobbles to prevent splaying of legs. A transphyseal bridging was performed by implanting bent cerclage wire to staple the proximal physis of the medial left tarsometatarsus. Despite intervention, the emu's left tarsometatarsus developed a rotational deformity. At 5-mo-old, a corrective osteotomy of the left tarsometatarsus was performed to allow continued ambulation on that limb while he was growing, despite a predicted need for a second osteotomy once the bird became mature. A custom-made 3D-printed guide system was designed from a computed tomography scan, which directed osteotomy position and reduction such that the final shape of the tarsometatarsus matched that of an unaffected emu. Once fragments were aligned, the tarsometasarsus was stabilized with external bone plates and screws laterally and medially. Post-operatively, this patient developed complications that involved swelling of the toes, necrosis of skin and bone along the dorsal aspect and associated nerve damage. Long-term wound care and weekly regional limb perfusions with intravenous ceftiofur was performed. Three months post osteotomy the patient developed progressively worsening pronation of the left foot. Custom orthotic shoes were constructed of 3D-printed plastics. This case report describes the challenges associated with ratite leg deformities, orthotic considerations, potential surgical interventions, and post-operative complications associated with this common presentation seen in emus.[1-5]

References

1. LaBonde J, Miller R, Michel C. Conservative orthotic management of angular limb deformities in ratites. *Proc Annu Conf Assoc Avian Vet.* 1994;143-145.

2. Greenacre C, Aron D, Ritchie B. Dome osteotomy for successful correction of angular limb deformities. *Proc Annu Conf Assoc Avian Vet.* 1994;39-43.

3. Langhofer M. Surgical correction of angular limb deformities in ratites. *Proc Annu Conf Assoc Avian Vet.* 1995;255.

4. VanDerHeyden N. Tibiotarsus derotation in emu chicks utilizing a K-E apparatus. *Proc Annu Conf Assoc Avian Vet.* 1994;259-264.

5. Zollinger T, Backues K, Burgos-Rodriguez A. Correction of angular limb deformity in two subspecies of flamingo (*Phoenicopterus ruber*) utilizing a transphyseal bridging technique. *J Zoo Wildl Med.* 2005;36(4):689-697.

Suspected Thromboembolic Event Resulting in Pelvic Limb Paralysis in a Spectacled Owl, Complicated by Interstitial Pneumonia, Iron Hepatopathy and Muscle Sarcocystosis

April Romagnano, PhD, DVM, DABVP (Avian Practice),
Kian Martin, DVM,
Julia Blackmore, DVM, DACVIM (Neurology),
Robert Ossiboff, DVM, PhD, DACVP,
M. Scott Echols, DVM, DABVP (Avian Practice),
Scott Martin, MS, DVM

Session #191

Affiliation: From Avian & Exotic Clinic of Palm City, 1508 SW Mapp Road, Palm City, FL 34990, USA (Romagnano, Martin K, Martin S), Veterinary Neurology of South Florida, 2227 S Kanner Hwy, Stuart, FL 34994, USA (Blackmore), Diagnostic Laboratory Department of Comparative Diagnostic and Population Medicine, University of Florida, 2015 SW 16th Ave, Building 1017, Room V2-186, Gainesville, FL 32610, USA (Ossiboff), Parrish Creek Veterinary Hospital & Diagnostic Center, 86 N 70 W St., Centerville, UT 84014, USA (Echols).

Abstract: An 8-y-old female spectacled owl had an episode that appeared to be a seizure. On examination, the owl was bright, alert, and responsive; it was thin and down with bilateral pelvic limb paralysis. The owl's pelvic limbs were paralyzed and stiff with tightly clenched feet, toes and talons. The right wing was weak and drooping slightly. Diagnostics included repeated bloodwork, polymerase chain reaction, zinc and lead testing, radiographs, computed tomography scan without contrast and a neurology consult. Supportive treatment was provided. After the bird died unexpectedly, necropsy and histopathology revealed interstitial pneumonia, along with sarcocystosis and iron hepatopathy. A discussion of this case, including diagnostics, treatment, and findings is provided.

Introduction

An 8-y-old female spectacled owl lived in large indoor-outdoor enclosures in South Florida. The bird's diet consisted of whole frozen thawed rodents and occasional fresh meat supplemented with raptor vitamins. The owl had an episode that was caught on camera, which appeared to be a seizure. On examination, the owl was bright, alert, and responsive (BAR). It was thin, and down with bilateral pelvic limb paralysis. The owl's pelvic limbs were paralyzed and stiff with tightly clenched feet, toes and talons. The right wing was weak and drooping slightly.

Diagnostics

Diagnostics included repeated bloodwork, polymerase chain reaction, zinc and lead testing, radiographs, computed tomography (CT) scan without contrast and a neurology consult. Bloodwork was within normal limits with no hint of inflammation and showed normal WBCs, EPHs, liver and kidney function, repeated negative aspergillosis testing and no heavy metals. Sarcocystosis blood testing validated for psittacine birds was negative. There was a low PCV on the second bloodwork, which improved by the third. PCRs were negative for

Haemoproteus, *Leucocytozoan*, *Mycobacterium avian*+ complex, Eastern and Western equine encephalitis, *Malassezia*, toxoplasmosis and *Chlamydia*. PCRs were positive for *Candida* genus+*albicans* and *Aspergillus* genus but no clinical or radiographic signs of these disease states were noted. Fecal floats were negative, but fecal ELISA was positive for *Giardia*.

Treatment and Follow-Up

The owl was placed in a warm oxygen incubator and hand fed. In addition to supportive treatment, antifungals, antibiotics, anti-inflammatories, gabapentin, and a short trial of levetiracetam were given. Initial, and multiple subsequent, hyperbaric oxygen treatments were performed, and the left leg returned to normal function and the bird's foot, toes and talons relaxed and opened up. The right wing became stronger, but the right leg remained paralyzed leading to the decision to do a CT scan and a neurologic consult.

CT scan

The CT scan, albeit without contrast, noted osteoporosis and a thin body condition. The scan ruled out vertebral and skeletal fractures, intervertebral spacing issues, masses, and organ enlargement, corroborating radiographic findings.

Neurologic examination

On neurologic examination, the owl was BAR with intact cranial nerves. Although not confirmed via CT, the neurologist suspected a severe thromboembolic event had affected the right rear leg originating from the lumbar region and a lesser thromboembolic event affected the right wing originating from the thoracic region. The right leg was held in rigid extension and was still stiff with a tightly clenched foot, toes and talons. It had no superficial or deep pain response and no motor and no withdrawal response and its return to function was extremely guarded. The owl was able to right itself, but often toppled over to the right. The right wing had superficial and deep pain but had decreased motor and withdrawal response. The left leg had intact motor function and normal reflexes. The left wing was normal neurologically and functionally.

Subsequent Findings

Post-neurologic consult, the owl remained BAR and continued improving. Its appetite increased and it became more vocal, but the right leg remained in rigid extension with tightly clenched foot, toes and talons causing the bird to repeatedly topple over when trying to right itself.

Six days before the bird's unexpected death, a tiny amount of clear discharge was noted in its right naris. This minimal discharge was aspirated and aerobic, fungal and anerobic cultures were performed. The aerobic culture grew coagulase-negative *Staphylococcus* sp., the fungal culture grew *Candida parapsilosis*, and the anerobic culture was negative. At that time, repeated micro-self-aspirations were suspected to be occurring when the bird would topple over, trying to right itself. This was corroborated after necropsy and histopathology revealed interstitial pneumonia, along with sarcocystosis and iron hepatopathy.

Postmortem Findings

Necropsy revealed severely congested lungs filled with serosanguinous fluid. The trachea and bronchi contained minimal serosanguinous fluid. The heart's pericardial sac was pale, and the surface was intermittently white with a powdered sugar appearance. Mild renomegaly and hepatomegaly were noted, where one liver lobe had rounded edges. The spleen was normal in size but mildly mottled. Thyroids appeared to be slightly enlarged, rounded and pale yellow. Other organs were within normal limits.

Histopathology and electron microscopy revealed a severe interstitial pneumonia. The owl had tissue apicomplexan cysts (*Sarcocystis* sp.) within the skeletal muscle. The liver had evidence of severe and chronic degeneration associated with large amounts of intralesional iron.

Acknowledgments: The authors thank Carolyn Cray, PhD and the amazing staff of the Avian & Exotic Clinic of Palm City.

Section 5

Potpourri Session

Todd Driggers, DVM
Session Chair

Application of Ultrasonic Technology to Dissolve Solid Medications in Avian and Exotic Veterinary Practice

Pat Macwhirter, BVSC (Hons), PhD, FANZCVS,
Matthew Gosbell, BVSc, MANZCVS,
Madeleine Rowe, DVM, MANZCVS,
Wallace Tench

Session #170

Affiliation: From Melbourne Bird Veterinary Hospital, 1 George St, Scoresby, Victoria, Australia, 3179 (Macwhirter, Gosbell, Rowe), Medasyst Pty Ltd, PO Box 398, Ivanhoe, Victoria 3079 (Tench).

Abstract: Administration of solid medications to many avian and exotic animal patients (e.g., parrots, rodents, primates) can be difficult because of anatomic, behavioral and safety constraints.[1] Medical conditions such as weakness or oral, esophageal or upper gastrointestinal tract disorders can make administration of solid medications challenging.[2,3] Manual crushing of tablets may result in inaccurate, variable-sized particles that may block feeding tubes. Use of a compounding pharmacy to liquify solid medications can be costly and involve critical time delays, particularly in emergencies or in remote locations.[4] Based on the literature on enteric feeding[5] and work undertaken by a manufacturer that showed ultrasonic dissolution did not affect drug quality, we tested a benchtop, ultrasonic system (Medasyst Mycare™, Medasyst Pty Ltd., Ivanhoe Victoria 3079 Australia, https://medasyst.com/about/), originally designed for geriatric human use. The system has a label claim of delivering 98% of active ingredients of suitable medications in a liquid form. It can be used to prepare volumes of 10-50 ml of liquid medication. Most tablets (e.g., colchicine, allopurinol, voriconazole) dissolved in a single 2-min cycle. Clinic staff or pet owners can operate the bench-top system with minimal training. Drug concentration can be varied to suit the requirements of individual patients. Masking flavors and proprietary excipients that improve stability (e.g., Ora-plus) are available. The system is not suitable for enteric-coated tablets but can be used with the contents of capsules (e.g., pimobendan). Work is ongoing to collate information for users, within legislative frameworks, on stability, drug interactions and suitable drugs and excipients for use with the system.

References

1. Hernandez-Divers S. Therapeutic techniques of birds. *Proc Annu Congr WSAVA.* 2002. Available at: https://www.vin.com/apputil/content/defaultadv1.aspx?pId=11147&id=3846189&print=1#:~:text=The%20most%20commonly%20used%20routes,intravenous%2C%20intramuscular%2C%20subcutaneous. Accessed January 23, 2024.

2. Fisichella PM, Carter SR, Robles LY. Presentation, diagnosis, and treatment of oesophageal motility disorders. *Dig Liver Dis.* 2012;44(1):1-7. doi: 10.1016/j.dld.2011.05.003.

3. Ganem D, Kistler A, DeRisi J. Achalasia and viral infection: new insights from veterinary medicine. *Sci Transl Med.* 2010;2(33)33. doi: 10.1126/scitranslmed.3000986.

4. Powers LV, Davidson G. Compounding and extralabel use of drugs in exotic animal medicine. *Vet Clin North Am Exot Anim Pract.* 2018;21:195-214. doi: 10.1016/j.cvex.2018.01.005.

5. Klang MG. Developing guidance for feeding tube administration of oral medications. *J Parenter Enteral Nutr.* 2023;47(4):519-540. doi:10.1002/jpen.2490.

The Science and Wonder of Birds

LoraKim Joyner, DVM, MPVM, MDiv

Session #124

Affiliation: From One Earth Conservation, 82-52 211th Street, Hollis Hills, NY 11427, USA.

Abstract: This session reviews the neurophysiologic and cultural aspects of how people relate and respond to nature, and specifically to birds. Recent studies show how immersion in nature can yield health benefits for humans, as can the practice of birding. Being intentionally mindful of birds, whether on a trail, in a conservation project, or in a clinic can also reap benefits not just for humans, but for the birds themselves. As people come to understand and connect to birds, they increase their resilience and health by being with birds, and in so doing, increase their capacity to care for themselves and others. Specific practices that help us connect to birds are explained and demonstrated, offering opportunities to grow multispecies, ecological, and spiritual intelligences. Growing our awareness and understanding of birds can take only minutes a day but conversely also involves a lifetime of commitment, exploration, and wonder, yielding benefits in both personal and professional settings.

Introduction

"The crisis is ultimately not precisely about biodiversity. It is about restructuring our relationship as human beings with the natural world, and birds are an excellent place to start."[1]

Not only has watching birds increased in recent years in many parts of the world, but so has the scientific understanding of how humans connect to nature and to birds, and how they benefit from these interactions. With intentionality, such as in a bird mindfulness practice, connections between humans and birds can be reinforced, not just resulting in greater health for humans and their societies, but also for birds and the entire biotic community.

Birds Are Good for Humans

Even if we aren't purposively seeking out birds to interact with, they are beneficial to humans in ways not directly tied to the ecological roles they fill to augment biodiversity and ecosystem functioning. For instance, we might not be consciously tracking birds, but subconsciously we are. Several studies have found that a higher density of birds or diversity of birds correlated with a higher degree of life satisfaction, lower depression, anxiety, and stress, and lower hospitalization rates due to anxiety and mood disorders.[2-9] One study found a positive correlation between listening to recorded bird songs and lower anxiety and paranoia in healthy patients.[10] Other studies have not resulted in findings that correlate bird richness to human well-being.[3,11] Studies reviewed used a variety of measuring tools for both birds and people and this might explain the differences in outcomes.[11]

Some people report that seeing wildlife, including birds, increases spiritual fulfillment and feelings of wellbeing.[12-15] Time spent in the outdoors provides benefits, such as decreased stress, increased creativity, kindness, and generosity.[16,17] According to the Attention Restoration Theory, people can concentrate better after spending time in nature, or even looking at scenes of nature.[10] Nature helps ease fatigue from informational processing.

Orienting your vision upwards, such as looking at the tops of trees or at birds flying, has been linked to greater mental integration through quieting the inner chatter, and dissolving constructed boundaries of self. Medical doctor, neuroscientist, Zen practitioner, and bird watcher Jim Austin has written much about this subject, including writing of the author's experience of a sudden moment of clarity when looking up to see and hear a flock of parrots suddenly burst loudly out of the forest tree tops.[18] Austin explains not just the neural basis for this human response, and practices to trigger these brain alterations, but also points out how in the Zen tradition there are multiple stories of people becoming enlightened after seeing or hearing a bird, such as in this story:

"The disciple was always complaining to his Master, 'You are hiding the final secret of Zen from me.' And he would not accept the Master's denials. One day they were walking in the hills when they heard a bird sing. 'Did you hear that bird sing?' said the Master. 'Yes,' said the disciple. 'Well, now you know that I have hidden nothing from you.' 'Yes.' said the disciple."[19]

Seeing birds might not lead to sudden clarity, but growing one's awareness and understanding of birds through intentional practices and with repetition can build up benefits over the long term. This can be from the general unfocused gaze, such as looking out and up, openly receptive to what comes, and also from the more cognitive-focused attempt to identify a bird, or understand its behavior. Both are mindfulness practices with resulting benefits.

Perceiving positive benefits, either consciously or subconsciously, has led to numerous birdwatching activities, which appear to be increasing, such as during the recent pandemic and afterwards.[20] Even if the definition of birding is limited to "the identification and observation of wild birds in their natural habitat as a recreation" bird watching would include some 45 million people in the United States alone.[21] If we include appreciating, focusing, caring for, and learning about birds, the numbers are much greater and include those who say they love birds, have bird art, have homed a bird, watch bird documentaries, read bird poetry, swerve their car to get a better look when an unusual bird flies overhead, or break when another crosses in front of the car. Over 57 million U.S. households maintained bird feeders in their yards, spending more than $4 billion annually on bird food.[22] There are also numerous organizations with professionals concentrating on birds, including veterinary clinics, bird clubs, nonprofits, parks, reserves, universities, and government agencies. That's a lot of people thinking about and benefiting from birds, who in turn could be a resource for care and conservation actions directed towards wildlife.

Humans Are Good for Birds, Except When They Aren't

As veterinarians involved with wildlife and specifically with birds, we know how hard many humans work to care for birds. We also know that at the same time, humans have tremendously negatively impacted birds and bird populations. In North America, there are 3 billion fewer birds in 2020 and one in four breeding birds has been lost since 1972.[23] Seventy species have lost two-thirds of their populations in this time, with a prediction of 50% more to be lost in the next 50 y. Half of the world's bird species are in decline with one in eight threatened with extinction. The main threats include habitat destruction and degradation from agricultural expansion and deforestation, invasive species, and human hunting and trapping. These threats also affect mammal, reptile, and amphibian species, and overall ecosystem health.

Intentional actions directed towards birds can also have had adverse impacts on birds. The wildlife trade, both legal and illegal, has caused many bird populations to plummet and individuals to be harmed, as has hunting, habitat degradation and loss, climate change, and conflict.[24] The presence of humans, such as in hiking and wildlife viewing, such as birding, can be a source of disturbance that can harm birds.[25,26]

The Intersection of Humans and Birds

Birds and humans have influenced each other in many ways, and in more nuanced and deeper ways other than simply having a harmful or beneficial relationship. For instance, who we think of as a self is actually a mesh of neural wiring created by relationships. Infants have been studied thoroughly in this regard and their emotional and mental development depends on those with whom they relate in their environment. They are most strongly impacted by mothers and parents, but also by animals and trees. The child's sense of self, called the ecological self, depends on who else is in their environment, interactions and relationships.

In the book, The Parrot in the Mirror, Antone Martinho-Truswell, a behavioral ecologist, said that the challenges in life on earth, indeed the ache of physical suffering and death, caused birds and humans to develop similar traits in a process called convergent evolution.[27] We both are very social, smart, live long lives, move around more in the day and raise our young more like birds than like mammals. In another text, Avian Illuminations, Boria Sax writes how humans' close relationship with birds shaped our culture and human identity.[1] With one example after another, he shows how intimately our bonds with birds are bound up in the matrix of ideas, practices, fears and hopes that make up human civilization. He concludes that a world without birds would effectively mean the end of humankind, for they not only tell us how we got to be where we are, but what we might yet become.

Given the great potential of both the harm and benefit of human-bird interactions, and the melding of human and avian lives, it behooves those whose vocation and avocation is caring for birds, to augment as much as possible awareness of these relationships and the beneficial aspects of these relationships. The intentional practice to connect and increase benefit is a means to increase mindfulness of which some major components are briefly described below.

Multispecies Intelligence

One way to increase awareness and understanding, and hence deepen connection beyond cultural constructs, is to grow multispecies intelligence (MI). MI is a form of social intelligence that extends beyond the human species. It is the ability to be curious and knowledgeable about the motivations, subjective experience, behavior, and welfare status of other species to maximum the benefit and minimize the harm for them.

To grow our MI, we need to be aware of the limitations of our conscious, culturally derived mind that separate humans through distinct boundaries from other species and from each other. We come by this naturally for humans evolved to see those of a possible competing group as distinct and different others. This in turn grows the strength of our own groups' relationships so that our family or small group is stronger and can protect more resources for own groups. One way to control the negative aspects of this othering and inherent tribalism is to develop our MI, which uses understanding of emotional intelligence, and animal communication and behavior across species lines for the mutual benefit of all. MI requires understanding species needs, behavior, motivations, and interconnecting relations with others and their habitat. In short, we ask, what is the individual feeling and needing?

We ask this by seeking to know the motivations for the behaviors, such as understanding their subjective experience (emotions and internal processing) and needs. This means employing what is known as critical anthropomorphism: Critical anthropomorphism means studying animal behavior by using the sentience of the observer to generate theories based on the knowledge of the species, its perceptual world, and ecological and evolutionary history.

By engaging in critical anthropomorphism, we avoid two errors on either end of the spectrum of multispecies understanding. One error is to say that other species are nothing like humans (anthropocentrism), and the other is to say they are exactly like us (uncritical anthropomorphism). Critical anthropomorphism means that we imagine what it is like to be in the shoes, paws, hooves, wings, claws, feet, and skin of another, and then check where we might have made either of the two types of errors. We put on our scientific lens, and ask, what is this individual feeling and needing? We put on our empathetic, embodied lens, and ask, what is this individual feeling and needing? We employ all the science and sensory and body resonance that is available to us, study, reflect, discuss, check our assumptions, and then ask: How might my perception of another lead to more harm than good?

There are many resources for growing our understanding of birds such as through the sciences of cognitive ethology and welfare science. In particular, the Five Domains Model which draws upon observations, cognitive ethology, and the general life sciences of ecology, physiology, and biology, can help inform us as to the status.[28] Intentional practices that meld science, observation, bodily immersion, research and study with the world of birds can grow our MI.

Ecological Intelligence

Ecological intelligence (EI) is the capacity to recognize the often-hidden web of connections between human activity and nature's systems, and the subtle complexities of their intersections, so that we may minimize harm and maximize flourishing for all. It is both an intellectual and an emotional expertise that strives for nourishing others as yourself, for caring for one is caring for all. EI is the awareness of the feelings and needs of all others and how they interact in the complex ecology of the web of life.

Interconnecting harm and benefit

Growing our EI means growing our awareness of interconnecting harm. By paying attention to birds, we discover the often-hidden interconnections of interdependent malfeasance, and this is often not enjoyable. Interdependent malfeasance is when one action or set of behaviors may also lead to a chain of harm that echoes through many lives and individuals. Opening ourselves to reality, we realize that birds near and far are in trouble. Species are declining at alarming rates and climate change produces extra stressors and challenges to which we are unsure the birds can adapt. The goal of birding though is not to experience shame, but to gain awareness about the mesh of interdependent malfeasance in which we all are stuck.

Growing our EI also means growing our awareness of interconnecting benefit. When one individual or species is nurtured, these actions positively impact a plethora of species and ecological niches, and this is known as interdependent beneficence. One example of this includes top carnivores that are involved in trophic cascades. Trophic cascades are the changes that occur in an ecosystem by adding, removing or changing the behavior of top predators in a food web, which then impacts other predators, herbivores, and plants. Another example is parrots serving as seed dispersers, pollinators, genetic linkers, seed facilitators for secondary dispersers, and plant protectors.[29] "Through both antagonistic and mutualistic interactions with their food plants, parrots can exert cascading effects on the plants' life cycles and shape the structure and functioning of ecosystems."[30]

Awareness of benefit and harm help us embrace reality

Birding helps us embrace reality, not just of the world of birds, but of all existence. In the movie, The Thin Red Line, the hero, a soldier in the South Pacific during World War II, muses as he discovers a dying parrot chick on the ground that had been blown out of her nest due to the bombs that were ensuring the mutual destruction of plant and animal life:

"One man looks at a dying bird and thinks there's nothing but answerable pain. Another man sees that same bird and feels the glory – feels something smiling through it."[31]

We embrace reality, not to torture ourselves, but to nurture ourselves and others. Our ecospiritual challenge is to walk a third way: not denying the reality we face, and not retreating into everyone-for-herself-survivalism. It is the path of open-eyed and open-eared awareness, and also the path of connection to both nature and neighbor, being not afraid to face reality and not avoiding needed knowledge because it's depressing and you'd rather not think about it. And at the same time not bunkering protectively. The ecospiritual challenge is to choose neither despair nor defense, but new community."

This new community is one where all belong and are connected and interdependent to each other, and to the whole. Birding brings us into a community and welcomes us to the family of life, which is a source of infinite wonder. Wonder is one way that we are able to face the harshness of this life, and it is a gift from the whole of earth that transcends our consciousness.

Wonder and awe

Awe is an emotional response to perceptually vast stimuli that transcend current frames of reference and often engenders feelings of fear and respect.[32] It is similar to wonder, which can be defined as a more reflective, curious approach to stimuli that is rare or unexpected. With both, our self-centered perspective shifts, resulting in connection where life's possibilities open before us. Wonder helps us engage with the world to live in ways that integrate the reality that beauty is ever present. It also helps us face the also true, but harsher reality of harm, illness, death, disappointment, and massive suffering. Without wonder, we risk closing off to life, living more shallow lives, less intimacy and vibrancy.

Wonder opens us up and can help us deepen our connections and relationships of all kinds. An experience, such as rafting, can cause one to feel more engaged and curious about the world, even a week later. Wonder also lifts depression, and decreases body inflammation. Wonder also helps our prosocial behaviors for we become more empathetic, humble, and generous.[32] When we have more empathy, others resonate with us better and we have improved relationships. Our self-identity moves from a separate self to being part of a whole, or the whole itself. By merely writing about awe, we become kinder, more compassionate, and this can extend to other species and the biotic community as a whole. The world needs all the help we can get and give, and two primary aspects of human nature we can nurture for this aim is wonder and its partner, empathy.

One way to have a wow experience is to get out into nature, where wonder can arise spontaneously and often. Nature is full of unexpected and surprising events that we cannot foresee, and this is good for us. James Austin, the neurologist previously mentioned, encourages us to have nature experiences because they help integrate our neurologic processing and contribute to mindfulness and living in the present moment with attention and gratitude.

Ecological empathy

Ecological empathy is our bodies and subconscious knowing that we belong, matter, and that we are heard and seen as we see and hear others. Higher avian diversity means humans score higher in life satisfaction. The same phenomenon happens when we are around water[33] as well as when we are around other animals and plants. More trees equate to greater human and societal wellbeing. People in hospitals heal faster if they have a window through which they can see trees. Healing still improves though at increasingly lower amounts, if there is a picture or realistic painting of trees, or even an impressionist rendering of a nature scene.[34,35] Studies have shown that the higher concentration of trees in an urban setting reduces stress, reduces crime rates, promotes healing, improves cognition and attention, contributes to happiness, improves mood, and improves health.

In the practice of Forest Bathing, where guides take people out to areas with high concentrations of trees, health, mental, and spiritual benefits are frequently recorded. Not all the mechanisms of this are known and certainly include the beneficial inhalation of tree pheromones known as terpenes. Being out in nature can also increase our body esteem.

Even if there won't be many birds, or perhaps just a low variety of urban birds, the fact that you move amongst trees and look up results in your subconscious receiving messages of belonging and connection. These benefits can then in turn be given back to a bruised and aching world.

Intentional practices that grow our understanding of the web of relationships along with bodily immersion into the world of birds can grow our ecological intelligence.

Spiritual Intelligence

There are many definitions and understandings of the word spiritual and what spiritual intelligence might entail.[36] For some, spiritual intelligence is the human capacity to be curious about the ultimate meaning of life and the interconnections between us and the world in which we live. For the sake of this paper, spiritual intelligence is defined as the ability to see beyond self-experience, or even human experience, stories, and projections, resulting in a greater sense of interconnection, acceptance, and compassion to not just life, but to all of existence. Spiritual intelligence is also referred to as existential intelligence, however it is defined or measured and there is a multitude of ways, it increases the psychological well-being of an individual and also results in increased longevity, improved emotional states, stronger immune systems, reduced risk of disease, and improved self-confidence and fitness. It can also possibly positively impact the well-being of society and the earth as spiritual intelligence is ultimately concerned with the well-being of the universe and all who live there.

Practices that support the profound interconnections within existence and between the self and others, are similar to those referred to as mindfulness practices, which help situate oneself in the present moment's reality. The reported health benefits of mindfulness practices to humans are many including relieving stress, pain, addiction, and depression.[37]

Intentional practices that help shift an individual or human-based frame of reference for relationships include any traditional mindfulness practices that are influenced by an understanding of multispecies and ecological intelligence. Specific practices based on birds will also serve, such as sitting or walking in nature in silence while attempting to ease the mind's inner chatter.

Birding For Life

A host of practices exist, based not just on multispecies, ecological, and spiritual intelligence, but also emotional and social intelligence. Emotional and social intelligence guide humans into knowing who they are, which is interrelated with and necessary to knowing other species. There is no sharp demarcation between the intentional growth of any of these intelligences, as what grows one intelligence will also augment the others. Taken as an integrated package, practices based on these five intelligences constitute a bird mindfulness practice that in some instances can yield immediate positive results with lasting impact. In general, however, neural pathways that impact our emotions, behavior and subconscious actions are slow to change and a lifetime of practice can continually improve the care we can experience and offer. In this way, a bird mindfulness practice is for life, and for all of life.

The author guides others to grow their natural intelligences with Birding for Life Walks.[38] During these bird walks, practices are experienced together with a group of people while looking for birds. Participants have also learned and experienced these practices in webinars, workshops, conferences, and books.[39,40] Though there has been no formal study investigating the short- or long-term impact of these specific bird walk practices, comments, appreciation, and anecdotal stories indicated that a bird mindfulness practice can improve the health and well-being of humans, as well as the biotic community.

Conclusions

With modifications depending on the individual and situation, birding for life practices can be incorporated into clinical veterinary and wildlife care situations, as well as into everyday life. Indeed, probably most of us have already experienced being with birds as a form of mindful attention that has improved our lives and increased our awareness of, connection to, and care for birds. By sharing and growing intentionally our bird mindfulness practices, we might also grow our veterinary practices with greater internal (inter- and intra-personal relationships to other humans and species) and external (societal and organizational) outcomes.

Examples of Bird Mindfulness Practices

Trans-species empathy: Being the bird

This practice is written so that it can apply to any animal or plant species, and so that you can do all the steps sequentially or even skip some. There are many variations on this exercise. For instance, one may choose an animal to empathize with that you can see currently, either inside or outside the home. One can choose an animal that they know personally or see frequently. As an extra challenge, one might also choose an animal species or individual with whom they are unfamiliar or with whom they experience uncomfortable feelings, such as strangeness, fear, irritation, or disgust. Philosopher Timothy Morton proposes that this strangeness reminds us that we are all interconnected and that "every life-form is familiar, since we are related to it."[41] Our relationship goes deeper than our differences.

You may follow these steps or go directly to step #5 which is the center and primary step of this exercise.

1. Think of an individual. Write what you know of the being. What is the species? Individual name? Gender? Age? Life stage (e.g., growing, juvenile, parent). Health status? If you can't think of an individual, choose a species you would like to get to know better or understand.

2. Observe them over a period of time and write what you see them do. Explain what you see as if you were a reporter with as little judgment or human projection as possible. In other words, don't try to interpret the behavior at this point. It may be easier to choose just a short period of time or one behavior for this exercise, although you might find it useful for your relationship to journal at some point about all behaviors you encounter.

3. Now guess what you imagine they are thinking and feeling, and how these are related to their needs. What is their arousal level: high, medium, or low? Are they moving towards pleasure or away from discomfort? As a journal exercise you can list your guesses here.

4. To help you understand what you observed, do some research on the species regarding behavior, communication, feelings, and thoughts. You may find it difficult to find information about emotions and thinking in nonhuman species. Did you discover any new feelings or thoughts that might exist in the individual you observed?

5. Now imagine that you are the animal. Get into their paws, scales, fur, or feathers for about 15 min. Pick an animal that is in your yard or along a walk or a hike. You can also watch a video or nature documentary. You become them and now are doing what you have observed them doing. As this animal, what are you thinking and feeling? For these 15 min, just be them without analyzing too much why they do what they do. After you are done, do a body scan. How has your body shifted being the animal? Ask yourself if you discovered anything new by pretending to be the animal? Share what you learned with another person and also invite them into this journal or imagination exercise.

6. To help you understand what you observed, do further research on the species regarding behavior, communication, feelings, and thoughts. Use the Five Domains Model as a guide to what another individual might be needing. After doing research, did you discover any new feelings or thoughts that occurred in the individual? If so, add them to your list.

7. Now looking over the list of feelings and thoughts, make a list of this individual's needs. Try to be as complete as possible as you go through the behaviors observed or if you have the time, what behaviors might exist in not just a normal day as this individual, but over a lifetime. How might these needs be different from another individual of the same species, or from the average needs of this species?

8. What feelings and needs arise in you when you consider the feelings and needs of this individual?

9. What have you discovered about this individual, this species, yourself, or life through this exercise? If you have discovered anything, what needs of yours or the individual does what you have learned meet, or not meet.

10. Go back and spend time connecting to the energy of the other being by reviewing their feelings and needs, and then do the same with yourself. Allow this to be a time of being and connecting to life, without thought of requests or demands.

11. Finally, consider possible actions or steps you might do, or ask of others, based on this multispecies empathy exercise.

12. Share what you have learned or experienced with others and invite them into the exercise.

Bird stories and a macaw meditation

Scientific writing and resources can help us interpret the behavior of others and gain a greater understanding of their daily lives. This scientific writing can also be told in a story form, such as in novels, film, poems, and children's books. One resource is the children's book, What Would the Parrot Say, which imagines what a scarlet macaw might say about being caged or trapped. The book ends with information about the bird and what the reader can do to help.[42]

There are also wildlife meditations that guide one into being the bird. One Earth Conservation has produced a guided meditation that helps one understand the life of a scarlet macaw.[43] While being anthropomorphic in style, the bird story is based upon thousands of hours of observing parrots by the author. The author developed this video to share what is known about parrots because not everyone has the opportunity to study and reflect on a parrot's life.

References

1. Sax B. *Avian Illuminations: A Cultural History of Birds.* Chicago, IL: Chicago University Press; 2021.

2. Buxton RT, Pearson AL, Lin H-Y, et al. Exploring the relationship between bird diversity and anxiety and mood hospitalization rates. *Geo.* 2023;10(2):e00127. doi: 10.1002/geo2.127.

3. Fisher JC, Bicknell JE, Irine KN, et al. Bird diversity and psychological wellbeing: a comparison of green and coast blue space in a neotropical city. *Sci Total Environ.* 2021;793:148653. doi: 10.1016/j.scitotenv.2021.148653.

4. Hammoud R, Tognin S, Burgess L, et al. Smartphone-based ecological momentary assessment reveals mental health benefits of birdlife. *Sci Rep.* 2022;12:17589. doi: 10.1038/s41598-022-20207-6.

5. Haupt A. Birdwatching has big mental-health benefits. Here's how to start. Time Magazine. Available at: https://time.com/6231886/birdwatching-mental-health/. Accessed February 13, 2024.

6. Hepburn L, Smith A, Zelenski J, Fahrig L. Bird diversity unconsciously increases people's satisfaction with where they live. *Land.* 2021;10:153. doi: 10.3390/land10020153.

7. Methorst J, Rehdanz K, Mueller T, et al. The importance of species diversity for human well-being in Europe. *Ecol Econ.* 2020;181: doi: 10.1016/j.ecolecon.2020.106917.

8. Strassmann JE. Slow birding: *The Art and Science of Enjoying the Birds in Your Own Backyard.* Los Angeles, CA: Tarcher Perigee; 2022.

9. Watts T. *Keep Looking Up: Your Guide to the Powerful Healing of Birdwatching.* New York, NY: Hay House; 2023.

10. Stobbe E, Sundermann J, Ascone L, Kühn S. Birdsongs alleviate anxiety and paranoia in healthy participants. *Sci Rep.* 2022;12:16414. doi: 10.1038/s41598-022-20841-0.

11. Xu W, Zheng D, Huang P, et al. Does bird diversity affect public mental health in urban mountain parks? A case study in Fuzhou City, China. *Int J Environ Res Public Health.* 2022;19(12):7029. doi: 10.3390/ijerph19127029.

12. Cox DTC. Doses of neighborhood nature: the benefits for mental health of living with nature. *BioScience.* 2017;67(2):147-155. doi: 10.1093/biosci/biw173.

13. Louv R. Our Wild Calling: *How Connecting with Animals Can Transform Our Lives – And Save Theirs.* Chapel Hill, NC: Algonquin Books; 2019.

14. Pizer A. The Psychological Benefits of Wildlife Viewing. WhereistheWildlife.com. Available at: https://www.linkedin.com/pulse/psychological-benefits-wildlife-viewing-andy-pizer/. Accessed February 13, 2024.

15. United Nations Environment Programme and the Secretariat of the Convention on the Conservation of Migratory Species of Wild Animals. Wildlife watching and tourism: a study on the benefits and risks of a fast growing tourism activity and its impacts on species. Available at: https://www.cms.int/sites/default/files/document/ScC14_Inf_08_Wildlife_Watching_E_0.pdf. Accessed February 13, 2024.

16. Sherwood H. Getting Back to Nature: How Forest Bathing Can Make Us Feel Better. The Guardian. Available at: https://www.theguardian.com/environment/2019/jun/08/forest-bathing-japanese-practice-in-west-wellbeing. Accessed February 13, 2024.

17. Weir K. Nurtured by nature. *Monit Psychol.* 2020;51(3):50.

18. Austin J. *Living Zen Remindfully.* Cambridge, MA: MIT Press; 2016.

19. de Mello A. *The Song of the Bird.* New York, NY: Doubleday; 1984.

20. MacLellan L. New data show that birding mania isn't just a lockdown fad. Quartz. Available at: https://qz.com/2024279/birdings-popularity-is-lasting-beyond-the-pandemic. Accessed February 13, 2024.

21. U.S. Department of the Interior, U.S. Fish and Wildlife Service, U.S. Department of Commerce, Bureau USC. 2016 National Survey of Fishing, Hunting, and Wildlife-Associated Recreation. Available at: https://www.census.gov/content/dam/Census/library/publications/2018/demo/fhw16-nat.pdf. Accessed February 13, 2024.

22. U.S. Fish and Wildlife Service. Addendum to the 2016 National Survey of Fishing, Hunting, and Wildlife-Associated Recreation Report 2016-2, Birding in the US. Report_2016_508. Available at: https://digitalmedia.fws.gov/digital/collection/document/id/2252/rec/2. Accessed February 13, 2024.

23. North American Bird Conservation Initiative. The state of the birds, United States of America, 2022. Available at: StateoftheBirds.org. Accessed February 13, 2024.

24. Joyner L. Guide to understanding and reducing harm to Suriname parrots. Available at: https://www.oneearthconservation.org/_files/ugd/d204d4_309fef0f70d4496893889975bc30b257.pdf. Accessed February 13, 2024.

25. Bjørvik LM, Dale S, Hermansen GH, et al. Bird flight initiation distances in relation to distance from human settlements in a Tanzanian floodplain habitat. *J Ornithol.* 2105;156:239-246. doi: 10.1007/s10336-014-1121-1.

26. Lee ATK, Marsden SJ, Tatum-Hume E, Brightsmith DJ. The effects of tourist and boat traffic on parrot geophagy in lowland Peru. *Biotropica.* 2017;49(5):716-725. doi: 10.1111/btp.12426.

27. Martinho-Truswell A. *The Parrot in the Mirror. How Evolving to Be Like Birds Makes Us Human.* Oxford, UK: Oxford University Press; 2022.

28. Mellor DJ, Beausoleil NJ. The 2020 five domains model: including human-animal interactions in assessments of animal welfare. *Animals.* 2022;10(10):1870.

29. Blanco G, Hiraldo F, Rojas A, et al. Parrots as key multilinkers in ecosystem structure and functioning. *Ecol Evol.* 2015;5(18):4141-4160.

30. Blanco G, Hiraldo F, Tella JL. Ecological functions of parrots: an integrative perspective from plant life cycle to ecosystem functioning. *Emu-Austral Ornithol.* 2018;118:36-49.

31. Twentieth Century Fox Home Entertainment. The thin red line. Beverley Hills, CA. 2001.

32. Piff PK, Dietze P, Fienberg M, et al. Awe, the small self, and prosocial behavior. *J Pers Soc Psychol. 2015;108*(6):883-899. Available at: doi: 10.1037/pspi0000018.

33. Nichols WJ. *Blue Mind.* New York, NY: Back Bay Books; 2015.

34. Ulrich RS, Gilpin L. Healing arts: nutrition for the soul. In: Frampton SB, Gilpin L, Patrick C, eds. *Putting Patients First: Designing and Practicing Patient-Centered Care.* San Francisco, CA: John Wiley & Sons; 2003:117-146.

35. Williams F. *The Nature Fix: Why Nature Makes Us Happier, Healthier, and More Creative.* W.W. Norton and Company: New York, NY; 2017.

36. Srivastava P. Spiritual intelligence: an overview. *Inter J Multidisc Res Dev.* 2016;3(3):224-227.

37. Creswell JD. Mindfulness interventions. *Annu Rev Psychol.* 2017;68(1):491-516.

38. One Earth Conservation. Birding for Life Walks. Available at: https://www.oneearthconservation.org/birding-for-life. Accessed February 13, 2024.

39. Joyner L, Koelln G. Nurturing discussions and practices. New York, NY: One Earth Conservation Publishing; 2017.

40. Joyner L. Birding for life. New York, NY: One Earth Conservation Publishing; In press.

41. Morton T. Humankind: Solidarity with Nonhuman People. Brooklyn, NY: Verso Books; 2019.

42. Koelln G. What would the parrot say? New York, NY: One Earth Conservation Publishing; 2021.

43. Joyner L. Scarlet macaw meditation. Available at: https://youtu.be/kja32cJV9rc?si=3yOSGO8LkDN9YGHk. Accessed February 13, 2024.

Effects of Sample Preparation on Artificial Intelligence-Assisted Assessment of the Avian Blood Smear

Kaori Davis, DVM, PhD, DACVP (Anatomic and Clinical Pathology),
Liam Cattell, MEng, DPhil,
Bianca Murphy, DVM, DABVP (Avian Practice),
Elyssa Diamaano,
Thanh Le, MS, PhD

Session #188

Affiliation: From Moichor Reference Labs, 1188 Harrison St., San Francisco, CA 94103, USA.

Abstract: The complete blood count (CBC) is a foundational assay to ascertain an individual's overall health status. Morphologic assessment of leukocyte populations further aids in interpreting the leukogram. Morphologic assessment is typically performed manually; however, manual morphologic assessment has limitations because it is time- and labor-intensive, making it available only in selected patients. Previously, we evaluated the effectiveness of cutting-edge artificial intelligence (AI) and computer vision techniques for analyzing cell morphology in 403 avian hematology samples of which the AI model was able to identify 97% of cases with at least one toxic heterophil or immature heterophil. In this review, slide preparation, staining, and cover-slipping technique were controlled within a single reference laboratory. In this year's study, we evaluated the impact of sample preparation, specifically, staining, and cover-slipping on the performance and effectiveness of leukocyte morphology identification in avian hematology samples. Blood smears from all animals were prepared by various veterinary professionals and scanned, using the Moichor Pathview in clinic device linked to the same AI model and methodology used in the previous study. Based on the information obtained in this review, lack of cover-slipping and poor stain quality resulted in reduced detection of white blood cells and cell morphology by our model. Multiple solutions to reduce variation in model performance based on slide preparation variables were evaluated. Style transfer technology was investigated as a solution since its applications in human histopathology have been promising.[1,2] Images were grouped into specific categories and the style transfer was applied with the goal of making the aberrant image as close as possible to images obtained from reference laboratory-prepared images. When then run through the same model post style transfer, leukocyte detection improved. The nature of these data prompted further investigations into the applications of style transfer in avian blood film evaluation and are discussed.

References

1. Nerrienet N, Peyret R, Sockeel M, Sockeel S. Standardized CycleGAN training for unsupervised stain adaptation in invasive carcinoma classification for breast histopathology. *J Med Imaging.* 2023;10.6:067502.

2. Zhu J-Y, Park T, Isola P, Efros AA. Unpaired image-to-image translation using cycle-consistent adversarial networks. *Proc Conf IEEE Comput Vis.* 2017.

Section 6

Technician Session

Jovanna Berls, CVA, LVT, FFCP (Veterinary-Avian);
Bri Bryan, RVT;
Kristina Palmer, RVT, VTS (Clinical Practice-Exotic Companion Animal)
Session Chairs

Clucks and Ducks: What Veterinary Technicians Should Know About Backyard Poultry Keeping

Jovanna Berls, CVA, LVT, FFCP (Veterinary-Avian)

Session #105

Affiliation: From ABC Animal & Bird Clinic, 11930 Highway 6 South, Sugar Land, TX 77498, USA.

Abstract: Chickens and other poultry birds increasingly are becoming beloved family pets, and that has meant a growing number of them are presenting for veterinary care. Backyard poultry share similarities with the parrots and ornamental birds traditionally kept as pets, including that many health issues can be prevented by excellent husbandry and nutrition. This session provides an overview of how veterinary technicians can assist owners of poultry birds in providing for the health and welfare of their pets. Topics covered include appropriate housing, biosecurity, issues related to pharmaceutical use in food animals, and how to advise clients about husbandry, nutrition, behavior, and vaccinations.

Introduction

Keeping poultry in suburban or even urban backyards somewhat fell out of favor in the 1950s, but the practice has enjoyed a resurgence recently. Where these birds most often were kept primarily for their ability to produce eggs and meat for their owners, these days they are likely to be considered beloved family pets as deserving of quality veterinary care as any dog or cat. Although poultry in general have a reputation for being tough, stoic, and able to fend for themselves, excellent husbandry is key to their health and welfare, similar to other exotic pets. Veterinary technicians can play an important role in educating poultry owners to provide that care.

Husbandry of Backyard Chickens

Biologic and physiologic parameters[1]

Heart rate (bpm): 220-360
Respiratory rate (breaths/min): 28-49
Body temperature (°C [°F]): 40.5-43 [104.9-109.4]
Typical adult weight (kg): 1.6-5.0, depending on breed
Lifespan (y): 3-8
Sexual maturity (wk): 22-24

Types, breeds, and varieties

Unlike the parrots that are commonly seen in veterinary practices, there are many breeds of chickens, but they are all the same species, *Gallus gallus domesticus*. The American Poultry Association classifies chicken breeds by size (i.e., standard or bantam) and classifies varieties by physical characteristics such as color and leg feathering. Note that there may be different breeds and classifications in other countries. Chickens of the same breed can be different colors, such as buff and black Orpington chickens. Many hybrids and mixed breeds also exist. It is not uncommon for clients not to know or care about the specific breed or origin of the chickens they own.[2]

In general, chickens have been bred according to their use: egg laying, meat production, dual-purpose (e.g., meat and eggs), or show and ornamental purposes. Some popular laying breeds include Americauna and Maran breeds that produce colored eggs. White leghorn chickens are known for their egg production. Meat breeds also lay eggs but often not as many; they are generally larger and more broad-breasted than laying breeds. Ornamental breeds, such as silkies and Polish crested chickens, are quite varied in their appearance. They often are good pets, and some will lay a fair number of eggs. Game bird breeds, such as Albany and lemon fowl, have long necks and legs, and were originally bred for cockfighting; roosters of these breeds should not be kept together, as they may fight.[2]

Housing

The essential characteristics of appropriate chicken housing are that it should be easy to clean, adequate in size, offer protection from inclement weather, and be able to keep out predators such as raccoons, raptors, or dogs.[3] Beyond that, a chicken coop can be as plain and simple or as elaborate and attractive as the owner wishes.

Ideally, the coop should have a concrete floor, have walls that are washable, and provide adequate insulation; the design should allow for good ventilation but should not be drafty. It should be located in an area with good drainage. It should be large enough to allow a minimum of 0.75-1 m^2 (8-10 ft^2) for each bird.[4] The floor should be covered with at least 20 cm (8 in) of a suitable litter, such as pine wood shavings, straw or shredded newspaper.[3,4]

Chickens are ground dwellers during the day, but adult chickens will roost off the ground at night, so perches will be needed. Each bird should have access to at least 15 cm (6 in) of perch space. These should be introduced once the chickens are about 6 wk old.[3] Poop boards can be placed underneath the perches to help keep the litter clean. Droppings can then be scraped off and disposed of regularly without changing out the litter.[4] Nest boxes should not be introduced until the chickens are about 17 wk of age; providing them at a younger age can contribute to improper use of them as litter boxes.[3] Next boxes should be positioned up off the floor in a quiet, dark area of the coop.[4]

Birds should have the ability to get out of the weather so they can regulate their body temperature, as extremes of heat or cold can be stressful and have an adverse effect on their health. A good general range for the temperature inside the coop is 10-24 °C (50-75 °F). Keep in mind that chickens and other poultry are susceptible to frostbite, especially on the feet and unfeathered areas of the head, such as combs and wattles.[3]

If the birds have access to an outside run, lighting inside the coop may not be necessary, depending on the design of the coop. If egg production is important, decreasing day length will necessitate indoor lighting through the autumn and winter months. For uninterrupted laying, the hens should have 16 hours of light and 8 hours of darkness.[3]

For predator and rodent control, all access points in the coop should be secured with wire mesh, but keep in mind that animals such as opossums or skunks may be able to reach through the wire to grab the birds. Holes in the mesh should be no larger than 2 cm (0.75 in) to keep out small paws. A wire mesh under the coop's floor can be used to protect against predators that may try to dig underneath to reach the birds.[4] A wire fence buried at least 8 inches deep can be used to enclose an outdoor run; wire netting over top of the fence may be needed to protect the birds from raptors. Feed should be kept in a rodent-proof container; a good practice is to only keep as much feed on hand as the birds can eat in 30 days, not only to discourage rodents but also to ensure the feed remains fresh and maintains its nutritional value. Feed that is allowed to get wet can become contaminated with toxic molds.[3,4]

Nutrition

Most research into nutrition for chickens and other poultry has been conducted using commercial birds, not backyard poultry. Thus, the exact nutrient needs of backyard flocks have not been studied extensively. Coupled with the tendency of backyard chicken owners to feed whatever is at hand or whatever the birds seem to like, a number of nutrition-related problems can occur.[5]

Ideally, all chickens should be fed a commercially formulated crumble or pelleted diet produced by a nationally recognized manufacturer. The bag should include a tag or label that specifies a mill date or expiration date, as well as a guaranteed analysis of the nutrients provided.[4]

One common nutritional problem is failure to feed chickens according to their life stage or gender.[5] Juvenile birds need a chick starter or grower diet that provides higher levels of protein (20%) and fat (4%). Adult laying hens have higher requirements for calcium, protein, and vitamins, so should be fed a diet specifically formulated for them. Laying hens also can be offered crushed oyster shell as a calcium supplement. Roosters and hens that are not laying, on the other hand, should be fed a maintenance diet, often labeled as a game bird diet, to avoid excessive levels of calcium. Commercial feeds may be supplemented with leafy greens and some fresh vegetables, such as corn, or tomatoes. Fruit, scratch grains, and dried meal worms will be readily accepted, but should be limited.[4] Note that feeding kitchen scraps to chickens is illegal in some locations.[6]

The form of the feed also may be important to ensure the birds are consuming the food effectively and efficiently. Feeding crumbles to young, growing birds may help ensure they are eating enough because the small particle size is easy for them to manage, while pellets reduce the tendency for older birds to eat selectively. Feeding pellets to chickens with elaborate head feathers, such as silkies, will help prevent accumulation of food around their faces.[6]

Chickens that are allowed to forage outside in a yard or run will snack on grass and insects and they benefit from engaging in this natural behavior. Make sure no chemicals, such as fertilizers or pesticides, are applied to the area to avoid exposure to toxins. Ingestion of long grass or grass clippings can result in a crop impaction. Scout the ground in the area carefully to remove any nails, screws, or other hardware that may be accidentally eaten; chickens are notorious for ingesting shiny objects.[4]

Fresh, clean water should be available at all times. Drinkers are available in a variety of types and sizes. Drinkers can be placed in a raised location to avoid having feed scratched into them, but attention should be paid to whether smaller birds can reach them easily.[6] Nipple drinkers should be placed at a height just above the birds' eye level.

A wide variety of feeders is available. Wastage can be reduced by positioning them so the birds must reach over the lip slightly to reach the feed. These heights may need to be adjusted as the birds grow.[3] Keeping feeders inside the coop at night can prevent rodents and wild birds from reaching the food. Alternatively, feeders that require birds to step on a treadle to reach the food also may prevent access by rodents and wild birds.[6]

Sanitation

The entire coop should be disinfected at least once a year to remove pathogens, ectoparasites, and dirty litter, but disinfecting too often can have adverse effects.[3] Most often recommended is a cleaning protocol that involves the following:

- removing all the birds,
- sweeping or blowing the ceiling, walls, and floor to remove dust and loose dirt,

- wet cleaning the house with a detergent,
- rinsing,
- allowing the entire house to dry thoroughly, and
- applying the disinfectant.[3]

Food and water containers should be thoroughly cleaned at least weekly.[6]

Husbandry of Backyard Waterfowl

Biologic and physiologic parameters

Ducks[1]

Heart rate (bpm): 180-230
Respiratory rate (breaths/min): 30-95
Body temperature (°C [°F]): 40.5-43 [104.9-109.4]
Typical adult weight (kg): 1.1 (female), 1.26 (male)
Lifespan (y): 10-15
Sexual maturity (y): 1

Geese[1]

Heart rate (bpm): varies depending on season and reproductive status
Respiratory rate (breaths/min): 5-15
Body temperature (°C [°F]): 40.5-43 [104.9-109.4]
Typical adult weight (kg): up to 10
Lifespan (y): up to 31
Sexual maturity (y): 2-3

Types, breeds, and varieties

All domestic ducks are descended from the mallard (*Anas platyrhynchos platyrhynchos*).[7] Duck breeds are classified by their purpose, as either meat-types, egg-types, or dual-purpose. Main breeds used for egg production are the khaki Campbell and the Indian runner, easily identified by an upright posture similar to that of a penguin.[8] The Muscovy duck (*Cairina moschata*) is a separate species.[7] According to some sources, the Muscovy duck is not technically a duck but instead is a South American waterfowl that has features of ducks, geese, and turkeys.[9]

Geese are classified by weight as light-bodied (2.27-5.45 kg), medium-bodied (5.45-7.27 kg), and heavy-bodied (7.27-9.1 kg).[8] They are descended either from the greylag (*Anser anser*) or the swan goose (*Anser cygnoides*).[7] African, Embden, and Toulouse geese are considered heavy breeds and are commonly kept in backyard flocks. Medium breeds are not as common in backyard flocks, but one medium breed, the Sebastopol goose, is appreciated for its curly, fluffy feathers. Light breeds generally are only seen in exhibition collections, as they are ornamental breeds. Geese should be kept in groups, in order to flock; a good ratio is one gander (male) for up to three geese (female).[8]

Waterfowl also can be classified according to important physical and behavioral characteristics. Dabblers are birds that feed mainly on the surface of water and have their legs placed more centrally under their bodies, which enables them to walk well on land. Most domestic ducks are dabblers. Divers, on the other hand, have legs placed more caudally, enabling them to easily plunge under water to find their food. These are generally wild ducks, as are many perching species, such as wood ducks (*Aix sponsa*) and black bellied-whistling ducks (*Dendrocygna autumnalis*) that spend time perching in trees. Muscovy ducks are perchers and have sharp claws to enable them to climb and grip branches. Grazers are primarily herbivorous geese that may spend more time foraging on land than in the water.[10]

Housing

Similar to chickens, waterfowl need shelter from inclement weather and predators, but also will need access to natural sunlight and fresh water to drink from and swim in. Geese need access to fresh grass for grazing.[10] All waterfowl need a soft substrate to protect their feet from injury; hard surfaces such as concrete can contribute to pododermatitis. Grass and dirt surfaces are ideal but will require some cleaning by re-sodding or turning over the dirt one to two times per month.[11] Nest boxes for waterfowl, unlike chickens, should be placed on the floor of the shelter.[3] Perches will not be needed for most ducks, but should be provided for Muscovy ducks.[12]

The shelter should provide at least 1.5-2.0 m (5-6 ft) of floor space per bird. One nest box for every three to four ducks also should be provided.[8]

After about 8 wk of age when they are fully feathered,[8] all waterfowl will need access to a pool or pond, even geese that may not use it as much as ducks. A common recommendation for the water's depth is at least 0.6 m (2 ft). Either natural or artificial (e.g., a plastic kiddie pool) water features can be used.[11] If the pool is large, aeration may be needed to keep the water from becoming stagnant, which will contribute to the growth of pathogens. Waterfowl droppings can be voluminous and thus regular cleaning may be needed, especially of artificial pools.[10] Netting over the pool or pond should be used to prevent access by predators or wild birds that may carry diseases such as avian influenza (AI). Muscovy ducks can fly, unlike some other species, and may need prevention from escape.[13] If it is not possible to enclose the birds' habitat, a secure pen should be provided to let the birds sleep comfortably at night and protect them. Covering plants can be provided to give waterfowl a place to hide. Hollow logs and other natural features can be placed in the enclosure for enrichment and aesthetics.[11]

Keeping ducks or other waterfowl solely indoors is not recommended.[11]

Nutrition

Feeds formulated for chickens are not appropriate for waterfowl. Both ducks and geese need good-quality commercial pelleted food formulated for their life stage, as well as grass for grazing. Ducklings should be fed a grower diet for about the first 24 wk of life to ensure adequate levels of niacin, an essential nutrient. Drakes and ducks that are not laying eggs should be fed a maintenance diet. A game bird diet that does not contain added calcium or protein can be used. Laying ducks should be fed a layer diet formulated for game birds. High-carbohydrate foods such as bread, crackers, or corn should not be fed;[11] ducks are highly susceptible to obesity and related disease, such as hepatic lipidosis. Limiting high-energy foods and encouraging exercise are recommended.[10]

Keep in mind that ducks will need feeders that are wide and deep enough to allow them to scoop up their food with their wider bills. A width of 15 cm per bird should be sufficient to allow birds to eat without being forced to compete with others in the flock.[12] Drinkers separate from the water in a pool or pond also should always be available.[14]

Geese are primarily herbivores; in the wild they feed mainly on grass, aquatic plants, and some cultivated grains. They should be fed commercial game bird feeds and have plenty of access to outdoor forage. Chopped dark leafy greens can be used to supplement the diet.[10]

Sanitation

A cleaning protocol for the shelter similar to that listed above for chickens will also suffice for waterfowl. Waterfowl will drink more water and excrete more urine than chickens, and their droppings contain up to 90% moisture,[13] so droppings should be removed from the bedding daily and fresh bedding supplied as needed to allow birds to scratch and dust bathe.[12]

Husbandry of Backyard Turkeys

Biologic and physiologic parameters[1]

Heart rate (bpm): 93-163
Respiratory rate (breaths/min): 28-49
Body temperature (°C [°F]): 40.5-43 [104.9-109.4]
Typical adult weight (kg): up to 20
Lifespan (y): up to 20
Sexual maturity (y): 1

Types and varieties

Turkeys (*Meleagris gallopavo*) are usually classified as varieties rather than breeds. Most backyard flocks are composed of either heritage turkeys, which originally derived from wild turkeys that were domesticated and bred for meat production, or commercial broad-breasted turkeys. The broad-breasted turkeys are raised to produce large amounts of meat, so much that their mobility may be impaired, and they have lost the ability to breed without assistance. There are six subspecies of wild turkeys that can be found in various regions of North America.[15]

Heritage turkeys include the Beltsville small white, black, bourbon red, bronze, Narragansett, royal palm, slate, and white Holland varieties.[15]

Housing

Adult turkeys are less dependent on indoor housing than chickens. A minimum requirement is a covered roosting space that provides shade and raised perches. The lowest roost should be at least 0.5 m (20 in) high. If roosts are to be the primary means of preventing predation, they should be approximately 2.5 m (7 ft) high.[15] Wood shavings should be used as litter, rather than sawdust, which the birds may mistake for food, or newspaper, which will be slippery when wet.[16] In winter months, a structure with walls will be needed to protect them from the weather. Fencing or netting will be required to provide protection from predators.[12]

Turkeys can be raised entirely indoors, but foraging and grazing are important natural behaviors, and the quality of their welfare is improved when the opportunity to range outdoors is provided.[12] The range area should be enclosed with a fence at least 2 m (6 ft) high to prevent escape, as turkeys can fly up to 3 m (9 ft) to reach a roost. Wing feathers can be clipped to reduce flight capability.[15]

Note that chickens and turkeys should not be kept together. Chickens may carry a nematode that can be a vector for the protozoal parasite *Histomonas meleagridis*, also known as blackhead disease. Turkeys are particularly susceptible to the parasite, which often is fatal.[16,17]

Nutrition

Feeders and drinkers similar to those for chickens are appropriate as long as sufficient space and capacity to accommodate larger birds are provided.[12]

As with other poultry, commercial pelleted foods formulated for the species and appropriate for the birds' life stage are recommended, along with foraging opportunities. If a starter food containing an antiparasitic medication for coccidiosis is used, it should be specifically formulated for turkeys; medicated chicken starter food should not be substituted because different species of the protozoan infect turkeys.[16]

Sanitation

Feeders and drinkers should be thoroughly cleaned daily.[12] Perches and nest boxes should be cleaned at least weekly, droppings should be removed, and soiled litter should be replaced regularly.[15] Disinfecting the turkey house can be performed using a protocol similar to that outlined for chicken coops above.

Biosecurity

An outbreak of AI in 2022 forced a renewed focus on keeping poultry safe from disease, particularly because a single bird contracting the AI virus is likely to result in the death of the entire flock. Educating clients in protection for their pets is an area where veterinary technicians can play an important role.

The following steps are recommended to reduce the risk of contracting and transmitting diseases.

- Restrict access to the property and the birds. Visitors who have birds of their own should not be allowed to come into contact with the flock. Take steps to prevent access by wild birds.
- Ensure clothing and hands are clean before working with the flock; change clothes and wash up again after finishing. Cages, feeders, drinkers, and all other equipment should be cleaned and disinfected before coming into contact with the birds; remove droppings before disinfecting equipment and enclosures.
- Quarantine any new birds for at least 30 days before allowing them in contact with the flock. Birds that travel to an exhibition should be quarantined at least 14 days before returning them to the flock.
- Do not share equipment with neighbors or other bird owners.
- Separate any sick birds from the rest of the flock immediately.
- Know the warning signs of infectious diseases, such as acute bird deaths; respiratory distress; watery or green diarrhea; swelling around the eyes, head, and neck; and discoloration of comb, wattles, or legs.[18]

Preventing exposure to rodents and insects is also important, as they can serve as vectors for poultry diseases. Free-living wild birds can serve as reservoirs for avian diseases, including AI and exotic Newcastle disease.

Wild birds, insects, and rodents will be attracted to available food sources; keeping food inside the coop, or feeding only early in the morning and in the evening, can reduce the ability of other animals to access feeders intended for the flock. If rodenticides or insecticides are needed, bait stations should be placed in an area inaccessible to the birds. Any dead rodents or insects should be removed and disposed of promptly.[18]

Use of Pharmaceuticals in Poultry

A detailed discussion of the use of pharmaceuticals in poultry is beyond the scope of this presentation. Veterinary technicians should be aware that, although clients may regard their backyard poultry birds as pets, they are food animals in a legal sense. In the United States and some other countries, chickens and turkeys, in particular, are classified as major food animals because there is the potential for meat, eggs, offal, or other by-products such as feathers or manure to directly or indirectly become part of the human food chain. As such, they are subject to regulations designed to protect against human exposure to drug residues that have been imposed by federal, state, and local jurisdictions.[19]

Most clients have no intention of eating their pets, but eggs produced by the pets commonly are consumed. Medications administered to the birds can remain present in poultry products, including eggs, at least for a time. Consequently, a withdrawal time during which eggs or other products from the treated bird cannot be used or consumed should be observed, and the client should be advised of the withdrawal interval. These withdrawal times vary depending on the drug used. The client should be further advised that the bird being treated should be separated from other birds in the flock unless the entire flock is being treated, and that drug residues can be present on or in the bird's litter, feeder, drinker, and any other equipment being used for its care. If the withdrawal time for the drug used is unknown, contact the Food Animal Residue Avoidance Databank (FARAD) (http://wwwfarad.org) for a recommendation. A similar organization, called cgFARAD, exists in Canada and can be accessed at https://cgfarad.usask.ca/language.php.[19] For some commonly used medications, such as carprofen and amoxicillin or clavulanic acid, the withdrawal period according to FARAD is 6-8 wk. Note that these restrictions also apply to presurgical medications, such as butorphanol and midazolam.[19]

In the United States, some medications commonly used for other species of companion birds, such as enrofloxacin and metronidazole, are prohibited under all circumstances in poultry birds. A list of other prohibited drugs is available on the FARAD website.[19]

It is not uncommon for clients to try over-the-counter medications, supplements, or home remedies before bringing their bird to a veterinary clinic. Many of these will do neither harm nor good, depending on the substance used and the nature of the bird's condition, but it is a good idea to check the ingredients of the product(s) to ensure its use will not interfere with any treatment prescribed by the veterinarian.

Other Applicable Regulations

When dealing with clients who own backyard poultry birds, it can be helpful to be familiar with local and regional regulations governing issues such as zoning laws, deed restrictions, and permits. These regulations vary widely from one jurisdiction to another.

Taking a History

Getting the appropriate information for a good history on a chicken, waterfowl, or turkey patient is somewhat different from that needed for other companion bird species. For example, with a few exceptions, it is generally not necessary to ask whether a parrot has been vaccinated, but that question should be posed to owners of poultry birds. Below is a list of topics that should be covered when taking a history for backyard poultry patients.

- General signalment, including age, breed (may not be known), and gender.
- Origin (e.g., breeder, feed store, flea market, adopted from a rescue).
- How long the bird has been in the client's possession.
- Purpose: Most chickens will be classified as companion layers. If used for meat or eggs, is the purpose for personal consumption or are eggs or meat sold or donated?
- If a layer, how many eggs are laid and how often? When was the last egg laid? (In a flock situation, it may not be clear whether a specific bird is laying.)
- What is the size of the flock on the property? Is the flock exclusively chickens, or does it include other species, such as turkeys or ducks?
- How much exposure does the flock have to wildlife, including wild birds?
- Details of the housing: Coop or pen? Size of the coop? What's the substrate in the coop? Do the birds have access to an outside run? If so, what is the substrate in the run? What kind of fencing is used to enclose the run? Do the birds forage outside? What kinds of plants or furnishings are in the run? Are nest boxes and perches provided?
- Nutrition: Brand and type (starter, grower, layer diet), and in what form (crumble, mash, or pellets)? Any medicated feeds used? Are table scraps, scratch grains, or treats used?
- Is there a supplemental calcium source, such as crushed oyster shell, available? Is grit supplied?
- Has the bird received any vaccinations for conditions such as Marek's disease or coccidiosis? (Note that many clients will not know about vaccinations unless they obtained the bird from a breeder. A vague or ambivalent answer likely means there have been no vaccinations.)
- What is the presenting complaint, and how long ago was it first noticed?
- Has the bird received any over the counter or home remedies aimed at treating the presenting complaint? If so, what response, if any, was noted?
- Is there any history of other medical or health issues?[20]

Conclusions

Backyard poultry present a different challenge for veterinary technicians, but one of the similarities they share with other companion bird species is a need for appropriate husbandry practices. Veterinary technicians can play a key role in educating clients to achieve positive outcomes for these patients.

References

1. Raftery A, Kischinovsky M. Clinical examination and emergency treatment. In: Poland G, Raftery A, eds. *BSAVA Manual of Backyard Poultry Medicine and Surgery.* Quedgeley, UK: British Small Animal Veterinary Association; 2019:62.

2. Gerhardt L, Greenacre C. Common breeds of backyard poultry. In: Greenacre C, Morishita T, eds. *Backyard Poultry Medicine and Surgery*. 2nd Ed. Hoboken, NJ: John Wiley & Sons; 2021:23-44.

3. Karcher D. Basic Housing and management. In: Greenacre C, Morishita T, eds. *Backyard Poultry Medicine and Surgery.* 2nd Ed. Hoboken, NJ: John Wiley & Sons; 2021:45-55.

4. Stern H, Goodman M. Caring for Backyard Chickens. Association of Avian Veterinarians. Available at: https://aav.org. Accessed January 10, 2024.

5. Applegate T, Fowler J. Backyard poultry nutrition. In: Greenacre C, Morishita T, eds. *Backyard Poultry Medicine and Surgery.* 2nd Ed. Hoboken, NJ: John Wiley & Sons; 2021:117-130.

6. Hodgkin H, Mackinson I. Nutrition. In: Poland G, Raftery A, eds. *BSAVA Manual of Backyard Poultry Medicine and Surgery.* Quedgeley, UK: British Small Animal Veterinary Association; 2019:39-44.

7. Roberts V. Origins and types of poultry. In: Poland G, Raftery A, eds. *BSAVA Manual of Backyard Poultry Medicine and Surgery.* Quedgeley, UK: British Small Animal Veterinary Association; 2019:4-5.

8. Morishita T. Waterfowl husbandry for the avian health professional. *Semin Avian Exot Pet Med.* 2004;13(4):184-190.

9. University of Kentucky. Duck. University of Kentucky Department of Animal and Food Sciences. Available at: https://afs.ca.uky.edu/poultry/duck#:~:text=Although%20we%20call%20the%20muscovy,and%20roosters%20like%20a%20chicken. Accessed January 11, 2024.

10. Echols M. Anseriforme husbandry and management. In: Greenacre C, Morishita T, eds. *Backyard Poultry Medicine and Surgery.* 2nd Ed. Hoboken, NJ: John Wiley & Sons; 2021:56-106.

11. Association of Avian Veterinarians. Caring for Ducks. Association of Avian Veterinarians. Available at: https://aav.org. Accessed January 11, 2024.

12. Colville-Hyde A, Poland G. Husbandry. In: Poland G, Raftery A, eds. *BSAVA Manual of Backyard Poultry Medicine and Surgery.* Quedgeley, UK: British Small Animal Veterinary Association; 2019:27-30.

13. Duck Housing and Management. Cornell University College of Veterinary Medicine. Available at: https://www.vet.cornell.edu/animal-health-diagnostic-center/programs/duck-research-lab/housing-and-management. Accessed January 11, 2024.

14. Sandhu T, Dean W. Basic Duck Care. Cornell University College of Veterinary Medicine. Available at: https://www.vet.cornell.edu/animal-health-diagnostic-center/programs/duck-research-lab/basic-duck-care. Accessed January 11, 2024.

15. Schrider D. *Storey's Guide to Raising Turkeys.* North Adams, MA: Storey; 2013:14-29:127.

16. University of New Hampshire. Raising Turkeys. University of New Hampshire Cooperative Extension. Available at: https://extension.unh.edu/resource/raising-turkeys. Accessed January 11, 2024.

17. Jackson R. Gastrointestinal disorders. In: Poland G, Raftery A, eds. *BSAVA Manual of Backyard Poultry Medicine and Surgery.* Quedgeley, UK: British Small Animal Veterinary Association; 2019:184.

18. Morishita T, Derksen T. Biosecurity. In: Greenacre C, Morishita T, eds. *Backyard Poultry Medicine and Surgery.* 2nd Ed. Hoboken, NJ: John Wiley & Sons; 2021:107-116.

19. Tell L, Marmulak T, Martin K. Regulatory considerations for medication use in poultry. In: Greenacre C, Morishita T, eds. *Backyard Poultry Medicine and Surgery.* 2nd Ed. Hoboken, NJ: John Wiley & Sons; 2021:535-561.

20. Carter M. Don't be a chicken! Basic chicken history 101. *Natl Assoc Vet Tech Am J.* 2023;6-8.

Intravenous and Intraosseous Catheter Placement for the Veterinary Technician

Bri Bryan, RVT

Session #182

Affiliation: From Avian & Exotic Animal Care, 8711 Fidelity Blvd. Raleigh, NC 27617, USA.

Abstract: Avian intravenous and intraosseous catheter placement can be challenging, but with the right knowledge and tools can be done safely and effectively. The variability in patient size, comorbidities, and species are just a few factors that play a role in the type and location of the peripheral indwelling catheter selected. In this lecture, anatomy, placement sites and techniques, maintenance, troubleshooting, and tips are covered.

Introduction

Placement of indwelling intravenous (IV) or intraosseous (IO) catheters in avian species can be challenging in part because of their anatomic and physiologic differences from their mammalian counterparts.[1,2] Sedation, analgesia, or general anesthesia is often required for successful placement. Hematoma formation, maintaining patency, patient comfort and mobility, tampering, and other complications need to be considered. Despite these challenges, fluid therapy for birds is very similar to that of other species. Shock, dehydration, anesthetic procedures, hypotension, renal disease, and hypovolemia, among other conditions, are all reasons to place and maintain a catheter in avian species and oftentimes will improve the overall outcome of the patient.[3]

Relevant Avian Anatomy

The avian respiratory tract varies greatly from mammalian and reptilian species. Birds have no diaphragm and instead rely on their ability to move their sternum for respiration. The lower part of the respiratory system is composed of air sacs. Most birds have a total of nine air sacs. Birds have light but aerodynamically strong bones that have developed for flight. Some bones are pneumatic in nature, meaning that they are filled with air. It is important to know that these pneumatic bones are involved in the avian respiratory system. Many long bones like the humerus and femur are often pneumatic and should be avoided for IV or IO catheter placement. This could drown and asphyxiate the patient when IO fluids are administered and enter the respiratory system.[1,2,4] Caution should be used when placing an IO catheter in an unfamiliar species. There is much species variation amongst the class Aves and some species may have more pneumatic bones than others. This is often seen in heavy-bodied birds and some seabirds like pelicans which have a pneumatic ulna.

Sites and Techniques

Right jugular vein

Sedation and/or general anesthesia is often required. The patient should be placed in right lateral recumbency. Alcohol or sterile saline should be applied to the neck to identify the apterylae, or featherless tract, and the right jugular vein. Once the vessel is identified, the restrainer may place gentle digital pressure at the base of the neck. A 26- to 22-gauge over-the-needle catheter is inserted in a cranial to caudal direction with a 15–30-degree angle of insertion and advanced into the vessel until a flash of blood is seen. The catheter is advanced off. The stylet until the catheter is flush with the skin. To avoid kinking, the catheter should extend in a caudal direction towards the thoracic inlet. Warmed sterile saline can be used to flush the catheter and ensure patency. A slip tip injection cap or slip tip t-port is attached to the catheter. Depending on the style of catheter, an absorbable or non-absorbable suture can be used to secure the catheter with butterfly wings using a simple interrupted suture pattern. If the catheter does *not* have butterfly wings, these can be made by sandwiching two small pieces of porous tape together around the catheter and suturing the tape to the skin in the same fashion mentioned above. Oftentimes, bandaging and/or an Elizabethan collar will need to be applied to avoid slippage or self-removal.[1,4,5]

Medial metatarsal vein

Sedation is often required. The patient may be restrained in several ways depending on mentation and/or adequate sedation level. Dorsal recumbency or towel restraint with the patient held upright are the most common forms of restraint. The vessel runs superficially along the medial aspect of the intertarsal joint. Once the vessel is identified, the restrainer may place gentle digital pressure along the cranial tibiotarsus. A 26- to 20-gauge over-the-needle catheter is inserted cranially with a 15-degree angle of insertion and advanced into the vessel once a flash of blood is seen. Warmed sterile saline can be used to flush the catheter and ensure patency. A slip tip injection cap or slip tip t-port is attached to the catheter. The catheter will be secured with porous tape in a fashion like any mammal species. In a conscious or mentally bright patient, further bandaging or splinting of the leg can be applied to avoid self-removal.[4,5]

Basilic or ulnar vein

Sedation and/or local anesthesia is required. The patient is placed in dorsal recumbency with the wing extended. The basilic vein is easily identified by applying alcohol or sterile saline over the humeroradial joint. Lidocaine can be injected just distal to the insertion site. Allow the bleb to dissipate before attempting catheterization. Once the vessel is identified, the restrainer may place gentle digital pressure over the region of the proximal humerus. A 26- to 20-gauge over-the-needle catheter is inserted with little to no angle and advanced into the vessel once a flash of blood is seen. Hematoma formation at this site is commonly reported. Warmed sterile saline can be used to flush the catheter and ensure patency. Blanching of the basilic vein can also be seen. A slip tip injection cap or slip tip t-port is attached to the catheter. Depending on the style of catheter, an absorbable or non-absorbable suture can be used to secure a catheter with butterfly wings using a simple interrupted suture pattern. If the catheter does not have butterfly wings, these can be made by sandwiching two small pieces of porous tape together around the catheter and suturing the tape to the skin in the same fashion mentioned above. In a conscious or mentally bright patient, a figure-of-eight style bandage with or without a body wrap *must* be applied to avoid slippage or self-removal.[1,4-6]

Proximal tibiotarsus

Sedation and analgesia are required; general anesthesia is strongly encouraged. The patient is placed in dorsal recumbency with the knee flexed. A small pilot hole is made in the skin medial or lateral to the cnemial crest over the proximal tibiotarsus. Take care to avoid the patellar ligament when placing proximal tibiotarsus IO catheters. Insert a 27- to 22-gauge standard hypodermic needle or spinal needle and apply a firm and steady twisting motion on the periosteum into the medullary cavity of the tibiotarsus distally. Remove spinal needle stylet and confirm correct placement by aspirating blood/bone marrow or flushing the catheter with warm sterile saline.[1] Two orthogonal radiographs are strongly encouraged to confirm proper placement. A slip tip injection cap or slip tip t-port is attached to the catheter. Butterfly wings are made by sandwiching two small pieces of porous tape together around the catheter and absorbable or non-absorbable suture can be used to secure the catheter by using a simple interrupted suture pattern. In a conscious or mentally bright patient, bandaging and/or an Elizabethan collar *must* be applied to avoid slippage or self-removal.[4-6]

Distal ulna

Sedation and analgesia are required; general anesthesia is strongly encouraged. The patient is placed in dorsal or lateral recumbency. The carpus is flexed and palpated to find the dorsal crest of the distal ulna. Proximally, insert a 27- to 22-gauge standard hypodermic needle or spinal needle and apply a firm and steady twisting motion on the periosteum into the medullary cavity of the ulna. Remove the spinal needle stylet and confirm correct placement by aspirating blood/bone marrow or flushing the catheter with warm sterile saline.[1] Blanching of the basilic vein can also be seen when flushing with large volumes. Two orthogonal radiographs are strongly encouraged to confirm proper placement. A slip tip injection cap or slip tip t-port is attached to the catheter. Butterfly wings are made by sandwiching two small pieces of porous tape together around the catheter and absorbable or non-absorbable suture can be used to secure the catheter by using a simple interrupted suture pattern. In a conscious or mentally bright patient, bandaging and/or an Elizabethan collar must be applied to avoid slippage or self-removal.[4-6]

Selecting the Optimal Route

IV or IO access can be lifesaving for critical and moribund patients, as well as patients undergoing surgical procedures. Several factors should be considered when selecting the appropriate type and location for IV or IO catheterization. Patient size and species are the most relevant factors to consider. IV catheterization is generally not ideal for birds conure-sized and smaller and an IO catheter may be selected instead.[2] One study showed that 50% of IO fluid administration is passed into central circulation within 30 seconds of administration.[4,7] Moderate to severe dehydration can affect peripheral vasoconstriction and in these cases, an IO catheter may be selected over an IV catheter. Large avian species over 300 g are more likely to tolerate IV catheterization in the basilic or medial metatarsal vein with appropriate bandaging. Fowl and waterfowl are less prone to self-removal and typically tolerate IV catheterization quite well.[2] IV catheterization is also ideal for patients undergoing surgical procedures that don't require long-term fluid therapy or vascular access post-operatively.

Non-Ideal Patients

Not every patient will be an ideal candidate for IV or IO catheter placement. Birds presenting with active respiratory distress, abdominal distention, or extreme debilitation may not be good candidates for IV or IO catheterization. Other factors to consider are healed fractures, osteomyelitis, and hypocalcemic or osteoporotic birds.[4]

IO catheter placement in laying hens or birds with hormone-producing tumors, may be extremely difficult due to polyostosis. This is a condition where the medullary bone density is dramatically increased due to changes in the bird's calcium metabolism. If an IO has been recently placed, a different site or location should be selected. IV catheterization is typically preferred in releasable wildlife candidates as IO catheterization can cause trauma or joint injury. When choosing a placement site, be aware that injections given in the caudal half of the body might go through the renal portal system. Columbiformes have a jugular plexus and jugular catheterization should be avoided in these species.[1]

Supplies

Below is a list of supplies needed.

- Dilute dawn dish soap in warm water
- Over-the-needle catheters (26- to 20-gauge)
- 27- to 20-gauge hypodermic needles
- Slip tip injection cap or slip tip t-port
- Johnson and Johnson 2-inch porous tape
- Lidocaine (2%)
- Alcohol or sterile sodium chloride (0.9%)
- Chlorhexidine gluconate solution (0.5%)
- Suture (taper)
- Vet wrap (1- or 2-inch)
- Elasticon
- Gauze
- Cotton-tipped applicators or cotton balls
- Bandaging material
- Gloves

Site Preparation

Feather plucking is indicated when placing an IO catheter in the proximal tibiotarsus and distal ulna.[1] Feather plucking is often not required for the basilic vein and medial metatarsal but can depend on the patient size and visibility of the vessel. Plucking feathers is painful and patient sedation and analgesia should be strongly considered. The apterylae makes visualizing the jugular vein relatively straightforward. The catheter site should be aseptically prepared with alcohol or sterile saline and chlorhexidine gluconate-saturated gauze squares. Cotton-tipped applicators or cotton balls may be selected instead of gauze squares based on patient size. Warmed sterile saline may be selected for small or obtunded species to avoid further heat loss.

Maintenance and Complications

IV and IO catheters should be maintained similarly to those in mammalian species. Indwelling catheters should be removed and placed in a different location after 72 hours.[4,6] Multiple attempts made placing an IO catheter at the same site are contraindicated as this can cause large holes and/or multiple holes in the periosteum. Medications and/or fluids can leak out of these and can cause discomfort of the surrounding tissue and bone.[5] Potential, but rare, complications described include compartment syndrome, osteomyelitis, iatrogenic fractures, infection at placement site, infusion of fluid or leakage of blood into adjacent air sacs, and fatal hemorrhage after self-removal of the catheter.[4,6] Great care should be used to ensure a patient is never left alone with an indwelling IV/IO catheter.[6]

Conclusions

Avian species require vascular access for many of the same reasons mammalian patients do. IV and IO catheterization of avian species can be accomplished with the correct knowledge of avian anatomy and physiology, placement locations and technique, and patient preparation.

References

1. Greenacre C, Garhardt L. Psittacine and passerine birds. In: Ballard B, Cheek R, eds. *Exotic Animal Medicine for the Veterinary Technician.* Wiley-Blackwell; 2017.

2. Girling S. An overview of avian therapeutics. In: Girling S, ed. *Veterinary Nursing of Exotic Pets.* Blackwell Publishing; 2013.

3. Pollock C. Intraosseous Catheter Placement in the Bird. Available at: LafeberVet. https://lafeber.com/vet/fluid-therapy-in-the-avian-patient. Accessed January 29, 2024.

4. Schnellbacher R, Beaufrere H. Catheterization and venipuncture. In: Graham JE, Doss GA, Beaufrere H, eds. *Exotic Animal Emergency and Critical Care Medicine.* Wiley-Blackwell; 2021:460-465.

5. Doneley B. Clinical techniques. In: Doneley B, ed. *Avian Medicine and Surgery in Practice Companion and Aviary Birds.* CRC Press; 2016.

6. Lichtenberger M, Lennox, A. Critical care. In: Speer BL, ed. *Current Therapy in Avian Medicine and Surgery*. Elsevier; 2016: 584-585.

7. Lamberski N, Daniel GB. Fluid dynamics of intraosseous fluid administration in birds. *J Zoo Wildl Med.* 1992;23(1): 47-54.

Dietary Conversion in Psittacine Birds

Bri Bryan, RVT

Session #184

Affiliation: From Avian & Exotic Animal Care, 8711 Fidelity Blvd. Raleigh, NC 27617, USA.

Abstract: Captive psittacine birds have specific nutrient requirements to help ensure a long and healthy life. Client education and a thorough understanding of avian husbandry is a vital part of the role of a veterinary technician. Topics include a brief outline of avian nutrition, evidence-based and anecdotal techniques for psittacine dietary conversion, and contraindications for dietary conversion.

Introduction

Psittacine birds have complex and specific nutrient requirements. Historically, our knowledge of avian nutrition is drawn from poultry research.[1] While great strides have been made in the last two decades, there are still many unknowns. With this in mind, general recommendations for most psittacine birds include a formulated diet. Current recommendations vary, but sources generally agree that 70-80% of a psittacine bird's diet should be made up of a high-quality, formulated diet. The remaining 15-20% of the diet should consist of fresh, dark green and yellow vegetables, including sweet potatoes, broccoli, and leafy greens, and small portions of low-sugar fruits. Quality seeds and nuts, such as almonds, pumpkin seeds, and Brazil nuts, can be offered in small quantities as a treat.[2,3]

Basic Nutrient Requirements

Macronutrients[4]

Below is a list of primary macronutrients required by birds in avian diets.

- Carbohydrates
- Protein and amino acids
- Lipids and fatty acids

Micronutrients[4]

Below are the primary micronutrients required by birds in avian diets.

- Vitamins
- Minerals

Why is a Nutritionally Balanced Diet Important?

Seed-based diets are high in fat and are nutritionally deficient in calcium, vitamins, and protein.[3] Seed-based diets have been linked to numerous medical conditions including hypovitaminosis A, egg binding, dystocia, nutritional secondary hyperparathyroidism, hypocalcemic seizures, renal disease, gout, feather picking, ophthalmic disease, and more.[1] Most captive psittacine birds also have a relatively sedentary lifestyle and a high-fat diet may lead to obesity, cardiac disease, atherosclerosis, fatty liver disease, hypercholesterolemia, hypertriglyceridemia, lipomas, arthritis, pododermatitis, and more.[1] In the wild, psittacine birds lead a more active lifestyle and spend many hours foraging for food items. Depending on the season, they may eat a wide variety of foods with many different colors, textures, and tastes. The diet of a psittacine cannot be completely recreated in a captive environment; however, a formulated diet seems to meet their needs.

Challenges to Dietary Conversion

Many owners are reluctant to engage with the dietary conversion topic for many reasons. They may have concerns over stressing their bird, financial constraints, and lack the proper knowledge of avian husbandry and nutrition. Seed-based diets are plentiful on nearly every shelf of local pet stores. Without proper knowledge, owners may be led to believe this is a complete and balanced diet for their pet parrot. In recent years, manufacturers have attempted to make seed-based diets more complete for psittacine birds by adding vitamin and mineral coatings and pellets to seed-based diets. According to the label, these appear to be a more nutritious option; however, this is not accounting for the fact that psittacine birds remove seed hulls and typically only consume the seed kernels. Many parrot species are also selective and despite having pellets mixed into their diet, will not actually consume them.[1] Some clients believe their bird will never successfully transition away from a seed-based diet and avoid transitioning altogether out of fear of stressing their bird.[1] It may take several attempts with different brands and different types of food to find a formulated diet that psittacine birds actually want to consume. Every bird is different, every client is different, and every household environment is different. While there is no one-size-fits-all, there are several evidence-based and anecdotal techniques for dietary conversion.

Techniques for Conversion

Evidence-based conversion techniques

One study published in 2022 outlined three techniques for successful psittacine dietary conversion in a non-laboratory home environment. The three techniques are described below.

- "Birdie's Choice" utilizes variety and choice as the main modality. This is a reward-based method in which three different pellets were offered on a tabletop and the birds were rewarded with non-food items for every positive interaction. This method is best utilized for ground-feeding species including budgerigars (*Melopsittacus undulatus*) or cockatiels (*Nymphicus hollandicus*).[1,5]

- "Slow and Steady" utilizes timed feedings at key points in the day. All food was removed from the cage the night before and the following morning only pellets were offered when the bird is hungriest. In the afternoon, a small serving of the previous diet was offered in a separate bowl if needed. The bird was left with only pellets overnight. This method is best utilized for foot-feeder species including blue-and-gold macaws (*Ara ararauna*), umbrella cockatoos (*Cacatua alba*), and Amazon parrots (*Amazona*).[1,5]

- "Tough Love" is best utilized for birds that are used to eating a wide variety of foods. Several food dishes containing pellets were placed around the bird's enclosure in locations the bird favors. A small dish of the previous diet was placed in the enclosure in the bird's least favorite location. Over time, the previous diet was reduced and then completely eliminated. This method is best utilized for more stubborn birds.[1,5]

Of all three conversion techniques studied, 90% of the 75 birds were converted to a pelleted-based diet successfully in under 30 days regardless of the method chosen.[1]

A separate study evaluated 14 cockatiels in a controlled laboratory setting that were successfully converted to a nutritionally balanced diet within 12 days. This study measured more of the physiologic parameters and effects of dietary transition rather than method of conversion. They found body weight was not significantly affected by dietary conversion and all cockatiels accepted one of two complete diets.[6]

Another study evaluated seven Amazon parrots in a controlled laboratory setting that were successfully converted to a formulated diet within 1 month. Authors of the study found several major dietary imbalances were corrected once converted to a nutritionally balanced diet.[7]

Anecdotal conversion techniques

- Many psittacine owners find success introducing new food items through bird bread. Fruits, vegetables, and formulated diets can be baked into bird bread and offered to parrots in addition to their new diet as the transition slowly occurs. There are many commercially available products such as Harrison's Bird Bread Mix, Original (Harrison's Bird Foods, Brentwood, TN, USA) or other homemade recipes available online.[8]

- Environmental changes are great enrichment for birds and encourage them to explore new areas or even food items in their enclosure. This can be accomplished by moving food bowls, providing new feeding surfaces on the bottom of the enclosure for ground-feeding species, and even moving the entire enclosure to a new area.[8]

- Certain formulated foods have more success as a transition diet such as Harrison's High Potency (Harrison's Bird Foods, Brentwood, TN, USA) or High Potency Pepper (Harrison's Bird Foods, Brentwood, TN, USA). These pellets have a higher fat content and can be more appealing to birds when initially transitioning.[8]

- Juice-soaked formulated diets have also anecdotally shown success for stubborn birds. The pellets are soaked in the bird's favorite fruit juice and offered. They should be removed and replaced frequently to avoid spoilage. Over time the amount of juice should be reduced until it is discontinued entirely.[9]

- Some avian species are willing to try new food items with the example of another. This can be accomplished in several ways. A mirror can be placed below the feeding dish so the bird will be tricked into eating in the company of another bird. Owners can also eat at the same time as their birds by sharing mealtime or showing them that they are also "eating" the formulated diet being offered. A recent study showed birds using a parrot-to-parrot video-calling system showed high engagement with bird-specific interactions and behavior learning.[10] Perhaps, dietary conversion could be taught through parrot-to-parrot video-calling; however, further studies would be needed.

- Perhaps the most tried and true method for dietary conversion for any species is a slow, gradual transition over 3-4 weeks, slowly increasing the amount of formulated diet offered and decreasing the amount of seed in the same bowl. Formulated diets should always be put on top of the seeds so the bird will have to become accustomed to the taste, texture, and smell as they eat.[9]
- Lafeber Company offers a resource for veterinarians to access a dietary conversion program and a free dietary conversion kit (Lafeber Company, Cornell, IL, USA). This kit provides instructions for dietary conversion and includes 5 different trial-sized portions of pellets and other formulated foods, including Avi-Cakes and Nutri-Berries to help get owners started.[11]

Monitoring

Regardless of the transition method selected, monitoring fecal droppings and weight is vitally important. The author recommends obtaining daily weights at the same time each day to ensure an accurate weight.[3] This may require some prophylactic training on the part of the owner before starting dietary conversion. Recommendations for pausing dietary conversion include losing more than 3% of the starting body weight in 1 week or more than 10% bodyweight throughout the conversion.[1,8] It is also important to notify owners that some birds will starve themselves by refusing new foods. If this occurs, an alternative method for dietary conversion should be considered. Fecal volume, frequency, consistency, and color change should be monitored. It should be noted that some fecal changes can be normal as formulated diets vary in moisture content and dyes or pigment, but a veterinarian should be consulted to ensure no concerning changes are occurring.[8]

Contraindications for Conversion

Dietary changes should not be attempted if there are environmental changes, illnesses, or new cage mates being introduced to the household. Hospitalization is also not the time to consider dietary changes.[4] Dietary conversion can be a stressful process. Some owners note increased screaming and aggression while undergoing changes.[1] With this in mind, birds should be examined by a veterinarian and deemed healthy before any dietary changes.

Conclusions

Converting psittacine patients to a nutritionally balanced diet requires patience and persistence on the part of the owner. Several options for successful dietary conversion exist, both from an evidence-based and anecdotal standpoint. Seed-based diets can lead to many long-term medical conditions. While our understanding of the exact needs of psittacine nutrition requires further studies, our current recommendations seem to meet their needs.

References

1. Cummings AM, Hess LR, Spielvogel CF, Kottwitz JJ. An evaluation of three diet conversion methods in psittacine birds converting from seed-based diets to pelleted diets. *J Avian Med Surg.* 2022;36(2):145-152. doi:10.1647/21-00025.

2. Koutsos E, Gelis S, Echols MS. Advancements in nutrition and nutritional therapy. In: Speer BL, ed. *Current Therapy in Avian Medicine and Surgery*. St. Louis, MO: Elsevier; 2016:142-148.

3. Greenacre C, Gerhardt L. Psittacine and passerine birds. In: Ballard B, Cheek R, eds. *Exotic Animal Medicine for the Veterinary Technician*. Wiley-Blackwell; 2017:49-50.

4. Girling S. Avian nutrition. In: Girling S, ed. *Veterinary Nursing of Exotic Pets.* Wiley-Blackwell; 2013:170-181.

5. Zupreem. Converting Your Bird to Healthy Pellets. Available at: https://zupreem.com/avian-conversion. Accessed January 29, 2024.

6. Foreman AL, Fallon JA, Moritz JS. Cockatiel transition from a seed-based to a complete diet. *J Avian Med Surg.* 2015;29(2):114-119. doi:10.1647/2013-069.

7. Brightsmith DJ. Nutritional levels of diets fed to captive Amazon parrots: does mixing seed, produce, and pellets provide a healthy diet? *J Avian Med Surg.* 2012;26(3):149-160. doi: 10.1647/2011-025R.1.

8. Harrison's Bird Foods. Quick Tips for Converting Your Bird to Harrison's Bird Foods. Available at: https://www.harrisonsbirdfoods.com/wp-content/uploads/2023/07/Quick-Tips.pdf. Accessed January 29, 2024.

9. Roudybush. Switching to Roudybush. Available at: https://roudybush.myshopify.com/pages/switching-to-roudybush. Accessed January 29, 2024.

10. Kleinberger R, Cunha J, Vemuri MM, Hirskyj-Douglas I. Birds of a feather video-flock together: design and evaluation of an agency-based parrot-to-parrot video-calling system for interspecies ethical enrichment. *Proc Conf CHI Human Fact Comput Syst.* 2023;534:1-16. doi:10.1145/3544548.3581166.

11. LafeberVet. Lafeber's New Dietary Conversion Kits. Available at: https://lafeber.com/vet/lafebers-new-diet-conversion-kits/. Accessed February 2, 2024.

Avian Influenza and the Sick Pet Chicken

Brynn McCleery, DVM, DABVP (Avian Practice)

Session #205

Affiliation: From Red Bank Veterinary Hospital, 197 Hance Avenue, Tinton Falls, NJ 07722, USA.

Abstract: This session briefly reviews the origin of the current high pathogenicity avian influenza (HPAI) outbreak and its significance to wildlife, poultry, and humans. Clinical signs of disease and testing are reviewed. The focus of the session is on in-hospital precautions to take when treating poultry during an HPAI outbreak.

Introduction

The United States and Europe are currently experiencing the deadliest outbreak of avian influenza virus (AIV) in history. This outbreak of H5N1 AIV began in 2020 and has been associated with an enormous loss of poultry, both commercial and backyard. This strain is especially transmissible and persistent in the environment and has also been associated with outbreaks and death of wild birds and increased spillover into mammals.

Avian Influenza Viruses

The avian influenza viruses (AIVs) are influenza type A viruses belonging to the family Orthomyxoviridae.[1,2] They are single-stranded, negative-sense ribonucleic acid (RNA) viruses with segmented genomes that code for 10 viral proteins.[2,3] Influenza type A viruses have more genetic variation and a broader host range compared to the other two types of influenza viruses.[3] Viral naming is based on the influenza type, place of isolation, strain, designation, and year of isolation. Further classification into subtypes is based on two surface glycoproteins: hemagglutinin (HA) and neuraminidase (NA).[2] HA mediates viral attachment and entry into the host cell by binding to sialic acid receptors on the host cell surface and is responsible for neutralizing host antibodies. There are 18 known HA subtypes and HA 1-16 are known to infect birds.[3,4] NA catalyzes the cleavage of virus particles from the sialic acid sites on the host cell, thereby allowing for spread of the virus within the host.[2,3] There are 11 NA subtypes; NA 1-9 are known to infect birds.[3,4] Many combinations of HA and NA proteins are possible to create different subtypes of AIV, and related viruses within a subtype are referred to as a lineage.

Genetic drift (e.g., mutations) and genetic shift (e.g., reassortment through exchange of gene segments) provide influenza viruses with a high capacity to change. Antigenic alterations, especially in the HA and NA sites, occur frequently and are a mechanism through which the virus adapts to the host and survives.[2] Mutations (i.e., antigenic drift) that avoid antibody neutralization are selected for over time. These types of changes occur yearly to every few years. Reassortment of the eight gene segments coding the viral proteins can lead to larger variations (i.e., antigenic shift) which may result in new viruses for which a host population has no immunity. This is the cause of influenza pandemics.[2]

Pathogenicity of the AIVs is classified as high (HPAI) or low (LPAI) based on the ability of the strain to produce disease and mortality in chickens, as well as a genetic sequence in HA.[1,3] Natural infections of LPAI typically cause no clinical signs or tissue lesions in wild birds, and LPAI infections in poultry are usually asymptomatic or cause mild symptoms.[1,3,4] H5 and H7 subtypes have the potential to mutate or evolve into HPAI viruses;

however, not all viruses of these subtypes cause HPAI.[1] HPAI are rarely isolated from wild birds and usually occur in wild birds in the vicinity of outbreaks of HPAI in poultry.[1] HPAI virus strains are theorized to occur only after the viruses have crossed from wild birds into poultry, where they mutate and adapt.[3,4]

Recent HPAI outbreaks

1996 outbreak

H5N1 HPAI (A/goose/Guangdong/1/1996) reinfected wild birds after spilling over into poultry. The outbreak was contained through culling all the poultry in Hong Kong, but the AIVs continued to circulate in healthy duck populations in the surrounding areas. This strain is now considered endemic in six countries (i.e., Bangladesh, China, Egypt, Indonesia, India, and Vietnam).[3,5]

2014-2015 outbreak

An outbreak of HPAI subtype H5Nx occurred in poultry in the United States and Canada. HPAI H5N8 spread through the flyways from Asia to North America. The outbreak affected 211 commercial flocks and 21 backyard flocks in 21 states, leading to $3.3 billion in economic losses and the loss of 50 million birds. Three mortality events were reported in wild birds, affecting snow geese and ringed-necked ducks, and multiple captive raptor deaths were reported. Subtypes involved included H5N8, H5N1, and H5N2.[6,7] It was last detected in poultry in June 2015.

2021 outbreak

The current 2021 H5N1 outbreak (variants in clade 2.3.4.4b) is a different strain than the 2014 outbreak. The infection traveled from western Europe to eastern Canada and has been detected in all four flyways with most early infections centered around the intersection of the Central and Mississippi migratory flyways. There has been wide geographic spread with all states in the United States affected except Hawaii and all continents affected except Antarctica and Australia. The current outbreak has been associated with a higher wild bird mortality than that of 2014, as well as with a wider diversity of species affected. Swans, geese, waterbirds (e.g., pelicans, mergansers), raptors, and scavengers are all considered highly susceptible. There had also been increased sporadic spillover into a variety of mammalian species (e.g., harbor seals, grey seals, foxes, skunks, bobcats, lynx, Eurasian otters, Eurasian polecats, European badger, domestic pigs, ferret, common raccoon dog, opossums, bears, farmed mink, sea lions, domestic cats).[4,5] Spillover into humans has been rare and associated with recent poultry exposure. Symptoms in humans have been nonexistent or very mild. This is the deadliest outbreak in U.S. and European history.[8] The virus has repeatedly made it through the summer season and may become endemic in wild birds.[5]

Natural and aberrant hosts

AIVs infect more than 100 different bird species worldwide and are endemic in some wild birds. Ducks (especially dabbling ducks such as mallards, pintails, and widgeons), swans, geese, terns, shorebirds (including storks, plovers, and sandpipers), and gulls are considered natural reservoirs.[3,4] Wild waterfowl and shorebirds may be infected with no symptoms and can spread the virus during migration.

Wild birds carrying LPAI can infect aberrant hosts that come into direct contact with infected birds or their droppings. Aberrant hosts include poultry (e.g., chickens, turkeys, pheasants, quail, domestic ducks, geese, and guinea fowl), pigs, wild mammals, and humans. After transmission to poultry, LPAI often loses its adaptation to wild birds and this helps prevent reinfection. With the current outbreak, H5 and H7 infections in poultry have been able to reinfect wild birds resulting in extensive geographical spread of the virus.[4]

Transmission and incubation

Viral particles are shed in secretions from the eyes, nose, and mouth, as well as in the urates and feces. Transmission is through the oral-fecal route and can occur from direct contact with infected birds, through contact with droppings of infected birds, or through fomites and mechanical transfer (e.g., contaminated clothing, equipment, truck tires).[1,2] Airborne transmission is not considered a significant route of transmission; however, droplet transmission is possible in close contact. HPAI is extremely infectious with rapid flock-to-flock spread reported. An incubation time of 1-8 days is commonly reported, but incubation times may extend to 14-21 days depending on the strain.

Clinical symptoms

LPAI infections may be asymptomatic in poultry or cause mild symptoms including the following:[1,3,4]

- Mild to moderate respiratory symptoms
- Fluffed feathers
- Reduced egg production
- Increased secondary infections
- Low mortality

HPAI is associated with high acute mortality (up to 100% in chickens, often within 48 hours). Symptoms include the following:[1,3]

- Reduced drinking (may be one of the first symptoms)
- Extreme depression, fluffed feathers, quiet mentation, lethargy
- Recumbent
- Unresponsive
- Severe respiratory symptoms: gaping, excessively watery and closed eyes, coughing, sneezing, gurgling, rattling, nasal discharge, dyspnea
- Gastrointestinal symptoms: severe diarrhea, discolored stool
- Swelling and purple discoloration of the head (comb, wattles, eyelids), neck, and legs
- Subcutaneous bleeding
- Cessation or marked reduction in egg production
- Irregular eggs (e.g., soft, thin-shelled, misshapen)
- Anorexia, reduced feed intake

- Neurologic symptoms: incoordination, head and body tremoring, drooping and/or paralyzed wings, hindlimb paresis or paralysis, twisted head and neck, swimming in circles

- Acute unexplained death

It is essential to become familiar with the clinical signs associated with avian influenza and to have avian influenza on the differential diagnosis list for birds presenting with the above symptoms. Unfortunately, LPAI and HPAI can be difficult to differentiate from other common avian diseases. HPAI should be on the differential diagnosis list for any acute unexplained poultry mortality.

Treating Poultry in the Veterinary Clinic: Mitigating Risk

While commercial poultry are typically treated on-site, backyard poultry are more likely to be brought to the veterinary clinic when ill. Due to the current HPAI outbreak, treating poultry in the veterinary clinic could lead to viral transmission between patients, especially since the current HPAI viral strain is quite persistent in the environment. It is also possible that staff members or clients could bring the virus into the building on clothing, shoes, or other equipment, especially if they have outdoor birds at home. To minimize risks to other avian and mammalian patients, as well as to staff members, a biosecurity protocol should be developed for all veterinary clinics treating poultry.

Anamnesis screening questions

Avian patients should be screened to determine their risk of exposure to HPAI. Hospitals should develop a set of questions that can be emailed, asked over the phone, or asked curbside before the patient enters the building. This will help keep high-risk patients out of common areas such as the waiting room. Potential questions may include the following:

- Do you live near a controlled area or recent outbreak?

- Do you live near any bodies of water or agricultural fields?

- How many birds in the flock are sick? What is the general flock health?

- Have there been any recent flock deaths?

- Do the birds have outdoor access?

- Is the outdoor access free range or contained? Are they supervised when free range?

- How much time is spent outdoors?

- Are there any wild bird feeders or other wild animal feeding stations?

- Does anyone in the home do any sport hunting of wild birds?

- Does anyone in the home work with or volunteer with any avian species?

- What are the symptoms prompting the visit?

Note: Waterfowl with outdoor exposure should be considered potentially infected with avian influenza even if asymptomatic.

Poultry and other avian species with outdoor access, wildlife exposure, compatible clinical signs, and from homes with other sick birds and/or acute flock deaths are considered high risk.

Separation

Approaches to separating the patient include the following:

- Avoid entrance to the hospital. If safe to do so, the bird can be triaged and/or examined outside in a temporary shelter or the owner's vehicle. A temporary triage station is especially recommended for veterinary clinics or wildlife rescues seeing a large number of wild birds, including waterfowl and wading birds. Home visits may also provide a solution for practices that have mobile veterinary capabilities.

- Limit client access to the building. If the screening questions identify a high-risk patient, the bird can be brought to a separate isolation/treatment area while the client remains in their car (similar to curbside policies enforced during COVID-19). This can help limit fomite transmission into the hospital.

- Schedule end-of-day examinations. Although not always possible, poultry examinations or appointments with high-risk patients (based on species or clinical signs) can be scheduled at the end of the day.

- Use separate treatment and housing areas. Restricted or isolation wards should be utilized for high-risk patients (based on history and clinical signs). These spaces should ideally have separate air space from areas where other birds may be examined or hospitalized.

- Limite mobile veterinarian visits: If possible, limit ambulatory calls to a single poultry site per day.

Limit staff risk

Approaches to limiting risk to staff include the following:

- Use basic hygiene for all avian cases including wearing protective outerwear (e.g., white coat, lab coat) and washing hands.

- Use adequate personal protective equipment (PPE) when examining high-risk or suspect cases (e.g., masks [N95 or more protective], nitrile gloves, disposable coveralls or gowns, eye protection, rubber boots, rubber boot covers or disposable shoe covers, hair covers). Double bag and dispose of disposable PPE after use. Non-disposable items should be laundered with bleach and dried with high heat (separate from other laundry) or properly disinfected.

- Limit the number of staff that handle high-risk patients. If possible, avoid having staff with birds at home (e.g., poultry, psittacine or passerine birds) from working with suspect AIV cases.

The current strain is considered a low zoonotic risk; however, the use of PPE will protect staff members and help limit transmission via fomites in the clinic.

Disinfect

Approaches to disinfecting include the following:

- Use disinfectants labeled to kill avian influenza virus on hospital areas and equipment coming into contact with birds suspected of avian influenza. Follow chemical labels for proper use and contact time.

- Use additional measures for mobile veterinarians. Wash vehicles before entering and when leaving areas with poultry. Change clothing and clean boots between each site.

Educate staff and clients

Stay up-to-date on outbreaks in your area through the Centers for Disease Control and Prevention (CDC) and U.S. Department of Agriculture (USDA) websites. Staff with poultry, outdoor birds, or indoor birds may want to consider changing clothing at work to limit exposure to hospitalized patients and to their birds at home. All staff in the clinic should understand basic biosecurity principles and should be trained with any new biosecurity protocols put in place to address the HPAI outbreak.

The USDA's Animal and Plant Health Inspection Service (APHIS) and the Association of Avian Veterinarians have excellent resources on biosecurity practices for backyard flocks and outdoor captive birds. Direct owners to the following resources to help reduce the risk of AIV transmission to their flock:

- www.aphis.usda.gov/animalhealth/defendtheflock

- https://www.aav.org/general/custom.asp?page=avian_influenza

Testing

HPAI is reportable in most countries. In the United States, poultry suspected of HPAI should be reported to the USDA APHIS, as well as to state animal health officials. State guidelines regarding testing and follow-up procedures may vary.

Reverse transcription polymerase chain reaction (RT-PCR) is typically recommended; however, agar gel immunodiffusion testing may also be performed for evaluation of AIVs. Appropriate samples include an oropharyngeal swab in gallinaceous poultry or a cloacal swab in waterfowl.[2] Serology is not recommended as mortality occurs acutely before antibody production. Post-mortem findings are highly variable and not pathognomonic for AIVs. Petechiae and hemorrhage are common, especially of the pancreas.

Testing of wild bird populations is performed for surveillance purposes only. Eradication of HPAI from wild birds is not being attempted. Preventive measures are instead being used to help limit spillover to domestic species from wildlife. If wild birds are found dead and brought to your clinic, the carcasses should be double bagged and kept outside if possible before collection for testing.

Vaccination

Vaccination is not allowed in the United States at this time due to impacts on international trade as vaccinated birds cannot be easily distinguished from infected birds through testing. Furthermore, vaccinated birds may still become infected and shed the virus. The USDA is currently trialing vaccines and policies may change in the future based on the current outbreak. As with other poultry vaccines, use in backyard flocks may be limited due to vaccine bottle size and price.

References

1. Alexander DJ. A review of avian influenza in different bird species. *Vet Microbiol.* 2000;74:3-13.

2. Thomas JK, Noppenberger J. Avian influenza: a review. *Am J Health Syst Pharm.* 2007;64:149-165.

3. Nunez IV, Ross TM. A review of H5Nx avian influenza viruses. *Ther Adv Vaccines Immunother.* 2019;7: 1-15.

4. Centers for Disease Control and Prevention (CDC). Technical report: highly pathogenic avian influenza A (H5N1) viruses. Available at: https://www.cdc.gov/flu/avianflu/spotlights/2022-2023/h5n1-technical-report.htm#print. Accessed March 25, 2023.

5. Adlhoch C, Fusaro A, Gonzalez JL, et al. Avian influenza overview March-June 2022. *EFSA J.* 2022;20(8):7415.

6. U.S. Department of Agriculture (USDA). Final report for the 2014-2015 outbreak of highly pathogenic avian influenza (HPAI) in the United States. Available at: https://www.aphis.usda.gov/animal_health/emergency_management/downloads/hpai/2015-hpai-final-report.pdf. Accessed March 13, 2024.

7. Greene JL. Update on the highly-pathogenic avian influenza outbreak of 2014-2015. Available at: https://nationalaglawcenter.org/wp-content/uploads/assets/crs/R44114.pdf. Accessed March 13, 2024.

8. Animal and Plant Health Inspection Service (APHIS). 2022-2024 confirmations of highly pathogenic avian influenza in commercial and backyard flocks. Available at: https://www.aphis.usda.gov/aphis/ourfocus/animalhealth/animal-disease-information/avian/avian-influenza/hpai-2022/2022-hpai-commercial-backyard-flocks. Accessed March 13, 2024.

Talking to Clients About Ultraviolet Light for Pet Birds

Jovanna Berls, CVA, LVT, FFCP (Veterinary-Avian)

Session #199

Affiliation: From ABC Animal & Bird Clinic, 11930 Highway 6 South, Sugar Land, TX 77498, USA.

Abstract: There is substantial research to show that ultraviolet light is beneficial to birds, helping to prevent or slow the progression of osteoporosis. Light in the ultraviolet-B (UVB) spectrum, in particular, is essential for calcium metabolism and good immune function. Providing UVB can be a challenge for many pet bird owners, who naturally have concerns about taking their pets outside and whether use of an ultraviolet lamp can be both safe and convenient. This presentation covers why ultraviolet light is recommended for pet birds, what safety issues may arise, and how to encourage and support clients who want to include it in their bird-care routine.

Introduction

Research has shown that full-spectrum lighting, especially ultraviolet-B (UVB) light in the 290-320 nm spectrum, is important for optimal health and welfare in pet birds because of the role it plays in synthesizing vitamin D for calcium metabolism. It stands to reason, then, that full-spectrum lighting is recommended for all pet birds. Providing that lighting is not as simple as it may seem at first glance, which raises questions and concerns for the bird owners. Many veterinary technicians are familiar with the need for full spectrum lighting in pet reptiles, but discussion of it in connection with pet birds has come into focus only relatively recently. This presentation is intended to give veterinary technicians tools for assisting clients in ensuring their birds have the light they need to thrive.

Why Ultraviolet Light is Needed

Birds need adequate access to vitamin D, an essential hormone, to metabolize and absorb the calcium and phosphorus needed for growth and development of bones and maintenance of bone density, as well as to support immune and neuromuscular function. Mobilization of stem cells to form osteoclasts, cells that are responsible for drawing calcium stores from bone, requires vitamin D. In conjunction with calcium, vitamin D has been shown to inhibit the development of common diseases such as osteoporosis, hypocalcemia, and hypovitaminosis D in poultry.[1] Research has shown that both vitamin D and calcium are needed to ensure reproductive success in birds.[1]

Of particular concern are conditions sometimes collectively known as metabolic bone disease, which involve loss of bone mass. These conditions include rickets, osteomalacia, and osteoporosis. These diseases often are associated with deformation or fracture of bones, as well as reproductive failure and dystocia related to thin-shelled eggs in female birds. Osteoporosis in birds resembles the age-related condition in humans and is exhibited as resorption of cortical bone, porous bones, joint disorders, and pain.[2]

A complete discussion of vitamin D synthesis in the body and its complex relationship with calcium metabolism is beyond the scope of this presentation. Vitamin D is acquired in the body in two ways: through diet via absorption in the gastrointestinal tract and through synthesis of 7-dehydrocholesterol to previtamin D_3 in the epidermis and the cornea after exposure to UVB light. Most pet birds fed commercially available pelleted diets will obtain some dietary vitamin D that will be absorbed from the intestine in their food; however, even dietary supplementation can be insufficient to prevent vitamin D deficiency and hypocalcemia.[3] Many pet birds are not fed (or do not eat) pelleted foods, leaving them susceptible to hypovitaminosis D and other related conditions. Those birds need another source of the vitamin because they are not consuming it in their diet. In addition, research has shown that the half-life of vitamin D_3 synthesized in the skin through UVB exposure is much longer than that of dietary vitamin D_3. Research in green iguanas (*Iguana iguana*) has shown that using the dietary route of acquisition of vitamin D may not provide sufficient levels unless it is given in such massive doses that toxicity could result, indicating that synthesis from the skin may be needed even if dietary supplies are provided. Studies in chickens have shown they have no dietary requirement for vitamin D if they are exposed to adequate levels of UVB light.[1]

An additional consideration is research showing that birds exposed to natural or artificial ultraviolet light may be motivated to engage in higher levels of activity, increasing the biomechanical load on the bones. This load bearing is considered crucial for normal bone development, and inactivity is known to be detrimental to bone mass and skeletal health.[2]

How to Provide UVB Lighting

The best source of exposure to UVB light is natural sunlight, which includes both UVB and ultraviolet A, which is believed to help birds select ripe foods and identify mates.[4,5] If this method is used, it is recommended to provide sunlight exposure for 20-30 min, 2-3 times/wk. Caution is required to prevent escape, attacks from predators such as raptors, exposure to diseases that may be carried by wild birds, and to protect pet birds from inclement weather conditions. Companion birds should be taken outdoors only if they are in a cage or wearing a harness, and they should never be left unattended. The ideal outdoor temperature is 18-29 °C (65-85 °F). Overheating may be exhibited by panting or wings held away from the body.[4]

Most pet birds will need indoor exposure to UVB at least in the winter, when the weather is not conducive to taking them outside. Birds that have been on a vitamin-deficient seed diet or who are chronic egg layers will particularly benefit from having UVB exposure year-round. Indoor exposure to UVB requires a special lamp that can be attached to the top of the bird's cage, hung from above the cage, or provided by a floor lamp tall enough to direct its light down from above the cage. Ideally, the lamp should be used a minimum of 3-4 hours per day to mimic mid-morning sunlight.[4]

Keep in mind that anything between the UVB lamp and the bird will decrease the amount of UVB the bird receives. In particular, placing a bird's cage near a window will not provide adequate exposure because the glass will either absorb or obstruct the radiation.[1]

How Much Exposure is Needed and How Much is Safe?

Different species of birds likely have different requirements for UVB exposure, but research in that area is currently lacking. It is important to keep in mind that overexposure can be harmful, leading to skin cancer and cataracts, similar to adverse effects noted in humans.[4] Birds have thinner skin and corneas than mammals, so they may be more sensitive to ultraviolet light than humans or other species; birds with unfeathered areas where direct exposure can occur, such as macaws, may be especially susceptible to overexposure. Generally, the UV index, or UVI, is used to measure the intensity of ultraviolet radiation in levels from 1-11; the level occurring in a client's local area is readily available on many weather apps accessible via smartphone or smartwatch. A UVI less than 5 is generally considered safe. Handheld ultraviolet meters also are available.[4]

The distance between the lamp and the bird's head is important to ensure adequate exposure and prevent overexposure, as ultraviolet radiation effectively extends only so far from the source. UVB lamps and bulbs vary greatly from one model to another, so it's important to note the package instructions to ensure the lamp is used at the recommended distance. In general, high output bulbs should be positioned no closer than 15 cm (6 inches), and low output bulbs should be no farther than 30 cm (12 inches), measured from the top of the bird's head at its highest perching location.[4]

The main adverse effect of overexposure is primary skin damage, which can be noted as sunburn or erythema. Chronic exposure can be a risk factor for skin cancer. Birds in the wild have the option to take shelter under leaves or other structures that block direct exposure to sunlight; similarly, companion birds should have the ability to avoid the light if they feel the need. Positioning the lamp to provide a gradient of exposure is recommended for that reason.[1] Note that the lamp should never be positioned at the side of the cage, as that can contribute to formation of cataracts.[4] While dietary oversupplementation can result in hypervitaminosis D, hypervitaminosis D directly related to exposure to UVB light has not been reported in any species.[1,6]

Any squinting, face rubbing, or redness of the skin around the eyes can signal overexposure. Discontinue use immediately if those signs are seen.[4]

Most bulbs should be replaced every 6 mo, even if the package label says they will last longer than that, because the amount of UV output will decrease, often significantly, with time, even if the bulb is still producing light.[1,4] Ideally, a solar meter can be used to check the lamp's output periodically.

An additional consideration is the lamp's flicker rate. Fluorescent lamps typically produce light in an on-and-off sequence at a rate too fast for the human eye to detect, commonly 100-120 Hz. Birds can see the lamp flickering at that rate, and some research suggests they may experience stress and behavioral changes, such as reduced eating and drinking or higher levels of preening, as a result of being able to perceive the flicker. One study showed that stress effects were lower with the use of high-frequency lamps (> 30 kHz, a rate that is imperceptible to animals).[7]

Discussing Ultraviolet Light with Clients

The first thing to keep in mind when discussing provision of ultraviolet light with clients is that the subject is complex. It may not be an easy or quick conversation. Offering a printed handout is a good idea to give the client a reference to take home; a useful handout is available on the Association of Avian Veterinarians website.

If the client prefers to expose the bird to natural sunlight, it may be helpful to discuss the outdoor environment to be used and the bird's normal behavior and demeanor. Is direct sunlight available, or are there trees or walls that will block or filter the sun? Will the client need to train the bird to accept a harness? If the bird will be in an outdoor cage, where will the cage be located and how will the bird be transported to the cage? Is there anything in the environment that may be stressful for the bird, such as loud noise or the presence of other animals? If the bird vocalizes, will neighbors object?

If using a UVB lamp, note that lamps vary greatly, often requiring some research before purchase; some models are primarily designed for use with reptiles, and clients may not initially recognize them as appropriate for their birds. Some suggested brands and types are listed on the Association of Avian Veterinarians website. Depending on the type and location of the bird's cage, the lamp may require some modification or effort to install it. For parrots, a method of protecting the electrical cord from inquisitive beaks will be needed. Advise the client to monitor the temperature in the cage to ensure the lamp is not overheating any areas. The client may need to think through the best time for the lamp to be turned on and off, which may require the use of a timer, depending on the client's daily routine.

Conclusions

Full-spectrum lighting including UVB light is important for the optimal health and welfare of pet birds. Although providing full-spectrum light initially can seem complex and inconvenient, once a workable plan is implemented, the needs of both the pet and the client can be met.

References

1. Watson M, Mitchell M. Vitamin D and ultraviolet B radiation considerations for exotic pets. *J Exot Pet Med.* 2014;23:369-379.

2. Echols M. Avian metabolic bone diseases. *Proc Annu Conf Assoc Avian Vet.* 2017;217-227.

3. Nightengale M, Stout R, Tully T. Plasma vitamin D (25-hydroxyvitamin D) levels in Hispaniolan Amazon parrots (*Amazona ventralis*) housed indoors over time. *Avian Dis.* 2022;66(2):148-154.

4. Association of Avian Veterinarians. Ultraviolet Lighting for Companion Birds: Benefits & Risks. Available at: https://www.aav.org. Accessed January 9, 2024.

5. West J, Tully T, Nevarez J, Stout R. Effects of fluorescent lighting versus sunlight exposure on calcium, magnesium, vitamin D, and feather destructive behavior in Hispaniolan Amazon parrots (*Amazona ventralis*). *J Avian Med Surg.* 2019;33(3):235-244.

6. Olds J, Burrough E, Madson D, et al. Clinical investigation into feed-related hypervitaminosis D in a captive flock of budgerigars (*Melopsittacus undulatus*): morbidity, mortalities, and pathologic lesions. *J Zoo Wildl Med.* 2015;46(1):9-17.

7. Evans J, Smith E, Bennett A, et al. Short-term physiological and behavioural effects of high- versus low-frequency fluorescent light on captive birds. *Anim Behav.* 2012;83(1):25-33.

Feeding the Pet Parrot

Jill K. Murray, RVT, RLATG

Session #185

Affiliation: From Lafeber Company, Cornell, IL 61319, USA.

Abstract: Feeding a pet parrot a nutritious diet is a complex journey, yet it is vital for optimal health and longevity. With nutritional deficiencies being a common primary or secondary presenting complaint, the veterinary health care team needs to be a knowledgeable resource for owners. Understanding the types of diets available, along with their advantages and disadvantages, better helps in our role to educate and guide owners in dietary selection for their pet parrots.

Introduction

Feeding a pet parrot a nutritionally complete and balanced diet is more complex than selecting a bag of bird food off the shelf. Not all foods are created equal, guidelines are open to interpretation, and the manner in which a bird consumes the food can result in nutritional deficiencies. When selecting a diet, one should consider the species of parrot, life-stage and health status, to promote optimal health, and meet energy requirements, while encouraging normal expression of foraging and enrichment behaviors. This requires feeding a multi-component diet of commercial feed in addition to fruits, vegetables, and whole foods. Understanding the different components of the diet is vital to providing individualized nutritional recommendations.

Most pet bird owners lack advanced knowledge to make an educated decision on the nutritional needs of their bird(s). The veterinary team should be actively educating and guiding pet owners in the development of an individualized feeding strategy for their parrot(s) because poor nutrition can greatly impact a bird's health and longevity.

Psittacine Nutrition

The primary function of eating for any species is to obtain nutrients and energy. Without these, survival is not feasible. In the wild, psittacine birds eat a variety of seeds, fruits, nuts, flowers, grasses, proteins and more. Seasonal changes can affect what the birds consume, since the availability of food items changes. Ideally, captive diets should reflect a bird's natural diet, although many limitations and challenges exist. Often, many components of the bird's natural diet are not readily available in our location and foraging constitutes only a small part of the bird's food intake in captivity.

Beyond providing nutrients to support the basic requirements to sustain life, nutrition is important to the physical makeup of the birds. Appropriate diet contributes to maintaining the bird's beak occlusion, healthy feathers, skin and nails. Unlike other orders of birds, diet does not affect feather coloration in psittacine species.[1] Nutritional deficiencies may constitute the primary cause of up to 90% of all clinical problems in pet birds presented to veterinarians.[2] Often, these deficiencies are subclinical, non-specific, and challenging to diagnose.

Commercial Diets

Commercial bird foods have seen many changes. Years of science, research and innovation have led to the development of more nutritionally complete commercial diets. The collaboration and hard work of leading experts in various facets of aviculture and medicine have contributed to an advanced understanding of psittacine nutrition, and the psychological and physical needs in relation to captive diets.

Commercial diet manufacturers cannot make a label claim that the diet is complete, because no standardized definition for a complete psittacine diet exists. Commercial bird food labels will list that they are fortified, healthy and nutritional based on the manufacturer's recommendations and interpretations of the 1996 Recommendations for a Complete Diet for Galliformes from the Association of American Feed Control Official's (AAFCO) Panel of Nutrition Experts. These recommendations were updated in 1998 when another committee of experts focused on the guidelines for psittacine and passerine birds.[3] Many limitations with these guidelines still exist, including the unique nutritional needs of individual species, life-stages and a focus on gross energy instead of metabolizable energy needs.

Commercial bird foods can be broken down into the following three types.

Seed diets

Seed diets most commonly provide a mix of whole seeds, grains and extruded pellets, with occasional dried fruit pieces. A vitamin, mineral, and amino acid mixture is sprayed onto the components as they pass through the production line, resulting in a nutritional and fortified diet as fed, but not necessarily as consumed. The bird must hull the seeds and thus the nutrient-enhanced coating is not ingested since the hull is tossed away by the bird that consumes only the internal kernel. The mixed diet also allows for selective consumption, with a bird's preference for oilseeds (e.g., pumpkin and sunflower) that are high in fats, predisposing them to nutritional deficiencies, obesity and resulting health consequences.

Seed mixtures provide great foraging and enrichment opportunities since they have various tastes, shapes and textures. They also promote natural feeding behaviors and use of the beak and feet in the prehension and manipulation of foodstuffs.

When feeding a mixed diet, segregation of different-sized and weighted ingredients can occur. Periodic mixing, to redistribute the components is recommended to maintain nutritional composition and distribution.

Pelleted diets

Pelleted diets are made with ground seeds and grains, supplemented with vitamins, minerals and amino acids, to create a uniform pellet, complying with the current AAFCO recommendations. Pellets are not cooked, but heat, moisture and grinding are used to create the longer fiber chain of the pellet, with maximum nutrient value. The uniform appearance of pelleted diets reduces the bird's selective consumption and subsequent risks of nutritional deficiencies from an imbalanced diet.

Extruded diets

Most of the commercial pelleted diets available are extruded diets. These diets are cooked and shaped into cereal pieces. Like pellets, these are made of ground seeds, grains and added nutrients, that are then cooked with high temperatures, pressure, and steam to take on the unique shape of the molds. Once set in the molds, the moisture is then rapidly extruded, rupturing the cells, resulting in shorter fiber chain, cooked diets. Many of these diets are colored using dyes for bright colors as a marketing tool, although natural diets are also available.

Pelleted and extruded pellet diets do pose some disadvantages to feeding. The standardized presentation reduces natural feeding behaviors, smaller sizes reduce use of the feet to prehend and manipulate the food, while reducing sensory experiences of taste and texture. They can also result in boredom that could be redirected into behavioral disorders.

Fruits, vegetables, whole foods and treats

In addition to the commercial diet, it is advised to round out the diet with a variety of fruits, vegetables, proteins, and other whole food items. The percentage of these will be variable based on many exogenous factors. When possible, sourcing foods indigenous to the species' natural habitat should be selected. Lists of safe and toxic foods for psittacine birds are readily available. Providing a variety of foods with varying colors, tastes and textures is important. It is also important, especially in warmer climates, to remove uneaten foods, before spoiling or bacterial growth begins.

Many owners also like to cook for and include their birds in activities that involve foodstuffs. They may enjoy sharing their meal or a snack with their birds, and this is an important time for bonding and engagement. Helping owners select healthy foods they can prepare and share in moderation with their birds, without the addition of seasonings, sauces and additives, can help to foster the human-animal bond while providing sensory enrichment for the bird and encouraging the birds to try new foods.

Storage

Improper storage of commercial diets can greatly affect nutritional composition. In addition to following the manufacturer's recommendations for best before and use within dates, improper storage conditions can lead to rapid degradation of food and its nutritional value.

Exposure to light, air and extreme temperatures can negatively affect the nutritional integrity of the food. Resulting in the reduced nutritional value of vitamins and oxidation of the polyunsaturated fats.[4] High humidity, temperatures and moisture can lead to spoilage, bacterial contamination, and the development of mycotoxins. While prolonged storage may promote the development of grain mites.

With commercial feeds, purchasing smaller quantities, monitoring best-before dates and employing proper storage will help to reduce the risks of nutritional degradation, spoilage, rodents and vermin. In most cases, keeping the feed in the original packaging is recommended by the manufacturer. If necessary, a tightly sealed, light-resistant container, stored in a dry, temperature-controlled environment is also appropriate. Should the feed become contaminated or concerns about its storage conditions arise, it is best to discard the remaining diet. Rotating feed bags, and not refilling containers without emptying, washing, and allowing the container to dry completely before refilling, can help to ensure the integrity of the diets.

Feeding Strategies

Pet psittacine birds pose numerous challenges in feeding that include selective consumption and neophobia. Despite being offered a nutritional multi-component diet, the bird may not consume the vital components in the ratios necessary or at all. This leads to a deficient intake of micro and macronutrients and excess intake of fats and calories. With reduced captive enrichment leading to boredom, unnecessary increased intake and behavioral disorders may occur.

Appropriate feeding regimens for a pet parrot will have variability due to many factors; age, health status of the bird, current diet, past food exposure, and other exogenous variables. By meal feeding smaller portions, adequate to their metabolic needs, owners can reduce food waste, closely monitor birds consumption and identify changes. The result is hopefully earlier detection of concerns that could be indicative of, or resulting from, illness or disease.

Dietary Conversion

Dietary conversion is individualized and must be addressed and addressed individually with the bird and the owners. Some birds will easily accept and try new things, whereas others will exhibit neophobic behaviors and avoid new or unfamiliar foods. Many factors can affect their intake preferences. It is common for birds eating seed mix diets to not readily accept a pelleted-type diet. Choosing a diet that reduces selective consumption while resembling their current diet can provide a more rewarding transition. Having knowledge of the species' natural feeding behaviors and incorporating this into a conversion program may also aid in dietary conversion efforts.

Disclosure Statement: J. Murray is a paid consultant for Lafeber Company.

References

1. Koski MA. Dermatologic diseases in psittacine birds: an investigational approach. *Semin Avian Exot Pet.* 2002;11:105-124.

2. Harrison GJ. Twenty years of progress in pet bird nutrition. *J Am Vet Med Assoc.* 1998;212(8):1226-1230.

3. Association of American Feed Control Officials. Nutrition expert panel review: new rules for feeding pet birds. *Feed Manag.* 1998;49(2):23-25.

4. Bavelaar FJ, Beynen AC. Assessment of commercial feeds for parrots. *Tijdschr Diergeneeskd.* 2003;128(23):726-734.

Section 7

Fundamentals Masterclasses

Armando Burgos, DVM, DABVP (Avian Practice);
Julianne Richard, DVM, MPH, CertAqV
Session Chairs

Introduction to Avian Cardiology

Brenna Fitzgerald, DVM, DABVP (Avian Practice)

Session #204

Affiliation: From Colorado Exotic Animal Hospital, 7120 E Hampden Ave #101, Denver, CO 80224, USA.

Abstract: Cardiovascular disease commonly occurs in companion birds and poses a serious threat to the quality of life and longevity of many avian species. Practitioners with a foundational understanding of relevant anatomy and physiology, heightened awareness of risk factors, recognition of clinical disease states most commonly encountered in practice (including atherosclerosis, congestive heart failure, and pericardial disease and effusion), and an understanding of the essential elements of a diagnostic workup are well-positioned to make accurate diagnoses and implement effective treatment strategies for these patients.

Introduction

Cardiovascular disease has traditionally been thought to be a rare occurrence in companion birds, but is in fact frequently encountered in practice and poses a serious threat to the quality of life and longevity of many avian species. Successful intervention requires a foundational understanding of relevant anatomy and physiology, heightened awareness of risk factors and recognition of clinical disease states, including atherosclerosis, congestive heart failure, and pericardial disease and effusion. Successful intervention also requires accurate and timely diagnosis and innovative treatment approaches, which should incorporate both available information specific to birds and components of therapy in human and small animal cardiovascular medicine. These proceedings review avian cardiovascular anatomy, recognized disease states, diagnostic options, and treatment approaches.

Anatomy and Physiology

The high-performance cardiovascular system of birds has evolved to meet the high aerobic demands of flight and other intense activity by providing adequate oxygen transport with maximal efficiency. The avian heart is 4-chambered, with left and right atria and left and right ventricles. The left ventricle perfuses the systemic circulation, and the right ventricle perfuses the pulmonary circulation. As resistance to blood flow is lower in the pulmonary circulation than in the systemic circulation, the right ventricle is smaller and the myocardium is thinner than in the left ventricle. The left ventricular wall is two to three times thicker than the right ventricular wall and can generate four to five times the systolic pressure. The cone-shaped left ventricle extends to the cardiac apex, and its right wall forms the interventricular septum. The right ventricle is a crescent-shaped cavity that wraps around at least one-half of the left ventricle. Its free wall is continuous with the outer wall of the left ventricle and it does not extend to the cardiac apex. Paired cranial venae cavae and the single caudal vena cava empty into the right atrium, while the paired pulmonary veins enter the left atrium. The left atrioventricular (AV) (mitral) valve is membranous, tricuspid, and is attached to papillary muscles via chordae tendineae. These are lacking in the right AV valve, which consists of only a spiral muscular flap. The pulmonary and aortic valves are each composed of three semilunar cusps. The ascending aorta extends from the left ventricle and unlike in mammals, curves to the right. It gives rise to the coronary arteries and paired brachiocephalic trunks, each dividing into common carotid and subclavian arteries, then continues caudally as the descending aorta, giving off the coeliac and cranial mesenteric arteries as its first major branches. The pulmonary trunk arises from the right ventricle and splits into right and left pulmonary arteries.

The heart is enclosed within a fibrous, relatively non-compliant pericardial sac, which normally contains a very small volume of serous fluid. The outer fibrous layer of the pericardium is continuous with the surface of the great vessels and attachments exist between the pericardium and the sternal plate, liver, adjacent air sacs and, indirectly, to the vertebral column. The heart is thereby well anchored within the cranioventral coelom, its apex partially enclosed within the lobes of the liver.

Adjacent respiratory structures include the paired, non-expansile lungs, resting dorsal to the heart, and the large, complex clavicular air sac which lies within the thoracic inlet and surrounds the heart base and great vessels. The position of these structures relative to the heart has great clinical relevance given the potential for cardiovascular involvement in cases of primary respiratory disease.

The cardiac conduction system of birds bears a close resemblance to that of mammals, consisting of the sino-atrial (SA) node, atrioventricular (AV) node, the bundle of His, and its bundle branches; however, there are three bundle branches (left, right, and middle) as well as an AV ring of Purkinje fibers surrounding the right AV opening that is unique to birds. A figure-of-8 is formed as the middle bundle branch passes around the aorta and joins the AV ring. Left and right bundle branches emerge from the interventricular septum to supply the left and right ventricles, following the tracts of the coronary arteries into the myocardium (periarterial Purkinje fibers).

Vascular microanatomy

Structurally the arteries can be classified into the elastic arteries and the muscular. The walls of elastic arteries are composed of three layers: the tunica intima, which is a single-cell layer of endothelial cells adjacent to the arterial lumen; the tunica media, which is composed of concentric layers of collagen and elastic fibers composed of smooth muscle cells and elastin; and the tunica adventitia, which is an outer layer of collagen fibers and connective tissue. Muscular arteries have the same three tunicae, but the media is only composed of smooth muscle cells and elastin and the adventitia is much thicker and has less compliance.

Hemodynamics

Systemic arterial blood pressure is a function of cardiac output and the resistance of the arterial system; the latter is partly determined by afterload. Afterload is determined by 1) impedance of the proximal aorta itself, 2) compliance (distensibility) of the elastic arteries, and 3) peripheral resistance to blood flow afforded primarily by small arteries and arterioles. The characteristic impedance of the proximal aorta increases with increased aortic stiffness and decreased radius, resulting in greater left ventricular afterload. Decreased compliance of elastic arteries, as well as increases in peripheral resistance (occurring with decreases in vessel radius), further contribute to afterload.

Vasoconstriction (and venoconstriction) are modulated by sympathetic tone and circulating catecholamines (EPI and NE acting via α-adrenergic receptors). Vasoconstriction is also mediated by activation of the renin-angiotensin-aldosterone system (RAAS), arginine vasotocin (AVT, the homolog of antidiuretic hormone in mammals), and endothelin-1.

Cardiovascular Disease States

Atherosclerosis, stroke, and intermittent claudication

Atherosclerosis is a chronic inflammatory and degenerative disease of the arterial wall wherein the lumen narrows by progressive accumulation of inflammatory cells, fat, cholesterol, calcium, and cellular debris that forms fibrofatty atheromatous plaques within the intima. Potential complications include stenosis, ischemia, thrombosis, hemorrhage, and aneurysm. Atherosclerosis is likely an underlying factor in the majority of non-infectious cardiovascular diseases diagnosed in pet birds and is undoubtedly the most common lesion of the cardiovascular system identified post-mortem in companion psittacine birds.

In psittacine birds, lesions are characterized by the accumulation of lipoproteins, macrophages called foam cells once they have internalized lipoproteins and stored cholesterol, necrotic debris, and fibrous tissue in the intima and luminal side of the vascular media. Lesions can be classified into seven types (I-VII); types I and II are considered early lesions, type III intermediate, and types IV-VII advanced. Lesion type VI is distinguished from IV and V by the additional complication of fissure, hematoma, or thrombosis, and type VII are calcific lesions characterized by large calcium plaques and osseous metaplasia. Clinical signs are attributed to advanced lesions, while early and intermediate lesions are generally subclinical.

Atherosclerotic lesions are most frequently recognized in the ascending aorta, brachiocephalic trunks, and pulmonary arteries. Although not as often appreciated, lesions also appear in peripheral vasculature, most notably the coronary arteries, carotid arteries, and descending aorta, as well as the subclavian arteries, coeliac artery, and ischiatic arteries. Coronary and aortic aneurysms have been known to arise secondary to atherosclerosis owing to decreased compliance of the affected vessels compared to normal elastic arteries.

Unlike humans, recognizable clinical disease in birds is primarily the product of progressive, flow-limiting arterial stenosis rather than thrombosis and hemorrhage of disrupted plaques that result in thromboembolism and acute arterial obstruction. Acute myocardial infarction is rendered less likely by the differences in avian coronary vasculature and greater collateral circulation compared to humans. In one study, advanced lesions were commonly characterized by severe arterial stenosis and occlusion, but atherothrombotic lesions, aortic dissection and rupture, and lesional hemorrhage were identified rarely (i.e., in 1.9% of lesions).

Patients with clinically significant atherosclerosis most often present for falling or collapse, frequently accompanied by transient or persistent weakness and dysfunction of one or more limbs, and with or without altered mentation. Extension and rigidity of one leg and sometimes the ipsilateral wing, clenching of the toes, ataxia or difficulty perching, and tremors or convulsions may be reported. These signs are considered most consistent with stroke or intermittent claudication (the latter defined in human medicine as exercise-induced intermittent weakness and pain in the pelvic limbs, resolving with rest), but rarely is this confirmed diagnostically. Some of these stroke-like events can be difficult to distinguish from seizure activity, and indeed seizures can occur subsequent to ischemic stroke. Signs may be acute and transient, or there may be persistent neurologic abnormities, including reduced mentation, blindness, seizures, vestibular signs, paresis of one or both pelvic limbs, and ataxia. Alternatively, sudden death may occur as a result of a cerebral infarct, aneurysmal rupture, or a lethal cardiac arrhythmia secondary to myocardial ischemia.

Reported prevalence of atherosclerosis in psittacine birds varies widely (i.e., 1.9 to 91.8%), with the highest recently reported among Amazons and African grey parrots (*Amazona* spp. and *Psittacus erithacus*, respectively). A significantly higher prevalence of advanced lesions (types IV-VI) was found for grey parrots, Amazon parrots, and cockatiels (*Nymphicus hollandicus*); cockatoos (*Cacatua* spp.) and macaws (*Ara* spp.) appeared less susceptible. There was a positive association between advanced atherosclerosis and increasing age, female sex, reproductive disease (predominately female), hepatic disease, and concurrent myocardial fibrosis. These findings are subjectively consistent with the author's experience in practice, where severe atherosclerosis is most often encountered in grey parrots and Amazon parrots over 20 years of age, followed by cockatiels, lovebirds (*Agapornis* spp.), Eclectus parrots (*Eclectus roratus*), cockatoos, and macaws.

Other risk factors for the development of atherosclerosis include high-calorie and fat diets, dyslipidemia (e.g., hypercholesterolemia), and limited physical activity. The predisposition of female psittacine birds to development of atherosclerosis likely relates to the profound effects of estrogens on lipid, protein, and calcium metabolism in reproductively active hens; these changes promote atherogenesis. Diet-induced atherosclerosis has been demonstrated experimentally in budgerigars (*Melopsittacus undulatus*) and Quaker parrots (*Myiopsitta monachus*). In the author's experience, companion birds ultimately diagnosed with advanced atherosclerosis often have a long-term history of a high-calorie and fat diet, regular provision of animal-based products (e.g., chicken bones, meat, and cheese), clinical obesity, and/or chronic reproductive activity (i.e., female birds).

Heart failure

Heart failure occurs when cardiac output becomes inadequate to maintain arterial blood pressure manifested clinically by low output signs of lethargy, depression, and weakness and when the heart is unable to empty the venous reservoirs manifested by vascular congestion and transudation of fluid within tissues and body cavities (i.e., congestive signs). In the case of right-sided congestive heart failure (CHF), peripheral venous congestion, hepatic congestion, ascites, and pericardial effusion are often present. Pulmonary edema and congestion of the pulmonary veins occur with left-sided congestive heart failure, and a combination of signs may be seen with biventricular failure. Compared to mammals, birds are thought to have a greater propensity for developing pulmonary hypertension and right-sided, rather than left-sided congestive heart failure owing to the morphology of the right AV valve, less deformable nucleated erythrocytes, and the rigid, non-distensible lungs which limit the ability of the blood capillaries to expand and accommodate greater blood flow.

CHF is not a primary disease in itself, but an ultimate consequence of structural or functional abnormalities of the cardiovascular or pulmonary systems, compounded by the chronic effects of compensatory mechanisms. Not all cardiovascular disease necessarily leads to CHF, but it is a frequent clinical end-point encountered in companion birds.

Heart failure can result from primary myocardial failure, ventricular pressure overload, volume overload, conduction disturbances, or diastolic dysfunction. Myocardial failure can occur with either primary disease of the myocardium or secondary to chronic pressure or volume overload. Pressure overload results from outflow obstruction (e.g., from congenital or acquired valvular stenosis, narrowing of the aorta). The affected ventricle(s) ultimately undergoes concentric hypertrophy, in which the ventricular wall thickens with a corresponding decrease in chamber volume. With time, myocardial fibrosis develops and impairs both systolic and diastolic function. As myocardial failure progresses, the ventricle ultimately dilates. Volume overload results from valvular insufficiency (e.g., endocardiosis, vegetative endocarditis) and abnormal communications between chambers (e.g., ventricular septal defect). The result is eccentric hypertrophy of the affected ventricle(s), characterized by chamber enlargement and increased myocardial mass with little increase in wall thickness. With the progression of hypertrophy, eventual myocardial failure ensues, with further chamber dilatation. Diastolic failure results from decreased ventricular compliance, abnormal myocardial relaxation, and cardiac tamponade or constrictive pericardial disease.

Dilative and hypertrophic cardiomyopathy have been reported in psittacine birds, with dilative cardiomyopathy generally considered idiopathic. Hypertrophic cardiomyopathy most often develops secondary to the arterial luminal stenosis and decreased compliance characteristic of atherosclerosis, or to pulmonary hypertension related to atherosclerosis, chronic pulmonary disease, or left-sided CHF.

Myocardial changes consistent with ischemic injury, including myocardial necrosis, inflammation, degeneration, and fibrosis, have been associated with atherosclerosis of the coronary arteries in psittacine birds. These changes have the potential to produce arrhythmias, particularly considering the periarterial course of Purkinje fibers through the myocardium. Arrhythmias could result in sudden death while systolic and diastolic myocardial dysfunction may ultimately culminate in CHF. Myocardial disease may also be of infectious, metabolic, nutritional, neoplastic, or toxic origin. Myocarditis can be caused by a variety of infectious agents, including bacterial, fungal, and viral pathogens, as well as protozoal and helminth parasites.

There are few published studies examining the prevalence and etiology of cardiac disease in psittacine birds and a relatively small number of case reports. The reported prevalence of gross lesions ranges from 9.7% to 36%; in one study, histologic changes in the heart and great vessels were identified in 99% of birds necropsied. In the author's practice, cardiac disease most consistently takes the form of CHF subsequent to advanced atherosclerotic disease, and as such, the represented age and species groups mirror those with a higher prevalence of atherosclerosis. Cases of valvular insufficiency (including vegetative endocarditis), chronic pulmonary disease with probable hypertension, neoplasms or fungal granulomas encroaching on the heart base and great vessels, infectious myocarditis, cardiac arrhythmias, and congenital defects are encountered less frequently.

Pericardial disease and effusion

Pericardial effusion is characterized by an inappropriate accumulation of fluid within the pericardial sac. It can develop due to inflammatory changes of the pericardium (i.e., pericarditis), trauma or aneurismal rupture, cardiac or pericardial neoplasia, or as a manifestation of metabolic derangements, toxic insult, or, commonly, right-sided CHF. Infectious pericarditis and associated effusion can be a manifestation of systemic disease or occur as an extension of epicarditis, myocarditis, or inflammatory or infectious conditions affecting other adjacent tissues, including the air sacs, lungs, and liver. Bacterial, fungal, viral, and parasitic agents have been implicated. Visceral gout is a common non-infectious cause of pericarditis and pericardial effusion.

Cardiac tamponade is defined as compression of the heart by a restrictive pericardium or accumulation of pericardial contents, resulting in impairment of ventricular filling (diastolic dysfunction) and subsequent decrease in stroke volume and cardiac output. Cardiac tamponade results in right-sided CHF in advance of left-sided failure. Cardiac tamponade may arise by constriction with a fibrotic pericardium subsequent to chronic, fibrinous pericarditis and adhesion to the epicardium. When pericardial effusion is the cause of compression, the clinical severity of cardiac tamponade depends in large part on the rate at which the effusion develops. As the pericardium is relatively non-compliant, acute accumulations of even small fluid volumes can have profound hemodynamic effects. Conversely, a more chronic onset of pericardial effusion allows time for expansion of the pericardial sac and compensatory physiologic mechanisms, including increased blood volume and heart rate via activation of the RAAS and sympathetic stimulation.

Diagnosis

History, clinical signs, and physical examination findings

Diagnosis of cardiovascular disease begins with careful history collection and thorough physical examination, to be performed in a manner in keeping with patient stability. Examination findings, as well as relative predisposition to cardiovascular disease given the species, age, gender, lifestyle, and diet should be taken under consideration when formulating initial differential diagnoses and prioritizing diagnostic maneuvers. Clinical presentations of cardiovascular disease share many commonalities, particularly as more than one condition can exist concurrently, but there are distinguishing features of some disease states that warrant mention. Many cases are likely subclinical and go unrecognized for many years until well advanced. Clinical signs can have a subtle, insidious onset, with owners reporting progressively declining activity level, reduced appetite, and waning interest in household activities, toys, and vocalizations that were historically of interest to the bird. In some cases, no premonitory signs are appreciated prior to sudden death. Presenting complaints common to most forms of cardiovascular disease are lethargy, weakness, reduced appetite, respiratory distress, and exercise intolerance. Physical examination findings may include cachexia, tachypnea, dyspnea, harsh lung sounds, cyanosis, tachycardia, arrhythmia, systolic murmur, coelomic distension and ascites, peripheral edema, and altered mentation.

Atherosclerosis

Patients with clinically significant atherosclerosis most often present for falling or collapse, frequently accompanied by transient or persistent weakness and dysfunction of one or more limbs, and with or without altered mentation. Extension and rigidity of one leg (and sometimes the ipsilateral wing), clenching of the toes, ataxia or difficulty perching, and tremors or convulsions may be reported. These signs are considered most consistent with stroke, but rarely is this confirmed diagnostically. These events can be difficult to distinguish from seizure activity, and indeed seizures can occur subsequent to ischemic stroke. Signs may be acute and transient, or there may be persistent neurologic abnormalities, including reduced mentation, blindness, seizures, vestibular signs, paresis of one or both pelvic limbs, and ataxia. In cases with concurrent congestive heart failure (CHF), dyspnea and exercise intolerance may be part of the history and appreciable upon physical examination along with other congestive signs.

Congestive heart failure

Birds with CHF present most consistently with respiratory embarrassment, owing to either intracoelomic air sac compression caused by ascites and organomegaly, pulmonary edema, or a combination of the two. Coelomic distention is among the congestive signs appreciable upon physical examination in cases of right-sided or biventricular failure. Peripheral venous congestion and edema may also be noted. In the case of cardiogenic pulmonary edema, increased respiratory noise is typically heard upon auscultation of the lung fields in a calm bird, but characteristic crackles and wheezes are generally absent except in severe cases, and cough is exceptionally rare.

Clinical pathology

Clinical pathology, overall, has a low sensitivity for specific cardiovascular disease. Abnormal findings are more often attributable to secondary effects on other organ systems, underlying primary disease processes, or to concurrent, unrelated disease. For example, elevations in uric acid and bile acids reflect renal and hepatic functional impairment, respectively. These may occur as a consequence of renal hypoperfusion or hepatic congestion secondary to heart failure. Given the association between dyslipidemia and lipid-accumulation disorders, including atherosclerosis, in some species of captive psittacine birds, the potential use of plasma lipids as biomarkers for cardiovascular disease and risk assessment in these birds has been the subject of much research. Select diagnostic tests may have value as part of identifying risk factors in individual pet birds, but plasma lipid elevations are not diagnostic of atherosclerotic disease. Currently, plasma total cholesterol and total triglycerides are the most clinically appropriate and useful tests for risk factor assessment and monitoring responses to dietary and lifestyle modifications. Hematologic changes, including elevated total white blood cell count and shifts in the leukocyte differential are seen with many forms of inflammatory or infectious disease, as well as with stress, and can be seen in patients with cardiovascular disease.

Radiography

Radiographic imaging is generally an insensitive method to identify cardiovascular disease, but allows assessment of the size, contour, and radiodensity of the cardiac silhouette and great vessels, including the aorta, brachiocephalic trunks, and pulmonary arteries and veins. Increased radiodensity and tortuosity of the great vessels are suggestive, but not conclusive, of atherosclerosis. Assessment of increased opacity is highly subjective and subject to variability in radiographic technique, though focal and linear mineralization associated with one or more vessels is highly suggestive of advanced, calcified atherosclerotic lesions.

Enlargement of the cardiac silhouette can be seen with pericardial effusion, myocardial hypertrophy or inflammation, chamber dilatation, cardiac or pericardial masses, and aneurismal dilatation. Confirmation and differentiation between possible etiologies requires echocardiography. The size of the cardiac silhouette on the ventrodorsal view can be quantified by taking standardized measurements and calculating ratios relative to skeletal landmarks. In addition to enlargement of the cardiac silhouette, radiographic abnormalities that may be seen in cases of CHF include an enlarged hepatic silhouette and caudodorsal displacement of the adjacent gastrointestinal tract, increased radiodensity of the pulmonary parenchyma, intracoelomic air sac compression, or loss of coelomic detail due to ascites. The hepatic silhouette may appear enlarged due to hepatomegaly, perihepatic effusion (i.e., fluid accumulation in the hepato-peritoneal cavities), or cardiomegaly or pericardial effusion, considering that the cardiac apex is positioned between the liver lobes. Increased pulmonary density, particularly of the reticular pattern, may be seen with cardiogenic pulmonary edema or with primary disease of the pulmonary parenchyma (inflammatory or neoplastic infiltration, fibrosis). The clavicular air sac normally provides the contrast necessary to visualize the heart base and great vessels. When this space is lost or diminished, as occurs with space-occupying masses or accumulation of exudate within the air sac, visibility of the great vessels is obscured.

Ultrasonography

Coelomic ultrasound allows the identification of cardiac abnormalities, pericardial effusion, ascites, and hepatic parenchymal changes that cannot be distinguished radiographically beyond enlargements of the cardiac and hepatic silhouettes. It can most often be performed in the awake or sedated patient and represents a good option for those too compromised to withstand extended restraint or general anesthesia. Birds can be allowed to stand or be held in an upright position to reduce positional respiratory or circulatory compromise.

Instrumentation and technique, anatomic limitations and approaches

Two-dimensional (B-mode) echocardiography is a well-established technique to assess morphologic and functional status in birds, but M-mode is not useful because only longitudinal and semitransverse views are attainable in most birds. Minimal technical requirements of the ultrasound system include a frame rate of 100 frames per second, a high-frequency probe ($\geq$ 7.5 MHz) with a small coupling surface (microconvex array, phased array, or one end of a linear array), Doppler function, and capability to record still images and video clips for later review. Simultaneous ECG facilitates interpretation of images in relation to the cardiac cycle, but is not essential as end-diastolic and end-systolic frames can be visually selected.

Cardiac views are constrained by the situation of the heart in an indentation of the sternum and surrounded by the clavicular and cranial thoracic air sacs. Given these limitations, two approaches are possible for transcoelomic ultrasound: ventromedian and parasternal. The ventromedian approach is the most commonly used for psittacine and raptorial birds. The probe is placed just caudal to the sternum where a normally featherless area (apterium) lies along the ventral midline between adjacent feather tracts in most birds (excluding waterfowl). The beam plane is directed craniodorsally and the liver serves as an acoustic window, avoiding the air sacs laterally and the sternum ventrally.

The heart can be visualized in two longitudinal views, horizontal (four-chamber view) and vertical (two-chamber view), the latter obtained by rotating the probe 90°. The horizontal view allows visualization of the left and right ventricles, interventricular septum, left and right atria, left atrioventricular (AV) valve, muscular right AV valve, and aortic valve, though it is difficult to distinguish the borders of the atria from the surrounding tissue. The right ventricle appears triangular in shape, tapering toward the cardiac apex. The vertical view permits visualization of the left atrium, left ventricle, and left AV valve; a marginal sliver of the right ventricle may also be visible adjacent to the sternum. The hepatic parenchyma should appear homogenous and of moderate echogenicity, with a fine, granular texture. Hepatic vessels appear as anechoic channels passing through the parenchyma. Although views are limited for transcoelomic echocardiography, acoustic windows are greatly improved and visibility is facilitated by cardiomegaly, hepatomegaly, ascitic fluid, and pericardial effusion, all frequently present in cases of CHF.

Subjective assessments

Subjective assessments of chamber size, wall thickness, ventricular contractility, diastolic function, and valvular morphology and function can be made. Chamber dilatation, myocardial hypertrophy, valvular insufficiency and anomalies (to include vegetations), cardiac masses, septal defects, some aneurysms, pericardial effusion, ascites, and hepatic venous congestion can be identified by ultrasound assisted by color Doppler. Ascitic fluid can be collected via ultrasound-guided coelomocentesis and pericardial fluid by pericardiocentesis. Hyperechogenicity of the walls of the ascending aorta is suggestive of mineralization that can accompany advanced atherosclerosis.

Advanced imaging

Advanced imaging options for diagnosis of cardiovascular disease and associated conditions include computed tomography (CT), CT angiography (CTA), and magnetic resonance imaging (MRI). Traditional radiographic imaging allows evaluation of the avian cardiovascular system through visualization of the cardiac silhouette, some major vessels, and the pulmonary parenchyma; however, visibility is limited by superimposition with other structures and greatly influenced by patient positioning. CT, particularly with contrast enhancement, offers numerous advantages over traditional radiography in visualizing cardiovascular and pulmonary structures. Although MRI has traditionally been considered more sensitive, cerebral complications of atherosclerotic disease, such as ischemic and hemorrhagic strokes, can be diagnosed by CTA.

Treatment

Although limited pharmacologic data in avian species complicates effective medical management of cardiovascular disease, treatment approaches that have been reported, utilized by the author, or show promise for future application are discussed.

Atherosclerotic disease

Treatment of atherosclerosis involves both controlling risk factors and managing sequellae, including peripheral hypoperfusion, ischemic stroke, and CHF. Atherosclerotic lesions cannot be resolved, but diet, husbandry, and lifestyle changes may help to prevent, slow progression, or decrease the size of lesions. Peripheral vasodilation may have symptomatic benefits by promoting improved peripheral perfusion and reducing afterload by decreasing vascular resistance. Isoxsuprine is a peripheral vasodilator that causes vascular smooth muscle relaxation predominately through α-adrenoceptor blockade. Its empirical use in birds with clinical atherosclerotic disease has resulted in apparent symptomatic improvement, including reduced frequency and severity of stroke-like events and neurologic signs. The author has used isoxsuprine at a dose and frequency of 10-15 mg/kg PO q12h in numerous cases of clinical, presumed atherosclerosis some of which were later confirmed at necropsy and by histopathology.

Angiotensin-converting enzyme (ACE) inhibitors, including enalapril and benazepril, effect vasodilation by blocking the formation of angiotensin II (AII). AII promotes vasoconstriction and venoconstriction by mediating the release of catecholamines, which act on vascular smooth muscle via α-adrenergic receptors. Although the relative vasodilatory effect of an ACE inhibitor compared to isoxsuprine in birds is not known, it is conceivable that the two used in combination would have synergistic effects: an ACE inhibitor by limiting α-adrenoreceptor stimulation and isoxsuprine by α-adrenoreceptor antagonism. The author has treated patients with clinical (presumed) atherosclerotic disease using enalapril at a dose and frequency of 1.25 to 5 mg/kg PO q8-12h. In more severe cases this is paired with isoxsuprine.

Sildenafil may have merit in cases with suspected atherosclerosis of the pulmonary arteries, potentially in combination with other vasodilators. The author has treated patients with suspected pulmonary artery disease and/or pulmonary hypertension using sildenafil at a dose and frequency of 1-11 mg/kg PO q8-12h.

In mammals, pentoxifylline promotes the passage of erythrocytes through damaged microvasculature by increasing their flexibility. It may have value in improving peripheral perfusion in birds with atherosclerotic disease, including those with concurrent polycythemia. The author has used it for these indications in numerous psittacine patients (15-25 mg/kg PO q8-12h).

Supportive care and husbandry considerations

Patients with signs of stroke may have marked neurologic deficits including reduced mentation, limb paresis, and ataxia that prevent normal eating and drinking and impair mobility. They may experience seizures or suffer injuries from falls. Supportive care measures include fluid and nutritional support, analgesia, anticonvulsant therapy when needed, and management of secondary conditions such as trauma and aspiration pneumonia. Exercise restriction should be part of the longer-term treatment plan, as well as appropriate housing modifications to accommodate and protect birds with persistent deficits. Operant learning methods should be employed to facilitate low-stress medication administration. Managements that may also have preventative value, particularly for at-risk species, include moderation of dietary calories and fat to prevent and resolve obesity, avoidance of dietary sources of cholesterol, and control of female reproductive activity. It is possible that increasing opportunities for exercise beginning early in life, generally lacking for captive birds, may have preventive value.

Congestive heart failure

Appropriate treatment requires that known or hypothesized underlying cause(s) be addressed, if possible. The mainstays of treatment of CHF in small animal medicine, namely diuretics, ACE inhibitors, and positive inotropes, can be applied to treatment of the condition in birds. Beta-blockers, considered part of standard treatment in humans, may also have potential application. The goal of initial treatment is to stabilize the patient presenting with acute or decompensated failure, followed by the design of a longer-term management strategy. In the author's experience, this has met with highly variable success, but some patients have been maintained in stable condition for up to 7 y.

Diuretics

Marked reduction of excessive circulating plasma volume (hypervolemia), edema, and effusion is an immediate treatment priority accomplished through the use of diuretics, principally furosemide. In the author's experience, a dose range of 1-5 mg/kg IM has been most efficacious when stabilizing psittacine patients with severely decompensated disease. An initial frequency of 2 hours is usually followed by a shift to 6- to 12-hour intervals. There have been isolated cases with severely decompensated disease in which higher doses (7-11 mg/kg IM q12h) were required to gain control of congestive signs. In patients whose disease is less severe and advanced, a lower initial dose and frequency (1-5 mg/kg IM q12h) may be adequate. Once the patient has been stabilized, furosemide can be administered orally, with the dose increased at least 2-fold (presuming oral bioavailability is 60-75% as in humans) at a frequency of 8-12 hours. The ultimate goal for long-term management is to identify the lowest dose that controls congestive signs. Adjustments will be needed over time as the disease condition changes or progresses. Patients with more severe, advanced disease may require a higher dose and frequency (30-70 mg/kg PO q6-12h) to maintain control of congestive signs. Spironolactone may be used concurrently with furosemide when congestive signs cannot be controlled with furosemide and an ACE inhibitor alone. Aside from its diuretic effects, spironolactone is thought to prevent or decrease myocardial fibrosis in humans and counteract the myriad deleterious effects of aldosterone in CHF. Thus far, reports on its use or efficacy in avian species are scarce. In the author's experience, doses between 1-13 mg/kg PO q8-12h have been well tolerated in numerous CHF patients.

ACE inhibitors

ACE inhibitors are an essential component of long-term medical management of CHF by blunting the effects of the renin-angiotensin-aldosterone system (RAAS). By interfering with angiotensin II (AII) formation and limiting aldosterone production, ACEI promote vasodilation and reduce sodium and water retention, thereby decreasing total peripheral resistance and pulmonary vascular resistance (afterload) and circulating volume (preload), allowing an increase in cardiac output. Enalapril is thought to attenuate myocardial remodeling in humans and dogs. Enalapril has been the most commonly used ACE inhibitor in birds and empirical evidence suggests it is both safe and efficacious. With inclusion of enalapril in the treatment regime, increased quality of life and longevity have been reported in birds with severe cardiac pathology. The author has found that when edema and effusion are mild, they can in some cases be resolved with enalapril alone without concurrent use of a diuretic. Pharmacokinetic data supports a dose of at least 1.25 mg/kg PO q8-12h in pigeons and Amazon parrots, although the half-life was shorter and lower maximum plasma concentrations were reached in the latter species. Therefore, the author has used higher doses (4-11 mg/kg PO q8-12h) in psittacine birds with appreciable symptomatic benefit and no apparent adverse effects.

Other vasodilators

Sildenafil is a selective pulmonary vasodilator that reduces pulmonary vascular resistance and right ventricular afterload. Compared to mammals, birds are thought to have a greater propensity for developing pulmonary hypertension and right-sided, rather than left-sided CHF. The inclusion of sildenafil should be considered in treatment of CHF patients with cor pulmonale. The author has used sildenafil for this indication in psittacine birds at dose and frequency of 1-11 mg/kg PO q8-12h.

Positive inotropes

Positive inotropes are used to enhance myocardial contractility and are appropriate for treatment of heart failure due to systolic dysfunction. They are contraindicated in cases of hypertrophic cardiomyopathy, where diastolic dysfunction is the primary problem, and of outflow obstruction (e.g., aortic stenosis). Likewise, it should be questioned whether their use is appropriate in birds with heart failure secondary to atherosclerotic disease and luminal stenosis of major arteries. In those cases with severe systolic dysfunction; however, it is fair to consider their inclusion in the treatment regime, at least in the short term, to stabilize the patient. Pimobendan enhances myocardial contractility without increasing myocardial oxygen consumption. Pharmacokinetic data suggests higher doses may be needed in psittacine species, relative to humans and dogs, to achieve therapeutic levels. A pharmacokinetic study in Hispaniolan Amazon parrots (*Amazona ventralis*) demonstrated that a single oral dose of 10 mg/kg is required to achieve a peak plasma comparable to levels considered therapeutic in dogs and humans. The author has incorporated pimobendan (6-20 mg/kg PO q8-12h) into the treatment regime for the management of CHF in many companion birds.

Supportive care and husbandry considerations

General supportive principles apply to the avian heart failure patient, namely rest, minimizing stress with judicious, limited handling and restraint, and provision of supplemental oxygen when appropriate. Oxygen supplementation is indicated for patients with pulmonary edema, while physical fluid removal is more efficacious to stabilize those with air sac compression and cardiac tamponade. Patients may be presented in a severely debilitated state. In-patient supportive care measures may include fluid and nutritional support, but parenteral fluid administration must be carefully questioned. Maintaining the delicate balance between management of hypervolemia and concurrent dehydration and renal dysfunction is a profound clinical challenge, requiring close monitoring for changes in clinical status and adjustment of the treatment plan accordingly. Longer-term husbandry considerations, including dietary changes that should also entail sodium restriction, and lifestyle and housing modifications, mirror those discussed for atherosclerosis.

Pericardial effusion and cardiac tamponade

Treatment is based first on removal of the fluid, and second on treatment of the underlying etiology of fluid accumulation. Fluid removal can be accomplished either by ultrasound-guided or endoscopic pericardiocentesis or by endoscopic or surgical fenestration of the pericardium. Diuretics are contraindicated in cardiac tamponade because they decrease the preload necessary to maintain ventricular filling pressure and cardiac output; however, furosemide with or without an ACE inhibitor may be useful to reduce pericardial fluid of low volume related to CHF.

References

1. Beaufrere H, Nevarez J, Gaschen L, et al. Diagnosis of presumed acute ischemic stroke and associated seizure management in a Congo African grey parrot. *J Am Vet Med Assoc.* 2011;239(1):122-128.

2. Beaufrere H, Holder KA, Bauer R, et al. Intermittent claudication-like syndrome secondary to atherosclerosis in a yellow-naped Amazon parrot (*Amazona ochrocephala auropalliata*). *J Avian Med Surg*. 2011;25(4): 266-276.

3. Beaufrere H. Avian atherosclerosis: parrots and beyond. *J Exot Pet Med.* 2013;22(4):336-347.

4. Beaufrere H, Nevarez JG, Holder K, et al. Characterization and classification of psittacine atherosclerotic lesions by histopathology, digital image analysis, transmission and scanning electron microscopy. *Avian Pathol.* 2011;40(5):531-544.

5. Beaufrere H, Ammersbach M, Reavill DR, et al. Prevalence of and risk factors associated with atherosclerosis in psittacine birds. *J Am Vet Med Assoc.* 2013;242(12):1696-1704.

6. Fitzgerald BC. Cardiovascular diseases in pet birds: therapeutic options and challenges. *Vet Clin North Am Exot Anim Pract.* 2022;25(2):469-501.

7. Fitzgerald BC, Dias S, and Martorell J. Cardiovascular drugs in avian, small mammal, and reptile medicine. *Vet Clin North Am Exot Anim Pract.* 2018;21(2):399-442.

8. Fitzgerald BC, Beaufrere H. Cardiology. In: Speer BL, ed. *Current Therapy in Avian Medicine and Surgery.* St. Louis, MO: Elsevier; 2016:252-328.

9. Guzman DS, Beaufrere H, KuKanich B, et al. Pharmacokinetics of single oral dose of pimobendan in Hispaniolan Amazon parrots (*Amazona ventralis*). *J Avian Med Surg.* 2014;28(2):95-101.

10. Powers LV. Pulmonary arterial hypertension in companion birds. *Proc Annu Cof Assoc Avian Med.* 2014: 295-299.

Pet Poultry Basics

Marcie Logsdon, DVM

Session #167

Affiliation: From Washington State University Veterinary Teaching Hospital, 205 Ott Rd, Pullman, WA 99164, USA.

Abstract: Backyard poultry continue to increase in popularity in the United States, and many owners are now seeking veterinary care for their backyard flock. These small populations of production birds require special considerations for treatment compared to traditional pet birds. Biosecurity measures, disease risk and food-animal status must all be taken into account when treating these patients.

Get to Know Your Patient

When a chicken presents for treatment, one of the first things that needs to be determined is what role that chicken fills in its household. Is it primarily a pet, or is it primarily kept for production? In other words, do the owners want a chicken, or this particular chicken? Knowing what the owner's goals are will help streamline diagnostic and treatment recommendations.[1,2] Do we want to put this chicken with an obstructive breathing pattern through the CT scanner to evaluate its air passages, or do we want to prioritize whole flock health and euthanize it on intake for necropsy and histopathology to expedite a final diagnosis?

Food-Animal Status and Medication Considerations

Chickens (and turkeys) are classified as one of the four major food-animal species in the United States by the Food and Drug Administration (FDA). Therefore, regardless of the owner's personal view or intended use of their chicken, medication administration must follow FDA guidelines. Unfortunately, the list of approved medications for chickens is relatively short, and most target whole flock management. Most medications needed for acute, individual-level treatments are therefore used in an extra-label drug use (ELDU) manner. Non-prohibited drugs may be used in an extra-label manner when there is no suitable approved medication available. Any of the following situations are considered ELDU in food animal species: administration of a medication at a different dose, volume per injection site, interval, or route; for a different indication or duration; or to a different species than what is listed on the label. However, if a veterinarian prescribes a medication to be used in an ELDU manner, they are responsible for obtaining and providing appropriate withdrawal recommendations to the owner. The Food Animal Residue Avoidance Databank (FARAD) is a U.S. Department of Agriculture-administered program that will provide ELDU withdrawal medications to veterinarians upon request. Turnaround time is usually 1-2 business days. Further information is available at www.FARAD.org.

If a medication necessitates an egg discard period, the owners either need to discard eggs from the whole flock, or isolate the treated bird to distinguish which eggs were laid by it. Mixed breed flocks are popular, so sometimes an individual hen's eggs can be distinguished by color. For example, if a flock consists of mostly Rhode Island Red hens laying brown eggs, but has a single Americana that lays blue eggs.

Medications commonly used in avian practice that are prohibited for extra-label use in major food-animal species in the United States include fluoroquinolone class antibiotics (enrofloxacin, marbofloxacin, and ofloxacin), nitroimidazoles (metronidazole), and any medication that is imported under the FDA Personal Importation of Drugs guidelines. This is the regulatory framework that allows U.S. veterinarians to import deslorelin implants for use in pet birds.[3] In addition, any medication indexed with Minor Use and Minor Species Act (MUMS) designation cannot be used extra label at all (e.g., the 4.7-mg deslorelin acetate implants marketed for treatment of adrenal disease in ferrets). Because of this, there is currently no legal framework to allow deslorelin acetate implants in chickens in the United States. This is unfortunate, as deslorelin implants can be an effective treatment for many reproductive disorders.

Common Disorders

Reproductive disease

In this author's experience, the top three reasons for backyard chickens to present are:

1. Reproductive disease.
2. Reproductive disease.
3. Reproductive disease.

Commercial chicken breeds have been selected for high egg production and therefore the reproductive tract is a common system for diseases. Common reproductive disorders include primary egg binding (most common in young pullets coming in to lay for the first time), cloacal prolapse, salpingitis, neoplasia of the ovary or oviduct, and egg yolk peritonitis.[1,2] The prognosis depends on chronicity and the underlying disease process. Salpingohysterectomy can be a successful treatment if the underlying disorder is not neoplastic in nature and if the patient is treated before the formation of extensive adhesions. It is important to note that many cases of reproductive disease will first present for difficulty in breathing or crop stasis. This is due to compression of the air sacs and/or the gastrointestinal tract (GI) tract by fluid, mass or effect in the coelomic cavity.

Marek's disease

Marek's disease is an oncogenic herpesvirus that is widespread amongst poultry in the United States. It is a double-enveloped DNA virus that is very persistent in the environment. It is shed in feather dust and once present in a flock, it is almost impossible to eliminate from a standard coop. The classic presentation of Marek's disease is a unilateral leg lameness in a young bird (i.e., < 1 y old) that does not respond to pain management and rest; however, it can cause lymphoma in older birds or present as motility dysfunction in the GI tract. Definitive diagnosis is through necropsy or histopathology and sample submission of a section of the sciatic nerve is recommended if Marek's disease is suspected. There is also a PCR test available from the University of Georgia Diagnostic Lab. Remember, Marek's disease is a reportable disease in some states. The best prevention of Marek's disease is to acquire only vaccinated chicks. Marek's disease is transmitted horizontally, but not vertically. So if an owner of an infected flock wants to keep chicks, they can be kept safe from disease through artificial incubation away from the flock and vaccination upon hatch. Remember, like all vaccines, protection is not instantaneous so vaccinated chicks must be quarantined for at least 7 days before exposure.

Zinc toxicity

Zinc toxicity is common in free-ranging poultry. Poultry will consume washers, bolts, nails, scrap wire, and even pennies in their search for grit. Chelation therapy with or without caudal sac ventriculotomy is the recommended treatment. While zinc toxicity is more common, this author removed a penny from the ventriculus of a duck with symptoms consistent with zinc toxicity, only to discover that the blood lead levels were elevated as well (i.e., levels came back after surgery). The duck recovered well with chelation therapy, but may not have needed the surgery to remove the penny.

Trauma

Trauma is a common presentation. Treatment is similar to other avian species. Traumatic fractures of the limbs are a common sequela to domestic predator attacks (dogs). These fractures are often open and infected. In addition to fracture treatment, a thorough assessment for puncture wounds and internal trauma should be made as well. Fracture-repair techniques are similar to other avian species; however, hens in full reproductive mode are likely to have very dense bones (i.e., physiologic hyperostosis) that can make pin placement more difficult. Degloving injuries are commonly caused by wild predator attacks (e.g., raccoons, raptors). While these injuries can look dire, they have an incredible capacity to heal as long as proper antimicrobial therapy and wound treatment are provided.

Nuisance noise

Nuisance noise has resulted in frequent requests to devocalize or caponize roosters in urban areas because their vocalizations have become problematic. The Association of Avian Veterinarians has a position statement against decrowing roosters. Caponization (i.e., castration of a rooster) in its traditional form was performed at a young age with no anesthesia or analgesia and is no longer considered humane. This procedure can be done from a lateral approach with full sterile technique and appropriate analgesia; however, it will not stop a rooster from crowing. Specialized collars have been created to halt crowing. The efficacy and ethical implications of these collars are still debatable.

Nutritional disorders

Nutritional disorders can best be prevented by providing a life-stage-specific crumble or pellet as the primary diet. Supplementation with reasonable amounts of vegetables and fruits, grazing, and insects is fine. Boutique and local diets are increasing in popularity. These diets are often sold as a mixture of whole grains and legumes. While they may be balanced when eaten in their entirety, chickens are notorious for picking out their favorite components which can lead to an unbalanced diet. If waterfowl are present in the flock, ensure the feed contains niacin. While chickens can synthesize niacin, waterfowl cannot.

Reportable diseases

While most chickens will present for individual-level health concerns, there are many reportable diseases of significant national economic consequence that can be seen in backyard flocks. The recent outbreak of highly pathogenic avian influenza has brought the biosecurity risk of backyard poultry flocks into the spotlight. The index for suspicion of a contagious disease increases in birds with exposure to wild birds or other flocks, and when more than one bird in a flock is sick. These are two pieces of information that should be obtained before a client brings their bird physically into the hospital. It can be obtained over the phone when the appointment is made or clients can be instructed to wait in the parking lot until a basic risk assessment of the patient has been obtained. If multiple birds are affected, or there is respiratory disease or neurologic disease with no obvious non-infectious cause, then this author recommends consulting with the state veterinary office. In our experience, they have always been incredibly helpful in guiding diagnostic and risk management decisions.

References

1. Greenacre CB, Morishita TY, eds. *Backyard Poultry Medicine and Surgery: A Guide for Veterinary Practitioners.* 2nd Ed. Hoboken, NJ: Wiley-Blackwell; 2021:343-367.

2. Damerow G. *Storey's Guide to Raising Chickens: Breed Selection, Facilities, Feeding, Health Care, Managing Layers & Meat Birds.* 4th Ed. North Adams, MA: Storey Publishing; 2017.

3. Food and Drug Administration. Regulatory Procedures Manual. October 2020. Chapter 9: Import operations and actions. Section 2: Coverage of personal importations Available at: https://www.fda.gov/media/71776/download. Accessed February 2, 2024.

Choosing Biochemical Analytes for (Mainly) Parrots

J. Jill Heatley, DVM, MS, DABVP (Avian Practice), DABVP (Reptile and Amphibian Practice), Dipl ACZM

Session #131

Affiliation: From the Department of Small Animal Clinical Sciences, College of Veterinary Medicine and Biomedical Sciences, Texas A&M University, College Station, TX 77843, USA.

Abstract: The choice and interpretation of biochemical analytes for parrots is made challenging and exciting by the wide variety of lifestyles, physiology and diversity of the many species often included within this category. In this comparative review, I delve into some commonly used biochemical analytes that differ based on species and assay and I also review some new information from the literature that holds promise for diagnostic use in these species. While a clinical pathology finding should not create, but confirm, diagnoses for all species, the exotic animal diagnostic clinician and veterinarian often struggle with what is normal and if a parrot believed healthy is in fact, suffering from occult disease. The stoic nature of many of these undomesticated prey species challenges us to correctly classify these patients to achieve the best outcome for the animal and owner. Biochemical analytes, which vary greatly in their diagnostic ability include mammalian-based enzymes, total protein, albumin and globulin (and multiple methods of assay), and bile acids (and a variety of assays). In comparison, many chemical assays, such as the electrolytes, venous blood gases, lactate, glucose and multiple acute phase proteins hold promise as being more widely applicable and therefore able to discern health and disease in parrots and other bird species.

Tissue Enzymes

Tissue enzyme assessment for parrots has not been performed. This lack of assessment is an amazing disservice to parrots because we all know they are not chickens, nor are they ducks, raptors, or pigeons. The cost for tissue enzyme studies is not a lot greater than that for a biochemistry panel. We based an assessment of parrot enzymes on changes observed in other studies and on what is known about tissue enzymes in other species (Table 1).

Table 1. Variability in the presence of avian tissue and plasma enzymes.[a]

Avian species	Enzyme[b]	Tissue	Clinical activity findings
Barn owl (*Tyto alba*) Screech owl (*Otus asio*) Black-crowned night heron (*Nycticorax nycticorax*) Canvasback duck (*Aythya valisineria*) Redhead duck (*Aythya americana*)	ALT, AST, ALP, CPK, LDH	Plasma, liver, kidney, muscle	CPK is best for muscle. Interspecific enzyme variation. All enzymes found in all tissues studied.
Black duck (*Anas rubripes*) Mallard duck (*Anas platyrhynchos*)	ALT, AST, ALP, CPK, LDH, GGT, OCT	Plasma, liver, kidney	OCT, AST activity in liver and kidney. ALT, CPK, ALP, GGT high in kidney. LDH high in liver. Plasma GGT in one mallard.
Turkey (*Meleagris gallapavo*) Adult male	AST, ALT, ALP, LDH, CK	Liver, heart, brain, lung, proventriculus, duodenum, breast muscle	All enzymes in every tissue. Highest AST, ALT in liver, then kidney. Low AST, ALT in lungs. ALP high kidney, duodenum. Low ALP in brain, lung, proventriculus, muscle. High LDH, CK in heart, muscle; low in duodenum, proventriculus.
Racing pigeon (*Columba livia*)	ALT AST GLDH LDH GGT ALP CK	Heart, liver pectoral/ quadriceps, mms duodenum, kidney, brain	GLDH is the most liver-specific enzyme, in plasma acute severe liver damage, LDH, AST sensitive for liver damage, also in muscle. Differentiate muscle damage from liver damage with CK, specific and sensitive for muscle cell damage.

[a]Selected enzymes were evaluated for each species; unlisted enzymes for these species could still be of use, adapted and updated. Note the absence of parrot.

[b]ALT: alanine aminotransferase; AST: aspartate aminotransferase; ALP: alkaline phosphatase; CPK: creatine phosphokinase; LDH: lactate dehydrogenase; CK: creatine kinase; GGT: gamma-glutamyl transferase; OCT: ornithine-carbamyltransferase; GLDH: glutamate dehydrogenase.

Source: Heatley JJ, Russell KE, eds. *Exotic Animal Laboratory Diagnosis.* Ames, IA: John Wiley & Sons; 2020.

Bile Acids

Based on limited studies in parrots, approximately 80% of the liver must be inoperable for bile acids to increase. Bile acid investigations reported for birds including the closely related falcon species: peregrine falcons, Gyrfalcons, Saker falcons, and many parrots. An older retrospective evaluation of 442 samples suggested that increased enzyme activities correlated poorly with increased bile acid concentrations; increased bile acid concentrations had the highest association with confirmed hepatic disease, followed by high concentrations of alpha2-globulins, aspartate aminotransferase (AST), lactate dehydrogenase (LDH), and alpha1-globulins for parrots.

In germ-free chickens, only cholic, allocholic, and chenodeoxycholic acids occur; however, in conventionally raised chickens, other bile acids originate from the intestinal microbiome. Chenodeoxycholyltaurine and cholyltaurine are the main bile acids in chickens and turkeys, whereas chenodeoxycholyltaurine and phocaecholyltaurine predominate in ducks. Duck bile contains mostly taurochenodeoxycholic acid, followed by cholic, ursodeoxycholic, chenodeoxycholic, lithochilic, deoxycholic and taurocholic acids. In chickens, a high concentration of glycolithocholic acid, followed by taurocholic, lithocholic, taurolithocholic, chenodeoxycholic and cholic acids occur. Among wild birds, chenodeoxycholic acid was the primary bile acid, with cholic and allocholic acids dominating in carnivorous species.[2] Human colorimetric assays commonly detect five bile acids that account for about ~ 99% of bile acids in human biofluids: conjugates of cholic, chenodeoxycholic, deoxycholic and lithocholic acids. I cannot confirm which bile acids VSP colorimetric bile acid assay detects, but based on a literature review, it is likely all bile acids, except those lacking a 3-hydroxyl group or are sulfated or esterified at the 3 position. Radioimmunoassays (RIA) are now less common based on reagent and sample volume needed, colorimetric values have been reported for fewer avian species; RIA and colorimetric are not directly comparable (Table 2). To my knowledge, the colorimetrically determined bile acids of parrots have not specifically been reported but are presumed similar to those of granivorous birds. Be aware that use of the avian-reptile panel will only provide values should they be greater than 30; however, use of the "liver" rotor of the Abaxis system will provide a lower numerical value; it is likely that many species have a much lower normal value. The enzymatic assay measures a wider spectrum of bile acids, therefore, reference values tend to be higher than those for RIA. The RIA may be less likely to be affected by sample tissue such as lipemia, hemolysis, or increased LDH activity.[4] Postprandial increases in bile acid concentrations may occur in pigeons, mallards, and peregrine falcons. In most avian species suffering liver disease, anorexia is typically a concern, thus long-term fasting is likely unnecessary as elevations with hepatobiliary disease are typically much greater than postprandial elevations, and only a single nonfasted sample in birds is necessary for diagnosis

Take home messages

Use RIA for the colorimetric assay. Use the LIVER rotor to measure bile acids.

Table 2. Parrot bile acid ranges based on incomparable assays (μmol/L).

Species	Radioimmunoassay	Colorimetric assay
African grey parrot	18-71	12-96
Amazon parrot	19-144	33-154
Budgerigar	20-65	32-117
Cockatiel	25-85	15-139
Cockatoo	34-112	8-11
Conure	32-105	8-15
Eclectus parrot	30-110	8.8-9.8
Love bird	25-95	12-90
Senegal parrot	0-85	20-94

Common Blocks: Glucose, Electrolytes, Minerals

These assays are generally suitable for the determination of health. Limits to their use are only the unknown range of health or the limits of the analyte tolerated by the species, which could be much greater than reported humans or other mammals. Ionized calcium is an excellent example of advances in this area that demonstrate the very basic building blocks, used by all species in the same basic form. The challenge remains making published reference ranges accessible and having the analyzer and pertinent conditions of the assay and animal reported for comparison: sample collection, UVB light exposure, age, sex, and diet or any statistical patient or environmental conditions that were tested made no difference within the population tested. Venous blood gases, pH, vitamin A, vitamin D, trace elements and heavy metals are included in this category.

Proteins

Assays range from acceptable with analyzer-specific reference intervals, to completely incompatible with the reality of the patient (e.g., Bromocresol green in rabbits, turtles, VSP). Refractometry remains a simple, cost-effective assay in those species without a stress glucose level above 100, and a lack of lipemia and or vitellogenesis (e.g., male reptiles, nonreproductively active female reptiles, most rodents, but not birds). Although protein electrophoresis remains the gold standard, species-based differences, analyzer differences and the lack of absolute values (i.e., these are calculated from the BCG assay or a similar enzymatic reaction) make the basic proteins of albumin and globulins quite unreliable for these species. Comparison on the same analyzer from a previous value (i.e., trends) or from a health conspecific is recommended. Nonetheless, albumin is considered an excellent acute phase protein that decreases in parrots. Fibrinogen is also an acute phase protein in parrots as well as serum amyloid that increases during inflammation. The assay for fibrinogen is relatively simple and can be performed alongside the PCV/TS. These indicate acute inflammation as well if not better than the CBC. In practice, I also like a slide review to look for toxic change.

For the naturally increased glucose concentration in birds, it is likely that we can create a correction for total solids to get a better idea of blood protein from refractometry. Most colorimetric and other chemical or biochemical assays are invalid. EPH gives you an exact percentage of proteins, but the actual amounts provided are from an enzymatic assay, which is poorly validated for birds.

Lactate has received preliminary investigation in multiple avian species (Table 3). Higher values of lactate for birds that have poor exercise conditions or are recently active appear acceptable for these species. Reasons for lactate increase are classified as Type A hypoxia or shock and type B nonhypoxic and nonshock causes. In parrots it appears that relatively increased lactate may be induced by handling, which resolves rapidly after calming in healthy parrots. Lactate values have been published for flamingos and will soon be published for Quaker parrots, Hispaniolan Amazon parrots and multiple free-living parrot species.

Table 3. Comparative lactate values reported for various avian species.

Avian species	Lactate (mmol/L)	Notes
Boat-tailed grackle (*Quiscalus major*)	5.61	Immediately after release from mist net.
	2.62	Calm after release from mist net.
House sparrow (*Passer domesticus*)	4.77	Immediate post release from mist net.
	3.14	Calm after release from mist net.
Passerine bird	2.79	With midazolam sedation.
	4.39	Without midazolam sedation.
Passerine chick	5.04	Without sedation.
Mourning dove (*Zenaida macroura*)	7.72	Immediate post release from mist net, wing vein.
	4.83	Calm after release from mist net.

Disclosure: The author declares no conflict of interest and has nothing pertinent available to disclose regarding this manuscript.

References

1. Heatley JJ, Russell KE, eds. *Exotic Animal Laboratory Diagnosis.* Ames, IA: John Wiley & Sons; 2020.

2. Zaefarian F, Abdollahi MR, Cowieson A, Ravindran V. Avian liver: the forgotten organ. *Animals.* 2019;9(2):63.

3. Battison AL, Buckzowski S, Archer FJ. Plasma bile acid concentration in the cockatiel. *Can Vet J.* 1996;37(4):233.

4. Vergneau-Grosset C, Polley T, Holt DC, et al. Hematologic, plasma biochemical, and lipid panel reference intervals in orange-winged Amazon parrots (*Amazona amazonica*). *J Avian Med Surg.* 2016;30(4):335-344.

5. Alberton S, Vergneau-Grosset C, Summa N. Advances in exotic animal clinical pathology. *Vet Clin North Am Exot Anim Pract.* 2019;22(3):367-385.

6. Feltrer Y, Strike T, Routh A, et al. Point-of-care cardiac troponin I in non-domestic species: a feasibility study. *J Zoo Aquar Res.* 2016;4(2):99-103.

Back to the Basics: Clinically Relevant Avian Anatomy and Physiology

Marcie Logsdon, DVM

Session #163

Affiliation: From Washington State University Veterinary Teaching Hospital, 205 Ott Rd, Pullman, WA 99164, USA.

Abstract: There are an estimated 9,500-11,000 extant avian species in the world, almost twice the estimated number of mammal species. This amazing amount of diversity makes generalizations about birds difficult to make. Some anatomic differences between birds and mammals, of which practitioners should be aware, hold true for most avian species seen in practice. This paper highlights many of these important considerations.

Note: Adapted from: Faux CM, Logsdon ML. Avian clinical cases. In: Orsini JA, Grenager NS, de Lahunta A, eds. *Comparative Veterinary Anatomy: A Clinical Approach.* San Diego, CA: Elsevier Academic Press; 2022:1251-1416.

Introduction

All veterinarians know that a mammal is a mammal, but what if your patient is a bird? Birds have many evolutionary adaptations that set them apart from mammal patients. Current veterinary curriculums focus heavily on common companion animal species (e.g., dog, cat), livestock, and horses leaving new graduates at a loss regarding avian and exotic species. This is unfortunate, since according to a 2022 national pet survey, 9.9 million U.S. households own a pet bird,[1] which puts birds in fourth place, behind dogs, cats, and aquarium fish. Before providing care to any unfamiliar species, practitioners should research normal parameters, husbandry and diet requirements, and common ailments for that species. A full description of avian anatomy and physiology would fill several textbooks. Below are some practical anatomic and physiologic considerations that are relevant when treating pet birds.

Ocular Considerations

Avian eye shape is classified as flat, tubular, or globular. Compared to mammals, avian eyes are often much larger in relation to their skull. A bony ring, the scleral ossicles, is found around the eyes and helps provide structural support. This also limits the capacity for ocular movement and many species like owls must move their entire head to shift their field of view. As in mammals, eye position in the skull varies by species and natural history. Predatory birds like raptors generally have forward-facing eyes to provide improved binocular vision whereas prey species have eyes set on the sides or even the top of the head (e.g., snipes) to help see predators.

Due to the avian eye's large size and integral relationship with the skull, enucleation carries a high risk of morbidity and mortality. Alternatively, removal of the cornea and evisceration of the contents of the globe is more commonly performed.

As in many exotic species, a traditional menace response is unreliable. A practitioner can watch for accommodation of the pupil when an object approaches, which is more reliable in calm birds than stressed ones. Or the practitioner can perform a modified menace or startle response where an object such as the hand starts behind the eye outside of the field of vision and then passes rapidly forward into the field of vision. Most patients will react by moving their head away or turning to look at the hand.

Pupillary light response (PLR) is still a valuable assessment tool. Birds have striated muscle in their iris and can have voluntary control of the pupil. A stressed bird may remain mydriatic even in bright light. To assess a direct PLR, start in a semi-darkened room and apply a bright light. The neurologic pathway being tested during a PLR is still CN II afferent, CN III efferent. If the PLR is intact, the pupil should constrict initially; however, if the patient is stressed with a high sympathetic tone then the pupil may begin to dilate again despite the continued presence of the light. Birds lack a consensual PLR as the optic nerve has complete decussation at the optic chiasm. This transient anisocoria should only be appreciable when there is severe discrepancy in light levels between the two eyes (e.g., when performing a PLR exam); during general observations under ambient lighting, pupils should be symmetrical and anisocoria is a sign of pathology. Due to the presence of striated muscle in the iris, atropine is not an effective mydriatic in birds. Intracameral administration of neuromuscular blocking agents such as atracurium can be effective however they are impractical for routine examinations. Anesthesia-induced mydriasis is often the most available method, especially if anesthesia is necessary for any concurrent diagnostics or procedures.

Unlike primates, the avian lens does not have a UV filter, and bird eyes contain a fourth type of cone (tetrachromatic vision) allowing many species to see into the ultraviolet range. In contrast, humans have three types of cones (trichromatic vision). This ability is used to find food including everything from determining which fruits are ripe to seeing urine trails left by mice. Many species with no apparent sexual dimorphism to the human eye have very different color markings when the UV spectrum is visible. For birds to fully appreciate these colors in their environment, full-spectrum lighting is required.

Fundic exams are an essential part of any ocular exams, especially in wild avian patients with suspected trauma. Birds lack a tapetum, and the retina is avascular. In many nocturnal birds, the retina lacks pigment as well, allowing for visualization of the vessels of the choroid. Because the retina is avascular, trauma resulting in retinal detachment results in permanent loss of vision in the affected area, even if the retina reattaches. Depending on species, birds have between 0-3 fovea. The pectin is a structure found in avian and reptilian eyes. This structure extends from the retina and is thought to help with nutrient distribution in the vitreous chamber. Functionally for the avian practitioner, it serves as a good focal point when performing a retinal exam. Because it is a dimensional structure you can focus it in your ophthalmoscope and then use it as a reference point to bring the retina into focus. It is also a common location for ocular trauma.

The avian lens can develop nuclear sclerosis or cataracts. Traumatic cataract formation can be a sequala to head and ocular trauma. While they usually occur rapidly following trauma, they can have a delayed onset of up to a few weeks later. Phacoemulsification has been performed successfully on captive birds to help restore vision. Use of intraocular lenses has been reported,[2] but is complicated by high species variability and rapid accommodation potential of the avian lens. In addition, some bird species such as falcons, dive at such high speeds that a replacement lens would be unlikely to remain in place when submitted to that much gravitational force.

For further discussion on the avian eye, readers are encouraged to reference the works of Dr. Rudiger T. Korbel, whose lectures are frequently presented at annual conferences and are available at the Association of Avian Veterinarians (AAV) online continuing education materials.

Integument Considerations

Feathers are one of the defining characteristics of birds. Flight feathers, remiges and rectrices, attach directly to the periosteum of the bone. This stability allows for the transfer of force from the skeleton to the feathers during flight. How firm this attachment is depends on the species. Many prey species have developed to allow for easy sloughing of those feathers as a predator escape mechanism, which is very useful for getting away if a predator tries to catch the bird. It is less useful if a patient is poorly restrained and feathers are pulled accidentally during an exam. Most of these birds can regrow feathers without incident; however, in larger birds without this adaptation, traumatic pulling of the feather can result in trauma to the underlying periosteum, bone or feather follicle.

The uropygial or preen gland is located at the base of the tail and is present in most species. Some common pet species, such as Amazon parrots, lack a uropygial gland. If present, it should be evaluated at each physical exam for signs of infection or neoplasia. The uropygial gland is a common site for squamous cell carcinomas to appear. The oily secretions of the uropygial gland are important for waterproofing of feathers. In some species, the precursor to vitamin D_3 is sometimes present in these secretions. Once present on the feathers it is converted to D_3 by sun exposure and then consumed during subsequent preening.[3] The secretions also have antimicrobial and antifungal properties that act as a defensive barrier on skin and feathers.

Birds lack biliverdin reductase, the enzyme responsible for converting biliverdin in bruises to bilirubin. Therefore, bird bruises will appear green for an extended period during healing. This should not be confused with gangrene.

Aural Considerations

Birds lack pinnae at the opening of the ear canal. The external ear acoustic meatus is often covered by feathers. These feathers are called ear coverts or auriculars and lack barbs; these specialized feathers help sound pass through and clinically can be used to help locate the ear. The entrance to the ear can be located caudal to the angle of the jaw and caudoventral to the eye. Ear size is highly variable by species. Species that rely heavily on hearing for hunting often have very large ears. For example, in many owl species, the external acoustic meatus is so large that you can easily visualize the posterior aspect of the eye and the scleral ossicles. Owls also have asymmetrical ears, helping them to more precisely pinpoint sounds. This asymmetry is achieved by actual differences in the skull in some species (e.g., Northern saw-whet owls) and by soft tissue differences in others (e.g., common barn owl). The parabolic appearance of facial feathers on owls and other auditory hunters helps funnel sound to the ears.

Primary otitis is uncommon in adult companion birds. This author has seen rare cases of otitis of unknown etiology in Quaker parakeets. Myiasis and secondary otitis can occur in nestlings that are outdoors and can be a cause of vestibular signs. In addition, ears are closed at time of hatch and ascending otitis (from the choana) can occur in nestlings.

Cardiovascular Considerations

Birds have a highly efficient cardiovascular system designed to provide energy during the high metabolic demands of flight. Because of this oversized system, and the fact that flighted birds have many weight-saving adaptations such as hollow bones, the total estimated blood volume is higher in birds than mammals. Estimated blood volume in most birds is ~10% of body weight compared to 6-8% in mammals. An estimate for blood sampling volume that is generally considered safe in a healthy individual is up to 10% of blood volume. In a healthy bird, blood samples of up to 1% of body weight can be collected during a single session. Therefore, you can safely draw up to 0.3 ml of blood from a healthy 30-gram budgerigar.

The turnover rate of red blood cells in birds is more rapid than in mammals, as is replacement after loss. Birds are also more tolerant of acute blood loss. In one study, the LD50 for acute blood loss in mallard ducks was 60%.[4] In mammals, the LD50 is usually 30-40%.

Avian (and reptilian) red blood cells are nucleated. This means that standard automated hematology analyzers cannot be used. Differentials must either be performed manually from slides, or with the use of more recent artificial intelligence programs that have been developed to identify and count cells from images taken from slides. Nucleated red blood cells cannot be used as an indicator of regeneration; instead, degree of polychromasia can be used as a measurement of regenerative response.

Common blood draw locations are the right jugular, basilic or ulnar vein, and medial metatarsal vein. The right jugular is usually larger than the left. Conveniently, there is a region of apteryla (i.e., featherless region) overlying the jugular, which is readily visible under the skin.

A capillary refill time is usually inaccessible in the oral cavity of birds. One can be obtained from everting the mucosa of the cloaca. A more readily available option is to assess the venous refill time in the ulnar vein.

Upper Respiratory System Considerations

The avian trachea is comprised of complete, interlocking tracheal rings. This prevents tracheal narrowing under high inspiratory pressure. Avian tracheas are longer and wider than mammals of comparable size. This increases relative dead space and is overcome by large tidal volume. Because they lack the expandable dorsal membrane of mammal tracheas, inflatable endotracheal tubes are contraindicated. Inflation of the cuff can lead to ischemia and necrosis. Sterile lubricant can help form a seal when using a well-fitting non-cuffed endotracheal tube. Alternatively, Cole ET tubes (Avian/Small Exotic Cole Endotracheal Tube, Jorgensen Laboratories LLC, Loveland, CO) have an increase in diameter designed to sit against the glottis and prevent leaking of anesthetic gasses.

The syrinx is located proximal to the bifurcation of the trachea and is the location of sound production. Depending on the species the syrinx may be located proximal to the bifurcation (i.e., tracheal), at the bifurcation (i.e., tracheobronchial), or comprise of two syrinxes caudal to the bifurcation (i.e., bronchial). Birds with these paired syrinxes can produce two different sounds simultaneously. Because of the location of the syrinx, birds can still vocalize when intubated.

The cervicocephalic air sacs connect to the complex of sinuses in the skull and can be found along the dorsolateral aspect of the neck. These are not involved in respiration. They lie proximally to the jugular. If they are punctured during a jugular blood draw, hemorrhage into the cervicocephalic air sacs can result in epistaxis. In addition, chronic or severe sinusitis can cause a secondary hyperinflation of the cervicocephalic air sacs.

Lower Respiratory Tract Considerations

The avian respiratory system is exceptionally efficient. Lungs are the location of gas exchange. In birds they are non-elastic and fixed in place. Instead of dead-end alveoli, air is passed through small air capillaries and gas exchange happens in a cross-current manner to blood capillaries. This cross-current mechanism helps extract oxygen more efficiently from the inspired air and subsequently is more efficient at expelling CO_2 results in a higher CO_2 in expired air than in mammals.

Birds lack a diaphragm and must be able to expand their body wall to move air. Therefore, care must be taken during restraint to ensure that the body is not being compressed. Large air sacs extend throughout the caudal coelomic cavity. No gas exchange happens in the air sacs. On inspiration, air first bypasses the lungs and goes to the caudal air sacs. This means that in a clinical setting, cannulation of a caudal air sac is a way to bypass the trachea and provide air to the lungs. This can be used in emergency situations of tracheal obstruction, or for maintenance of anesthesia during procedures that require access to the trachea. The air sacs also provide a window for endoscopy; no insufflation is needed for visualization because the patient is doing that for you!

Gastrointestinal Considerations

The gastrointestinal tract from beak to cloaca varies depending on the species' diet. The crop is a diverticulum of the esophagus. Most pet species have a crop; however, it is notably lacking in waterfowl and owls. The primary function of the crop is storage of food; digestion happens later in the GI tract. Columbiformes have an especially well-developed crop. In addition to food storage, in these species it is the site of production of crop milk, a secretion used to feed young. Gavage feeding can be done into the crop with either a red rubber feeding tube or a feeding needle. A good guideline for volume is ~1 ml/30 g of bodyweight, or 25-30 ml/kg.

The avian stomach is divided into two compartments, the proventriculus and the ventriculus (i.e., gizzard). Enzymatic digestion occurs in the proventriculus, while mechanical digestion occurs in the ventriculus. In many species of granivores (e.g., chickens) the ventriculus is highly muscular with a thick koilin layer lining the inner surface. These species ingest grit, which is transiently stored in the ventriculus and helps with the mechanical grinding of food. A common misconception is that all birds require grit. While a small amount of grit in a parrot's diet may be innocuous, it is not required. They grind and crush their food with their beak. Carnivorous birds generally have a poorly defined distinction between the proventriculus and ventriculus as they rely more heavily on chemical digestion as opposed to mechanical. The pellet of indigestible material (i.e., cast) regurgitated by owls and many other birds of prey is produced in the ventriculus.

Musculoskeletal Considerations

In flighted birds, the skeletal system shows adaptations for flight. These adaptations decrease weight and increase structural stability. Distal limbs show a reduction in the number of bones. This is achieved evolutionarily by loss (e.g., digits in the thoracic limb) and fusion (e.g., tibia and first tarsal bone to form the tibiotarsus).

The number of cervical vertebrae ranges from 11-25 giving birds extreme flexibility in the neck. When at rest, most birds hold their cervical vertebra in an "S" position, concealing the true length of the neck. When administering oral medications or gavage feeding it is important to extend the neck to its full length. In most species, the thoracic vertebrae are fused to form the notarium. In species lacking this fusion, such as bald eagles, there is extensive connective tissue providing similar support. The lumbar vertebrae are fused to create the synsacrum. These fusions are necessary to provide the stability and frame for flight. Between the notarium and synsacrum are a small variable number of mobile vertebrae. This creates a vulnerable spot and is the most common location of spinal trauma in birds.

Extensions of the air sacs extend into the appendicular skeleton. Most commonly, the humerus and femur are pneumatic bones. These air sacs connect with the respiratory system. Care must be taken during the treatment of pneumatic bone fractures. If performing an open repair, lavage is contraindicated as fluid can enter the respiratory tract. In addition, inhalant anesthesia may leak from an open fracture and create a fire risk if cautery is used. Respiratory infection is a possible complication of these fractures. Subcutaneous emphysema is a common sequela but is usually self-limiting.

Reproductive Considerations

Male testes are internal and paired. They change in size seasonally and may be visible on the ventrodorsal view radiographically if the patient is in breeding season. Most birds lack a phallus; however, ducks have evolved a very extensive phallus. Clinically, this can sometimes become infected or necrotic if prolapsed and may necessitate amputation. The phallus has no connection to the urinary tract and so can be safely amputated if needed.

Most female birds have only a single oviduct and ovary, the left ("The left is left, and the right recedes."). The Kiwi has a only right ovary and oviduct, and many raptor species have two ovaries but only a left oviduct. The ovary is closely associated with the common iliac vein and surgical removal carries a risk of hemorrhage.

Females do not require access to a male or fertilization to form eggs. Unlike reptiles, which produce an entire clutch simultaneously, eggs develop one at a time in birds. There should never be two calcified eggs visible on radiographs at one time. Egg production starts with follicle development at the ovary. Once ovulation occurs, the mature follicle is essentially a fully formed yolk. The ovulated follicle is caught by the infundibulum. If the follicle is not captured by the infundibulum, due to disease or neoplasia, an egg-yolk peritonitis ensues. Once it passes the infundibulum it enters the magnum. This is the longest portion of the oviduct and is where the egg white and albumin are deposited. Next, it passes into the isthmus where the inner and outer shell membranes are formed. The uterus is where water is added to the albumin and the shell is calcified. The vagina is the final passage to the vent and is where shell pigment is added.

Renal Considerations

Avian kidneys have three subdivisions: the cranial, middle, and caudal divisions, and are non-mobile, situated within depressions along the ventral aspect of the fused caudal spinal vertebrae (i.e., synsacrum). The lumbosacral plexus, including the femoral and obturator nerves, exits the synsacrum dorsal to the cranial division of the kidney, while the sacral plexus runs through the parenchyma of the middle renal division itself. Therefore, any condition causing enlargement of the kidney (e.g., cystic kidney disease or renal neoplasia), or that applies pressure to the kidney (e.g., gonadal neoplasia) can compress the lumbosacral and sacral nerve plexuses and result in pain and dysfunction of the associated limb.

The primary end waste product of nitrogen metabolism in avian species is uric acid (i.e., uricotelism), as opposed to urea in most mammalian species (i.e., ureotelism). Uric acid is non-water soluble so very little water is needed for excretion. This mechanism minimizes water loss and may also be a weight-sparing adaptation for flight. The most important advantage of uricotelism, as opposed to ureotelism, may be in the storage of nitrogenous waste in eggs during incubation. In this environment, a buildup of a water-soluble nitrogenous waste product such as urea may be toxic to a developing embryo.

Cloaca Considerations

The cloaca is the shared termination of the urinary, reproductive, and gastrointestinal systems. It is comprised of three subdivisions: urodeum (urinary), proctodeum (reproductive) and coprodeum (gastrointestinal). The vent is the muscular sphincter visible externally. Vent tone can be affected by neurologic disfunction and is an important parameter to assess in cases of suspected spinal trauma. In reproductively active female birds, the vent will relax and stretch to allow passage of eggs.

Birds can retropulse secreted urine from the urodeum into the colon and ceca where water is reabsorbed. Because of this ability and since fecal matter and urinary waste are excreted simultaneously, urinalysis is of limited value.

References

1. American Pet Products Association. 2021-2022 National Pet Owner Survey. Available at: https://www.americanpetproducts.org/news/News-Public-Relations/pet-industry-market-size-trends-ownership-statistics. Accessed January 31, 2024.

2. Carter RT, Murphy CJ, Stuhr CM, Diehl KA. Bilateral phacoemulsification and intraocular lens implantation in a great horned owl. *J Am Vet Med Assoc.* 2007;230(4):559-561. doi: 10.2460/javma.230.4.559.

3. Uva B, Mandich A, Vallarino M. The site of 7-dehydrocholesterol ultraviolet photolysis in domestic fowls. *Acta Histochem.* 1983;73(2):175-80. doi: 10.1016/S0065-1281(83)80026-9. PMID: 6320573.

4. Lichtenberger M, Orcutt C, Cray C, et al. Comparison of fluid types for resuscitation after acute blood loss in mallard ducks (*Anas platyrhynchos*). *J Vet Emerg Crit Care.* 2009;19(5):467-472. doi: 10.1111/j.1476-4431.2009.00465.x. PMID: 19821888.

Avian Triage, Vascular Access and Fluid Resuscitation

Jennifer Graham, DVM, DABVP (Avian Practice), DABVP (Exotic Companion Mammal Practice), Dipl ACZM

Session #120

Affiliation: From Graham Veterinary Consulting, LLC, Madison, AL 35758, USA.

Abstract: Other than acute trauma cases, avian patients presenting on emergency often have significant underlying disease. The owner may not note any abnormalities until the bird has decompensated. The clinician must be prepared to intervene with life-saving measures when necessary. This presentation discusses triaging avian patients with a focus on vascular access and fluid resuscitation.

Note: Adapted from Graham J. Avian triage, vascular access, and fluid resuscitation. *Proc Annu Conf Int Vet Emerg Crit Care Soc.* 2022.

Triage and Examination

Before taking a complete history, determine if the bird needs immediate care. As in mammals, the ABCs (i.e., airway, breathing, cardiovascular status) should be immediately assessed in avian patients with trauma. Stabilization efforts should be made in situations of hemorrhage, respiratory distress, severe neurologic signs including seizures, open fractures, or if the bird is weak and/or on the bottom of the enclosure. If immediate care is needed, the bird should be placed in a warm and oxygenated incubator, ideally 29- 32 °C (85-90 °F). Supportive care measures including fluids, nutritional support, analgesics, or antibiotic therapy can stabilize the patient and allow for additional diagnostics and treatment later.

Prior to handling the bird, examine it in its cage. Evaluate how well the bird can move about the enclosure, body position, respiratory effort (e.g., open-beak breathing or tail bobbing can be signs of increased respiratory effort), if the feathers are fluffed, and overall level of responsiveness. Take note of color and consistency of droppings, any regurgitated material, and the presence of any blood or discharge.

Physical examination is best performed systematically, thoroughly, and in a timely fashion to minimize stress for the patient. Have all materials, including any needed for the exam or diagnostics, organized prior to handling the bird. Work in a small room with locking doors and covered windows to minimize the potential for additional trauma with any escapes. Darkening the room can reduce movement of particularly fractious patients to facilitate initial capture. A scale that weighs in 1 g increments should be used to determine the weight of the patient prior to or immediately following the exam. Some birds will willingly step onto a tared perch, but some birds may need to be placed in a small container for weighing.

Evaluation of the head should include examination of the external ear canals, nares, eyes and surrounding tissues, beak, and the oral cavity including choanal slit. The beak is a common sight for trauma and should be closely inspected for cracks or bleeding. Head trauma can result in ocular trauma, hemorrhage, or bruising. Palpate the crop at the base of the neck and note any distention or evidence of wounds. Body condition should be assessed by palpation of the pectoral muscling and palpation of the body for any excess fat stores. The coelom should be concave in psittacine species; any distention could be a sign of reproductive disease, obesity, organomegaly, or other masses. Wings and legs should be evaluated along with the overall quality of the skin and feathers. The presence of a wing or leg droop warrants further investigation. Evaluate the cloaca and perform a cloacal eversion in larger species to evaluate for any abnormalities including papillomas. Check feet for signs of pododermatitis or overgrown and injured nails. Auscultation of the heart is best performed over the keel while the respiratory tract can best be ausculted across the back. Note any heart murmur, arrhythmias, or increased respiratory sounds. Any crackles, wheezes, or clicking can indicate pathology as respiratory sounds are normally minimal. Note that there can be a normal clicking sound in some birds from the uncinate processes (i.e., projections off the caudal aspect of each rib) hitting one another; this is most common in cockatiels in the author's experience. Bite wounds can result in subcutaneous (SC) emphysema and a crackling sound may be detected if this is present along with crepitus when palpating the skin. Skin turgor can be assessed by gently pinching the eyelids; if the skin stays tented this can be a sign of dehydration. The oral cavity can be dry and pigmented in many species so evaluation of the cloacal mucosa can give a better idea of mucous membrane color and hydration. Perfusion status can be assessed by refill time of the basilic vein. Complete neurologic examination can be challenging in some patients but should be attempted for any bird with suspected neurologic disease or head trauma. At any point in time a stressed bird can be given a break or placed in oxygen with the remainder of the exam completed later. Evaluate the recovery of the bird and note excessive recovery time, dyspnea, or weakness. Overweight birds can easily get overheated and may display panting and holding the wings away from their body to facilitate cooling during recovery.

Vascular Access

Stabilization procedures including hemorrhage control, securing an airway, which may include air sac cannula placement, and bandage application to prevent further trauma may be needed in addition to fluid resuscitation. Although not appropriate for critical patients, SC fluids can be helpful for stable dehydrated birds or patients with minor blood loss, such as that associated with a broken blood feather or minor trauma. SC spaces are found in inguinal, axillary, and dorsal regions. The inguinal region is the preferred site for administration of SC fluids in the bird. The patient should be restrained upright or in dorsal recumbency (if no coelomic pathology) with the legs pulled down. A small amount of alcohol can be applied to the medial side of the most proximal portion of the leg where the top of the leg meets the body to part the feathers. Using a small gauge needle or butterfly catheter, administer fluids into the visible skin fold making sure a bubble forms, indicating the fluids are in the SC space. Although the same thing can be done in the axillary and dorsal regions, there is more risk with these fluid administration sites, and less fluid can be administered. Wing flapping is likely with fluids administered in the axillary region and dorsal administration has a higher risk of penetrating into an air sac or lungs. Care should be taken if administering fluids when penetrating bite wounds are present to avoid flooding the air sacs.

Intravenous catheterization

Intravenous (IV) catheters are more difficult to maintain than intraosseous (IO) catheters due to the thin skin of the bird. They are generally best tolerated by Galliformes, Anseriformes, and hooded raptors. Locations for IV catheters include the jugular vein, basilic (or ulnar) vein, or the medial metatarsal vein. The area for catheter placement is plucked and scrubbed. A 22- to 25-gauge needle can be used to perform a small dermal cut down to avoid catheter burring and bending. Pre-flushing the catheter (generally 24- to 26-gauge over-the-needle catheter) with heparinized saline can help prevent clot formation. The catheter is then guided into the vein and secured to the skin with glue or tape sutured to the skin. Inserting the catheter at a steep angle followed by flattening it against the skin surface parallel to the vein helps avoid penetrating the far wall of the vessel. A T-port or stop cock is attached to the catheter, after which the catheter can be flushed to ensure patency. Tegaderm can be used over the catheter to further secure it to the skin and feathers. A figure-of-8 bandage can be used to protect and secure basilic vein catheters. Tape guards on the bandage can be helpful to distract from biting on the bandage. A neck wrap can be used to help secure jugular catheters. Unless it is a long-legged bird > 300 g, it is difficult to use the medial metatarsal vein other than for anesthesia since securing this catheter in place is more challenging due to the distal location on the leg in most psittacine birds. The advantage of using this vessel is that the scales of the leg and feet help hold the catheter in place. Interdigital veins can be used in birds with large, webbed feet. A syringe cap can be used as a catheter guard. Using a syringe pump or fluid pump that can administer small volumes is ideal to control slow delivery of fluids.

Intraosseous catheterization

IO catheters can be advantageous to use in birds < 60 g, when IV catheterization is difficult, and to prevent hemorrhage if the catheter is removed by the bird. IO catheters should never be placed in pneumatic bones such as the humerus or femur to avoid the risk of air sacculitis, pneumonia, or asphyxiation due to fluid entering the respiratory tract. IO catheters can be placed either in the ulna or the tibiotarsal crest and are generally well tolerated by the patient. Fluid is rapidly absorbed into the systemic circulation when administered intraosseously, with over 50% of the fluid reaching the central circulation within 30 seconds. A major complication of IO catheters is osteomyelitis; therefore, sterile surgical preparation of the area is essential. Placement can be performed in an anesthetized bird or with a local block in severely debilitated birds. After the area is plucked and scrubbed, a local block of 2% lidocaine diluted with saline can be used to numb the area. The dose used can be up to 0.2 mg/100 g bird. Medium to large sized birds will use a 20- to 22-gauge, 4- to 5-cm (1.5- to 2.5-inch) spinal needle. A 25-gauge hypodermic needle can be used in smaller birds. A stylet can be made for the hypodermic needle using sterile cerclage wire or 3-0 to 6-0 stainless steel suture, thereby preventing obstruction of the needle lumen with cortical bone. The manus should be pronated while palpating the dorsal tubercle of the distal ulna as a landmark for ulnar IO catheter insertion. For tibiotarsal placement, the knee should be flexed while the cnemial crest is palpated; the needle should be inserted into the crest in a cranio-medial direction, avoiding the patellar tendon. A padded bandage or lateral splint can be used to stabilize the catheter. Compared to IV catheters, more fluid resistance is encountered with IO catheters, making small-volume syringes or continuous rate infusion with a small syringe pump preferable. With appropriate sterility and maintenance, IO catheters can be used for up to 3 days. IO catheters should not be used in osteopenic birds due to the risk of iatrogenic bone fracture.

Fluid Resuscitation

Water constitutes about 60-70% of a bird's weight with one-third constituting extracellular fluid and two-thirds as intracellular fluid. Blood volume varies by bird species, generally ranging between 5-10.5% of body weight. Water requirement is estimated between 50-100 ml/kg/d, with neonates and passerines having higher maintenance requirements. Fluid can be administered via oral, SC, IV, and IO routes. Oral fluid administration is generally not recommended for critical patients. SC fluids are appropriate for mildly dehydrated birds and as a vehicle for administration of tissue irritating drugs, such as enrofloxacin. No more than 10-30 ml/kg/site SC is recommended for fluids, and it is imperative that a fluid bleb is visualized under the skin during administration to avoid entering the coelomic cavity. IV or IO fluid administration is necessary for more dehydrated or hypovolemic patients, and when hypertonic, colloid, or > 2.5% dextrose fluids are administered. All fluids should be warmed to avian body temperature (38-40 °C, 100-104 °F) prior to administration.

Fluid therapy is a vital component of normalizing blood volume and restoring tissue perfusion in cases of shock. Crystalloids are the preferred treatment for avian dehydration. Most isotonic fluids in mammals are slightly hypotonic in birds because birds have a higher plasma osmolality. Plasmalyte-A 7.4 is the closest fluid to avian plasma in pH, electrolyte composition, and osmolality.

Hydration status is difficult to assess in birds when compared to mammals. Indirect blood pressure measurement is unreliable in small to medium sized birds. Therefore, fluid therapy is guided by physical examination unless an arterial catheter is in place or blood gas and electrolyte measurements are available to guide fluid choice. Some general guidelines for hydration assessment are: 5% dehydrated with history of fluid loss, skin slow to return when tented over keel or eyelids; 7-8% dehydrated with lethargy, dry and tacky mucous membranes, strings of thick saliva in the oral cavity; 10-12% dehydrated with skin very slow to return when tented over keel and eyelids, sunken eyes; 15% dehydrated with marked weakness and comatose. Signs of hypovolemia include altered consciousness, tachycardia, > 1-2 seconds ulnar vein refill time, weak pulse, hypothermia, and hypotension.

Fluid therapy is divided into resuscitation, rehydration, and maintenance. The fluid deficit equals the degree of dehydration (%) multiplied by the body weight in grams. Fluid deficit can be replaced over 6-8 hours in acute losses and over 12-24 hours in chronic losses in birds without cardiopulmonary compromise. Up to 50 ml/kg can be administered as an IV bolus. Recommended maintenance dose of fluids is 50-150 ml/kg/day (depending on species). If early decompensatory phase (blood loss greater than 25-30% of total blood volume), give crystalloid bolus (10 ml/kg) and HES bolus (hydroxyethyl starch, 3-5 ml/kg IV or IO). In decompensatory phase (LD_{50} of ducks was 60% of blood volume), give HSS bolus (hypertonic saline, 3-5 ml/kg IV or IO) over 10 min ± HES (3 ml/kg IV or IO); warm patient and administer crystalloids (10 ml/kg) and HES (3-5 ml/kg) bolus IV or IO. Repeat for 3-4 boluses or until blood pressure is greater than 90 mm/Hg. Atropine and epinephrine can be used if non-responsive. If PCV < 20% consider blood transfusion.

Blood transfusion

If PCV < 20% and blood loss is acute (i.e., patient not compensated), consider whole blood transfusion. Whole blood is typically used and ideally collected from the same species. Calculate the patient's total blood volume (8% of body weight) and replace 10% of the blood volume IV or IO. Alternatively, calculate 10% of the donor's body weight. Blood is typically administered over 1-4 hours using a pediatric microfilter.

References

1. Schnellbacher R, Beaufrere H. Catheterization and venipuncture. In: Graham J, Doss G, Beaufrere H, eds. *Exotic Animal Emergency and Critical Care Medicine.* Hoboken, NJ: Wiley-Blackwell; 2021:457-467.

2. Kanda I, Brandao J. Nutrition and fluid therapy. In: Graham J, Doss G, Beaufrere H, eds. *Exotic Animal Emergency and Critical Care Medicine.* Hoboken, NJ: Wiley-Blackwell; 2021:758-768.

3. Graham J. *Blackwell's Five-Minute Veterinary Consult: Avian.* Hoboken, NJ: Wiley-Blackwell; 2016.

4. Speer B. *Current Therapy in Avian Medicine and Surgery.* St. Louis, MO: Elsevier; 2016.

5. Carpenter J. *Exotic Animal Formulary.* 5th Ed. St. Louis, MO: Elsevier; 2018.

6. Harrison G, Lightfoot T, eds. *Clinical Avian Medicine.* Palm Beach, FL: Spix Publishing; 2006.

Cry Me a Liver: The Avian Liver in Health and Disease

Shangzhe Xie, BSc/BVMS, MVS (Conservation Medicine), PhD, DABVP (Avian Practice), DACCM

Session #203

Affiliation: From Mandai Wildlife Group, 80 Mandai Lake Road, Singapore 729826.

Abstract: The avian liver anatomy and physiology are comparable to the mammalian liver, with some individual species peculiarities. The range of diseases and abnormalities of the avian liver encountered in clinical practice is wide and covers the full spectrum of infectious and non-infectious etiologies. This masterclass covers the basic anatomy and physiology of the avian liver, disease syndromes associated with the avian liver, diagnostic modalities and their limitations for the avian liver, therapeutic options available for the avian liver, and case studies highlighting the complexity of avian liver diseases. These case studies cover a range of avian orders and etiologies, including mycobacteriosis, chlamydiosis, salmonellosis, hepatic lipidosis, hemochromatosis, and neoplasia, among others.

Liver Anatomy and Physiology

The avian liver is usually made up of the right and left lobes joined cranially, enclosed in a capsule of connective tissue. Blood supply is via the right and left hepatic arteries, arising from the coeliac artery, and hepatic portal veins, which drain blood from the proventriculus, ventriculus, duodenum, pancreas, intestines, and cloaca.[1] Terminal portal venules and arterioles empty into sinusoids between the hepatocytes. Hepatocytes absorb molecules from the blood in these sinusoids through pressure differences. Additionally, Kupffer cells collect particulate matter and microorganisms in the sinusoids through phagocytosis. The hepatic veins then drain blood away from the liver via the caudal vena cava.[1]

The hepatic portal triad, also known as the hepatic acinus, consists of the arteriole, portal venule and bile ductule, along with associated hepatocytes, bile canaliculi and sinusoids.[1] Bile produced by hepatocytes enters the bile canaliculi and then the ductules in the portal triad, which then empty into the interlobular ducts. These culminate in the common hepatoenteric duct emptying into the duodenum for avian species without a gall bladder; or, in avian species with a gall bladder, the hepatocystic duct followed by the gall bladder.[1]

The liver has several functions, including digestion via bile. Bile contains bile acids, which emulsify fat, and amylase. Due to the lack of biliverdin reductase and glucuronyl transferase in birds, the primary bile pigment is biliverdin, which is green. Hepatic enzymes facilitate glycogenesis, protein synthesis and lipogenesis. Hypoglycemia during fasting stimulates glucagon production, which activates liver enzymatic pathways involved in carbohydrate metabolism to produce glucose through glycogeolysis, gluconeogenesis and lipolysis.[1] The liver also processes both endogenous metabolites and exogenous chemicals. Hydrophobic, lipid-soluble molecules are converted by the liver to hydrophilic, water-soluble molecules and excreted in bile and urine, e.g. urea and uric acid. The liver is also the primary site of synthesis of essential proteins such as albumin, fibrinogen, prothrombin, and clotting factors.[1]

Diseases of the Liver

Liver disease in birds usually results in one or all of the following clinical signs:

- Green biliverdin-stained urates and urine due to increased levels of biliverdin in the blood due to cholestasis
- Weight loss or diarrhea due to maldigestion from decreased bile secretion into the intestine
- Weakness, depression, personality changes, behavioral disturbances, seizures, and/or paresis due to inadequate conversion of ammonia to uric acid and urea
- Coagulopathies due to inadequate production of clotting factors
- Anemia due to decreased synthesis of ceruloplasmin, leading to decreased iron mobilization
- Ascites due to decreased albumin and portal hypertension
- Feather and skin changes, overgrown beak and nails due to abnormal carbohydrate and fat metabolism
- Endotoxemia and bacteremia due to Kupffer cells not functioning
- Unexpected responses to medications due to decreased ability of the hepatocytes to metabolize chemicals
- Apparent pruritis or feather picking due to bile salts deposited in the skin or pain from the stretching of the liver
- Dyspnea due to compression of air sac space the enlarged liver or ascites

Liver disease can result from the full range of differential diagnoses, including congenital abnormalities, trauma, metabolic and nutritional disorders, toxins, parasites, infectious agents (e.g., bacterial, fungal, viral, chlamydial, mycobacterial, and rickettsial), and neoplasia. The more common diseases are discussed below.

Hepatic lipidosis

Hepatic lipidosis can be the result of consumption of excessive fatty acids, increased lipolysis due to diabetes mellitus or egg-laying activity, decreased fatty acid oxidation in the liver, or decreased ability of the liver to secrete processed fatty acids back into circulation. The diagnosis is confirmed histologically with the presence of vacuolation in hepatocytes. Treatment involves a reduction of dietary fat, an increase in dietary protein, provision of carbohydrates to replace fat as the main source of calories, and supplementation with vitamins E and B-complex.

Iron storage disease

There does not appear to be a consensus on terminology for iron storage disease. In some cases, hemosiderosis and hemochromatosis are used interchangeably to refer to iron storage disease, whereas, in others, hemosiderosis is used to refer to the stage before clinical signs are apparent and hemochromatosis after. To avoid such confusion, only the term iron storage disease will be used in this text. Iron storage disease refers to the excessive accumulation of iron within the liver, resulting in damage to hepatocytes due to excessive dietary iron intake or variations in iron metabolism.[2] Iron storage disease is commonly reported in Sturnidae (i.e., mynahs, starlings),

Paradisaeidae (i.e., birds of paradise), Ptyonorhynchidae (i.e., bowerbirds), Bucerotidae (i.e., hornbills) and Ramphastidae (i.e., toucans and toucanettes), but should be considered a potential differential diagnosis in all avian species.[2] Affected livers are enlarged and are usually golden-brown in color. Iron can be seen histologically in the hepatocytes and Kupffer cells, using Prussian-blue stain.[2] Hematology, biochemistry, diagnostic imaging, and iron serum testing generally are inadequate to confirm the diagnosis but may assist in evaluating concurrent pathologies.[2] Clinical management is dependent on the disease severity, and dietary review ensuring iron intake remains below 50 mg/kg of dry matter fed in susceptible species may be sufficient in mild cases.[2] Severe iron storage disease requires treatment with iron chelators or serial phlebotomy, and control of concurrent conditions.[2] Treatment of chronic cases can be challenging.

Parasites

Many parasites can infect the liver. Common parasites include systemic *Isospora*, primarily reported in passerine birds; *Plasmodium* (avian malaria), especially in penguins; *Leucocytozoon* especially in ducks and geese; *Histomonas* especially in peafowl and turkeys; *Trichomonas* especially when there is heavy infection; flukes of the family Dicrocoelidae; and migrating intestinal nematode larvae. Medications used to treat hepatic protozoan infections (e.g., metronidazole) are the same as those used for intestinal or systemic infections. Trematode infections (e.g., *Paratanaisia bragai* and *Bothriogaster variolaris*) are often difficult to diagnose antemortem. Treatment using medications such as praziquantel has been shown to reduce egg laying but may not eliminate the parasite itself.[1]

Bacteria

Salmonella spp., *E. coli*, *Pseudomonas* spp., *Yersinia* spp. and *Campylobacter* spp. are common isolates in avian cases of bacterial hepatitis. Affected livers are enlarged, with grey-white foci throughout the parenchyma. Multifocal hepatocyte necrosis with a heterophilic inflammatory response is usually seen. Bacteria are usually seen in macrophages and Kupffer cells. Endotoxins arising from bacterial enteritis can also enter the portal circulation, damaging the periportal hepatocytes.[1]

Avian chlamydiosis has traditionally been defined as the infectious, contagious, systemic and potentially fatal disease caused by the bacteria *Chlamydia psittaci*. Other *Chlamydia* species may also cause disease in birds (e.g., *C. avium*, *C. gallinacea*, *C. abortus*, *C. pecorum* and *C. trachomatis*).[3] The pathogen has been reported to occur in more than 460 avian species across 30 orders. The methods used for surveillance and/or confirmation of avian chlamydiosis evolved rapidly during the last decades, especially after molecular methods were developed. *C. psittaci* appears to possess cellular-level characteristics that make it remarkably efficient in disseminating within an individual host but also host-to-host spread, as well as adaptable in escaping host immunity.[3] Furthermore, the bacteria is capable of causing latent or asymptomatic carrier status in birds, and all these things suggest that pathogenesis is likely associated with co-evolved or adaptive host-pathogen interactions, which are in part genetically driven by patterns of microevolution.[3] Treatment usually involves the use of appropriate antibiotics (e.g. doxycycline, azithromycin).[4]

Avian mycobacteriosis is a disease caused by bacteria from the *Mycobacterium avium* complex (MAC). Mycobacteria can survive for long periods in the environment, particularly in soil and water. The proposed routes of transmission are through the ingestion of the pathogen in food and/or water, contact with infected soil, or inhalation of aerosolized pathogen.[5] Physical examination and conventional clinical diagnostics of diseased birds with avian mycobacteriosis are often non-specific and include weight loss, dullness, weakness, muscle and fat atrophy, and dull feathers. There may be focal granulomatous lesions in joints or soft tissue.[5] Post-mortem gross examination remains one of the most commonly used tools to preliminarily diagnose avian tuberculosis but it cannot be the sole modality. Gold standard confirmatory diagnosis still requires successful isolation of the *Mycobacterium* organism in specialized media. This can take up to 3 mo, and may produce false negative results due to sample variations. Molecular techniques are available as alternative confirmatory diagnostic methods. Several primer sets with differing target *Mycobacterium* species are available, and the use of each primer set should be adapted to the sample type and objective of testing.[6] In some situations, utilizing sets of clinical parameters (e.g., history of housing in aviaries with previous mycobacteriosis cases, poor body condition score, elevated white cell count) may be provided to diagnose mycobacteriosis with high specificity.[6] Some birds are relatively resistant to the disease and can act as asymptomatic hosts over long periods. Treatment options have been described in the literature and consist of prolonged courses of antibiotic combinations. Some of these antibiotics are last-resort human antibiotics and mycobacteriosis is potentially zoonotic, so the ethics of treating avian patients for mycobacteriosis should be considered and discussed before embarking on treatment.[6] Treatment is often pursued for individual cases with localized and mild signs of mycobacteriosis. Flock management may involve the removal of birds from others that are likely to be carriers of the pathogen, as well as having a focus on maintaining low to moderate stocking density, good hygiene practices and maintenance of high welfare with low-stress environments.[6]

Viruses

Several viruses can result in hepatic damage, including herpesvirus (Pacheco's disease virus, pigeon herpesvirus, owl herpesvirus, falcon herpesvirus, crane herpesvirus, stork herpesvirus, quail herpesvirus and finch herpesvirus); polyomavirus, primarily in parrots and finches; adenovirus; paramyxovirus; circovirus, especially in young birds; reovirus; hepadnavirus in ducks (duck viral hepatitis); and togavirus (Eastern equine encephalitis).[1] No specific treatment is currently found to be consistently effective and the focus of treatment is on supportive care.

Neoplasia

Hepatic neoplasia is usually primary, although metastatic neoplasia is also possible. Primary bile duct tumors are more common than hepatocellular tumors and are associated with Pacheco's disease in psittacine birds. Some neoplasms (e.g., lymphoma) may respond to chemotherapy using medications such as carboplatin and cisplatin.[1]

Diagnosis

Clinical pathology

Hematology does not provide any specific indication of liver disease in birds but may be useful in providing information regarding chronicity and differentiating between infectious and non-infectious causes.

Blood biochemistry may help differentiate between hepatocellular damage (AST, LDH, GLDH), function (plasma protein, bile acids), and cholestasis (GGT). CK is usually used to differentiate between muscle damage and hepatocellular damage, as many of the other enzymes (e.g., ALT and LDH) can be elevated with muscle damage as well. Given the differences in half-lives of these enzymes, at certain timepoints post-muscle damage, enzymes such as AST and LDH can remain high after CK has returned to normal range. In these cases, it is easy to misdiagnose muscle damage as hepatopathy. Cholesterol, triglycerides, VLDL, LDL and HDL may indicate hepatic lipidosis.[1]

Diagnostic imaging

Lateral and ventrodorsal radiographs of the coelomic cavity provide an indication of liver size and organ displacement. On the lateral view, the liver should not extend past the end of the sternum and the proventriculus should slope down toward the ventriculus at an angle of 30-45°. In some species, the liver shadow should not extend past a line joining the shoulder and acetabulum on the ventrodorsal view.[1] It is normal for there to be space between the heart and the liver in some avian species (e.g., hornbills) but not others (e.g., psittacine birds). There may be differences in normal liver size indicators in different avian species, so it is best to use radiographs from normal individuals of the same species as a comparison. For more detailed imaging of the liver, computed tomography is preferred.

Ultrasonography has several advantages over X-ray-based imaging modalities, in that it allows examination of the internal structure of the liver as well as examination of the nearby heart and pericardium, spleen and gastrointestinal tract. The presence of air sacs reduces the acoustic windows available for ultrasonography, so experience and practice are needed.

Once the organ of interest is confirmed to be the liver after utilizing the above imaging modalities, endoscopy or celiotomy for biopsy of the liver can help identify the etiology and pathophysiology of the disease process. Several approaches have been described: through an incision behind the caudal ribs on either side, through the caudal thoracic air sac, and through a ventral midline incision. The ventral midline approach for endoscopy allows more of the liver to be examined through one incision, compared to the lateral approach.[1] It is important to stabilize the patient before anesthesia to increase the chance of survival from the procedure. Where possible, cytology can be performed if gross lesions consistent with diagnoses associated with poor prognoses are found (e.g., acid-fast stain for mycobacteriosis or cytology to check for neoplastic cells).

Treatment

Treatment of an avian patient with confirmed liver disease should consist of the following actions.

Supportive care

Reverse clinical signs that may result in death and debilitation (e.g., dehydration, anorexia, weight loss, hypoglycemia, hepatic encephalopathy, anemia, septicemia, dyspnea, coagulopathy and diarrhea).

Treatment of the specific condition

These treatments are described above with a discussion of each condition.

Creation of an environment favorable for hepatocyte regeneration

Some medications that have been shown to stabilize hepatocyte cell membranes, maintain the acinar structure and minimize fibrosis will have a positive effect in mammalian patients with liver disease and they are used in the same way for avian patients. For example, ursodeoxycholic acid can decrease the detergent effects of the hydrophobic bile acids, thereby their reducing hepatotoxic effects. It may also increase the flow of bile from the liver, reducing the concentration of hydrophobic bile acids and therefore their toxic effects.

Colchicine is used as an antifibrotic drug that inhibits the synthesis and secretion of collagen by stellate cells and fibroblasts and may induce the production of collagenase.

Silibinin, the extract from the milk thistle, *Silybum marianum*, is frequently recommended as a therapy for avian disease. It is thought to have antioxidant and free-radical-scavenger effects as well as cell-membrane-stabilizing effects that prevent toxins from entering hepatocytes. In its original state, its water solubility and bioavailability are poor. Recent developments have improved both solubility and bioavailability but not all available commercial products utilize these advances. Care must be exercised in the selection of a suitable formulation, as formulations containing alcohol or other adjuvants may potentially cause harm.

Another hepatoprotective medication to consider in avian patients with liver disease is S-adenosyl-methionine (SAMe). SAMe is part of three major biochemical pathways in the body that serve to activate or excrete drugs; to maintain functional cell membranes; and for anti-inflammatory effects and cellular regeneration. The conversion of methionine to SAMe requires SAMe synthetase, an enzyme produced in the liver and decreased or deficient in the presence of liver disease. A lack of SAMe results in a deficiency of glutathione, important in many metabolic processes and cell detoxification. Exogenous SAMe (rather than methionine) requires supplementation in animals with liver disease and has been shown to increase liver and red cell glutathione levels and/or prevent its depletion.[1]

Conclusions

The liver is a major organ and it is subject to many different diseases. Due to the wide range of causes of disease and resulting clinical signs, biochemical and organ changes, definitive diagnoses can be difficult to achieve without a combination of clinical pathology and diagnostic imaging tests. Treatment can only be more targeted if a definitive diagnosis can be achieved, although supportive care and the creation of an environment favorable for hepatocyte regeneration can be undertaken regardless. Some diagnoses may not be treatable (e.g., mycobacteriosis and neoplasia) and early diagnosis will allow these patients to be put out of prolonged suffering rather than be put through treatment that will not change the outcome.

References

1. Doneley B. *Avian Medicine and Surgery in Practice: Companion and Aviary Birds*. 2nd Ed. Boca Raton, FL: CRC Press; 2016.

2. van Zanten T, Rasidi E, Xie S. Practical management of iron storage disease in captive birds. *Proc ExoticsCon*. 2022;63-73.

3. Xie S, Tomic DH, Chong SM, Lee J. Avian chlamydiosis. In: Fereidouni S, ed. *Ecology of Wild Bird Disease*. Boca Raton, FL: CRC Press; 2024.

4. Xie S. A Tale of two outbreaks: in-water doxycycline versus in-food azithromycin as means of aviary-wide treatment for *Chlamydia psittaci* in a large outdoor mixed species aviary. *Proc Annu Conf Am Assoc Zoo Vet*. 2021;115.

5. Xie S, Chong SM, Lee J. Avian mycobacteriosis. In: Fereidouni S, ed. *Ecology of Wild Bird Disease*. Boca Raton, FL: CRC Press; 2024.

6. Xie S. Avian mycobacteriosis: from individual to flock management. *Proc Annu Conf Assoc Avian Vet Virtual*. 2023.

Section 8

Clinical Practice Masterclasses

Karen Holmes, DVM;
Andrea Winkel, DVM, MPH
Session Chairs

Avian Renal System: Anatomy to Clinical Perspectives

Susan Orosz, DVM, PhD, DABVP (Avian Practice), DipECZM (Avian), and
M. Scott Echols, DVM, DABVP (Avian Practice)

Session #195

Affiliation: From Bird and Exotic Pet Wellness Center, 5166 Monroe St, Ste 306, Toledo, OH 43623, USA (Orosz), Medical Center for Birds, Oakley, CA 94561, USA (Echols), Parrish Creek Veterinary Hospital and Diagnostic Center, Centerville, UT 84014, USA (Echols), Scarlet Imaging, Murray, UT 84157, USA (Echols).

Note: Adapted and reprinted with permission from Orosz S, Echols MS. Avian renal system: anatomy to clinical perspectives. *Proc ExoticsCon Virtual.* 2020;132-140.

Abstract: When discussing the urinary system and its unique aspects in birds, the important consideration is how do birds osmoregulate? All animals must have adequate water in their extracellular fluid compartment to regulate cellular functions; however, animals live in environments ranging from deserts to aquatic environments, which include fresh to brackish to marine water. Each of those environments presents challenges that require a variety of strategies to maintain the appropriate internal milieu for life. To achieve water and ion homeostasis, animals range from osmoconformers to osmoregulators of their internal environment. Birds are osmoregulators. Birds fall between mammals and other non-mammalian vertebrates for the organs of osmoregulation. Many avian species have salt glands that perform important osmoregulatory functions, particularly in a marine environment. Like reptiles, birds utilize the lower portion of the gastrointestinal tract as an important component for osmoregulation. In addition, they have a unique kidney that has an important part in regulation. These organs are discussed from an anatomic and physiologic perspective for clinicians. Clinical aspects that relate to anatomy and physiology are included in this masterclass.

Kidney

External anatomy

The renal system consists of the kidneys, ureters and the urodeal portion of the cloaca. The kidneys of birds are fixed in ventral depressions of the synsacrum called the renal fossae, are symmetrical and are retroperitoneal. They extend from the proximal end of the synsacrum just distal to the caudal extent of the lungs to the cranial end of the synsacrum. There can be diverticula of the abdominal air sacs that extend between the kidneys and the pelvis.

Unlike mammalian kidneys, the kidneys of birds are not divided into lobes grossly but into divisions. Birds tend to have three divisions: cranial, middle and caudal divisions. The external iliac artery cranially and the ischiatic artery caudally run through the kidney so that it forms the three divisions. The portion of the kidney cranial to the external iliac artery is the cranial division, the portion between the two arteries is the middle and that portion caudal to the ischiatic artery is the caudal division. In chickens and many of the parrot species, these divisions are distinct. The middle division in passerines is not distinct as this portion blends with the cranial and more often the caudal division. In puffins, penguins and herons, the caudal divisions are fused in the midline.

The spinal nerves from the lumbar and sacral plexuses move through the substance of the kidney to innervate their respective muscles to the lower body wall and the leg or are sensory to that area of the body. This is important clinically, as swelling of the kidneys causes reduction in nerve function and the patient may present with inability to use the leg. This more often occurs in budgerigars and are presented to the veterinarian for a possible leg fracture.

In the domestic fowl, each division has on its surface slightly rounded projections that represent the renal lobules as they reach its surface. In general, the basic unit of the avian kidney is the lobule. They tend to be pear-shaped and wedged between the interlobular veins of the renal portal system. The wide, superficial portion represents the cortical region of the lobule and contains nephrons of both the cortical and medullary type nephrons.

The artery that supplies each lobule is found in its center with radiating branches. The drainage of the lobule consists of a vein that is intralobular like the artery. The collecting ducts of birds are found on the periphery of each lobule, which has been likened to the outer staves of a barrel. This is the opposite of mammals where the collecting ducts are intralobular and the arteries are interlobular.

Diagnostics

Characterizing renal disease can be challenging and is often best performed by using a combination of diagnostic methods. Some diagnostics rely in blood and urine chemistry values such as uric acid and urinary gamma-glutamyl transferase (GGT), respectively. A thorough understanding of avian renal anatomy is needed when considering radiology, ultrasound, endoscopy and advanced imaging (especially computed tomography [CT]).

A lateral view is the best method to radiographically view the kidneys. As viewed with a lateral radiograph, the absence of the normal dorsal diverticulum of the abdominal air sac (dorsal to the kidney and ventral to the synsacrum) may indicate renal enlargement. Also, *most* normal sized kidneys do not extend ventral to an imaginary horizontal line (which is parallel to the spine) that passes through the ventral border of the acetabulum. Improper positioning can artifactually change the appearance of this air-filled diverticulum. Because the renal silhouettes are superimposed on a lateral view of the abdomen, an oblique view may also be used to distinguish each kidney. Renal density and gross size changes may indicate renal disease.

A barium series on the ventrodorsal view may also be helpful in identifying severe renal swelling. In normal birds, the angle between the proventriculus and the ventriculus relative to the spine should be about 45 degrees. With massive renal enlargement and lateral displacement of the proventriculus, and sometimes ventriculus, this angle increases.

Due to the presence of surrounding air sacs (ventrally) and bone (dorsally and laterally), ultrasonographic imaging of normal avian kidneys is difficult. In one study of 386 mixed bird species that underwent ultrasonographic evaluation of the urogenital tract, abnormalities, such as renal cysts (6), cancer (12), and inflammatory nephromegaly (11) were identified in only 29 patients. The authors concluded that sonographic imaging of the normal kidney was not possible. Some disease conditions that either obliterate the air sacs or result in fluid accumulation in the coelomic cavity may actually improve renal ultrasonographic imaging. Renal tumors can be heterogenous, compared to homogeneous with normal tissue, and have no discernible border with the synsacrum dorsally as viewed with ultrasound. In these abnormal situations, ultrasonography can serve as a non-invasive and safe means to evaluate coelomic structures such as the kidneys. Also contrary to statements above, some skilled ultrasonographers can position birds such that the kidneys are visible via ultrasound even if there is no pathology present.

Intravenous excretory urography has been described in birds as a method to gain information on kidney size, shape, and function. Lumeij reports using organic iodine compounds given IV at 2 mg/kg in the basilic vein. The organic iodine can be visualized radiographically in the heart and pulmonary artery within 10 seconds and outlining the kidneys and ureters 20-50 seconds later. After 2-5 min, the cloaca will be outlined. This technique should not be used in birds with severe renal compromise.

Intravenous excretory urography may have some limited uses in a clinical setting as demonstrated in the case report below. Dennis and Bennett successfully used a water-soluble iodinated contrast agent (Renografin-76, Squib Diagnostics, Princeton, NJ, USA) to evaluate the ureters post-ureterotomy in a double-yellowheaded Amazon parrot (*Amazona ochrocephala*). The agent was dosed at 400 mg/kg and was given in the right medial metatarsal vein. Radiographic images were taken at 1, 2, 7, and 10 min post-injection. Ureter peristaltic movement and size were successfully evaluated using this technique.

Excretory urography combined with CT can provide significant information about the avian kidney. CT (computed tomography) is essentially a series of X-rays compiled to form 3D images. High-quality CT scans (generally 200-µm slice thickness or less) can provide unparalleled 3D information about kidneys in health and disease. MSE uses the following protocol for contrast CT studies in birds. First, a whole- body or region-of-interest (ROI) plain CT scan is performed. Second, 4-6 ml/kg of Isovue 370 IV is delivered over 2 min. Thirty seconds into the contrast perfusion, start the second whole-body or ROI CT scan. Third, complete a last whole-body or ROI scan immediately after completing the contrast perfusion.

This form of CT angiography can demonstrate not only anatomic features of the kidney but also provides information about renal contrast clearance. Tumors, cysts, vasculature and more can be highlighted by the contrast study giving anatomic information. In a normal (third) post-contrast CT scan, the ureters and cloaca fill with contrast. Lack of contrast filling in the ureters gives information about renal clearance and patency of the ureters.

When history, physical examination and/or laboratory abnormalities support the presence of renal disease, consider biopsy. Currently, the only means to definitively diagnose avian renal disease and specific pathologic patterns is with a kidney biopsy and histopathologic evaluation. A renal biopsy is most frequently performed during endoscopic examination of the coelomic cavity and specifically, kidneys. Alternatively, the kidneys can be approached dorsally by cutting through the synsacrum. This dorsal approach obviously requires cutting through bone and increases the risk of damaging the more dorsally located lumbar and sacral nerve plexuses. Before a renal biopsy is performed, the cost: benefit of the surgical procedure versus conservative therapy must be considered as many birds have compromised health, especially if they have kidney disease.

In a study of 89 free living birds of prey, 126 endoscopic coelomic approach renal biopsy samples (2 biopsies from 37 birds) using 1.8-mm biopsy forceps were taken. Post-biopsy hemorrhage averaged 67 seconds (10-172 seconds). The average biopsy was 2.2 mm long, 1.3 mm wide, and 1.0 mm deep. All samples contained proximal and distal tubules and 1 to 89 glomeruli, with most having 25-29 glomeruli per histologic slide. One hundred thirteen of 126 samples could be evaluated well or very well. Sixty-six samples revealed lesions including subcapsular bleeding (19/66), inflammation (16/66), cell casts (12/66), periodic acid Schiff positive reactions (8/66), and protein casts (6/66). Correlation between endoscopically visible change and histologic disease was 76.1 % (96/126). The cranial division was considered the best site to collect biopsy samples due to its size and visibility. The authors, Müller et al., noted it was possible to obtain specimens from the middle and caudal renal divisions in larger birds.

Renal histologic lesions are rarely pathognomonic for a specific disease process. Many different diseases cause similar renal lesions. Additionally, different pathologists may make differing morphologic diagnoses on the same renal tissue. The author (MSE) encourages veterinarians to work with a pathologist familiar with normal and abnormal avian histology. Often, it is the pathologist's interpretation of a renal biopsy combined with the attending veterinarian's case familiarity that enables both parties to make a definitive diagnosis or build a reasonable differential diagnoses list compatible with the kidney lesions noted.

Avian Nephron Types

As suggested, the avian kidney contains two basic types of nephrons—cortical, loopless or reptilian-type nephrons, and the medullary, looped or mammalian-type nephrons that contain medullary loops of Henle. The cortical type of nephron, as the name implies, is found in the cortical region only and does not have loops of Henle, hence the term loopless nephrons. They form the majority of the nephrons in the kidneys of birds that have been investigated. This is the type of nephron found in reptiles and the nephrons secrete primarily uric acid.

The medullary type has a loop of Henle like in mammals that has the potential for producing and, to a certain extent, concentrating urine. However, a recent study suggests a gradient of nephrons with the outermost representing cortical nephrons and the innermost nephrons (with the largest loops of Henle) representing the mammalian types. There are transitional ones with elongated looping intermediate segments that are not found within medullary cones or rays. For this reason, it has been proposed that a new terminology that should be used—loopless (LLN) for the old cortical nephrons without a loop of Henle, transitional (TN) for the intermediate forms, and looped (LN) for the ones with a loop. It has been estimated that the total nephron population consists of between 10–30% with LN. There is the suggestion that there are habitat related patterns to kidney structure–those species that live in an arid environment tend to have smaller kidneys, a larger medullary volume, and/or a smaller cortical volume.

Both types of nephrons have a renal corpuscle at their proximal ends. The corpuscle consists of a glomerular capsule that is indented by its tuft of capillaries, the glomerulus. The renal corpuscles lie about midway between the inter- and intralobular veins. All the nephrons have a proximal and distal portion to their nephrons, but only the mammalian type have the loop of Henle.

The cortical nephrons have their proximal tubule in the outer portion of the lobule and form approximately half the length of this nephron type. The middle segment of this nephron is poorly defined but is short and convoluted. The distal tubule of the reptilian nephron has compact convolutions near the intralobular veins.

The medullary nephrons have, as their intermediate segment, the loop of Henle, which is anatomically similar to the short mammalian nephron type with the proximal descending loop enlarging prior to the bend. Many species do not have a well-defined arrangement of their loops of Henle and collecting tubules because the components intermingle. However, passerines usually have their descending limbs just medial to an outer ring of perilobular collecting tubules. An inner ring is formed by the ascending limbs of the loops of Henle.

The outer ring of collecting tubules of each lobule consists of perilobular collecting tubules, while those that are part of the medullary region are termed medullary collecting tubules. Commonly, these medullary tubules combine to form a single larger vessel, a collecting duct. When several of these collecting ducts combine, they form a secondary branch of the ureter.

Birds, like mammals, have a juxtaglomerular apparatus. This consists of a macula densa, which is an epithelial thickening at the beginning of the distal convoluted tubule. This macula densa makes anatomic contact with the afferent arteriole through their secretory juxatglomerular cells, which secrete renin. The third component is the extraglomerular mesangium found between the macula and the specialized cells of the afferent arterioles. This mesangium of cells represent modified connective tissue cells.

Useful Blood Chemistry Parameters

Uric acid

Plasma uric acid can be useful as a screening tool for advanced renal disease. With the exception of gastrointestinal uricolysis, uric acid and its salts (urate) are the end product of nitrogen metabolism in birds. Elevated uric acid has been correlated with histologically confirmed severe renal disease in chickens (tubular nephrosis and interstitial nephritis). In a separate study involving dehydrated chickens, increased serum uric acid was associated with histologic renal lesions. Broilers given oosporein (renal toxin) developed visceral and/or articular gout, swollen pale kidneys, and had a 48% increase of uric acid over control birds. In a similar study with oosporein in turkey poults, intoxicated birds had dose-dependent increases in uric acid (over controls) ranging from 76% to 140%. Lumeij noted that fasting hyperuricemia (> 16.7 mg/dl [> 1,000 µmol/L]) in peregrine falcons (*Falco peregrinus*) indicates renal failure.

Uric acid is produced and secreted in the avian liver, kidney, and pancreas. Specifically, nephrogenic uric acid synthesis may increase when plasma purine precursors are elevated. These findings suggest the avian kidney has an important role in the synthesis, in addition to elimination, of uric acid, especially when increased precursors are available. Precursors, including body proteins degraded because of poor nutritional status, have been suggested as a cause of elevated uric acid and should be considered in birds with hyperuricemia. Hyperuricemia is defined as "any plasma uric acid concentration higher than the calculated limit of solubility of sodium urate in plasma."

Uric acid is mainly cleared via tubular secretion and is largely independent of glomerular filtration, water resorption, and urine flow rate. Blood uric acid levels are mildly affected by a bird's hydration status but rather reflect the functional capacity of the renal proximal tubules. However, in a study with dehydrated chickens, uric acid levels increased after 24-48 hours of water restriction, but only in those birds allowed free access to food. Serum uric acid levels actually dropped within 24 hours in birds denied food and water. It has been estimated that renal function must be below 30% of its original capacity before hyperuricemia develops; however, this statement has never been substantiated in birds. Suggested normal avian uric acid levels range from less than 1 to 10 mg/dl.

Uric acid represents 80% or more of the nitrogen excreted by birds. Therefore, a significant increase in the proportion of nitrogen excreted as uric acid is not likely, even with increased dietary protein consumption. At least in chickens, hyperuricemia is likely due to reduced renal tubular secretion of uric acid and not excessive production as can occur in humans. These findings imply that renal tubular diseases are likely responsible for hyperuricemia and uric acid abnormalities may not be evident until very high protein diets are fed. Specifically in chickens, dysfunctional proximal convoluted tubules result in reduced urate secretion and can lead to hyperuricemia if severe.

In chickens, the uric acid renal tubule transport system does not appear to become saturated until plasma uric acid levels exceed 60 mg/dl. Chickens genetically predisposed to hyperuricemia and fed high (60%) protein diets develop an elevated steady state of plasma uric acid (10-60 mg/dl) in order to excrete their daily loads of this by-product. The increased basal plasma uric acid made the affected chickens susceptible to articular gout formation. One group suggested that these chickens genetically predisposed to gout had a defective uric acid transport mechanism at the peritubular membrane.

In birds of prey, uric acid production is directly related to the amount of protein consumed and transient rises are noted following high protein meals. Peregrine falcons (*Falco peregrinus*) and red-tailed hawks (*Buteo jamaicensis*) are reported to have a significant post-prandial increase in plasma uric acid concentration (hyperuricemia) for up to 12 hours after ingesting a natural meal. The significant post-prandial uric acid increase noted in peregrine falcons was up to 32 mg/dl (reported as 1,881 μmol/L) between 3 and 8 hours after being fed. Lumeij stated that significant post-prandial increases in both urea and uric acid persist for up to 15 hours in peregrine falcons. It was not clear why these birds of prey did not develop gout lesions, but the authors recommended a 24-hour fast prior to evaluating serum uric acid in peregrine falcons. The authors further recommend that a 24-hour fast should be considered for all carnivorous avian species prior to blood uric acid testing. Almost identical findings of post-prandial hyperuricemia were noted in black-footed penguins (*Spheniscus demersus*) and represent another species that should be fasted before measuring uric acid levels.

Uric acid production following a high protein meal has been studied in various psittacine birds. In one study with African grey parrots (*Psittacus erithacus* sp.), plasma uric acid concentrations showed a positive correlation with dietary protein consumption. However, even though the fed protein level was as high as 30%, plasma uric acid levels remained within normal ranges. In cockatiels fed 11, 20, 35, and 70% protein for 11 months, serum uric acid increased linearly with dietary protein levels; however, the serum uric acid level was significantly greater only in birds fed 70% protein diets. Because no histologic or gross renal lesions were found at necropsy, the authors concluded that the rise of uric acid was related to dietary protein concentration and not kidney damage. Angel and Ballam found that feeding diets containing 13.5, 18.2, and 24.6% protein for up to 24 wk had no effect on serum uric acid levels in parakeets.

In consideration of the above-described causes of elevations in uric acid, this single biochemistry value can help identify significant renal disease. The author (MSE) prefers to repeat (fasting) uric acid levels on well-hydrated birds before a suggestion of renal disease is made. In birds with suspect renal disease that have a single laboratory value of hyperuricemia, the author will often give a total of 50-100 ml/kg SC, q24h to q12h of isotonic fluids for 2 days and then recheck the uric acid level. In the author's experience, birds with persistent hyperuricemia after fluid therapy and/or fasting have some form of renal disease.

Urea

Unlike mammals, urea in birds is produced only in small amounts (by renal mitochondrial breakdown of arginine) and does not serve as the end product of protein metabolism. Plasma urea in birds is excreted by glomerular filtration, and unlike uric acid, blood urea concentrations are more significantly affected by the bird's hydration status. During normal hydration, filtered urea is 100% excreted but is 99% reabsorbed in the tubules during dehydration. Plasma urea has also been shown to significantly increase in peregrine falcons for up to 15 hours post meal.

In studied cockatiels, serum urea levels increased linearly with dietary protein levels (11, 20, 35, and 70%). Separate studies involving the domestic fowl and pigeons demonstrated decreased urea elimination and/or increased blood urea levels (6.5- to 15.3-fold increase in pigeons) in dehydrated birds. Pegram et al. showed that plasma urea nitrogen "increased in a dose dependent fashion (in turkeys) at every level of dietary oosporein (nephrotoxin)." It should also be considered that these intoxicated turkey poults were also showing signs of dehydration. Lumeij has proposed that plasma urea is the single most useful indicator of prerenal (dehydration) causes of kidney failure in birds.

The urea/creatinine and urea/uric acid ratio can be used to better define pre- and post-renal azotemia. Because reabsorption of urea is disproportionally higher than both creatinine and uric acid, these ratios should be high during dehydration and ureteral obstruction. The formulas for these ratios are listed below:

Urea / creatinine = [urea (mmol/L) × 1,000] / creatinine (μmol/L)
Urea / uric acid = [urea (mmol/L) × 1,000] / uric acid (μmol/L)

Creatinine

Birds produce little creatinine from its precursor, creatine. Creatinine is eliminated by tubular secretion but clearance is variable. Clinically, creatinine may be elevated in pet birds by feeding high-protein diets. Lumeij showed that plasma creatinine will also increase significantly in dehydrated pigeons. The relationship between creatine and creatinine in birds with renal disease is poorly understood and differentiation does not appear to be useful clinically.

Urine GGT

Elevated urine gamma-glutamyl transferase (GGT) suggests early renal tubular damage in mammals. The same principle has been applied to birds and can be easily evaluated. First, collect the urine portion of avian droppings in a hematocrit tube: suck up the sample via capillary action and fill tube as much as possible minimizing fecal and urate contamination. Second, run GGT (just as for blood) on the collected fluid. Third, interpret the values as follows:

0-6 U/L = normal
6-20 U/L suspicious for early renal tubular damage**
> 20 U/L more supportive of early renal tubular damage**

**coelomic inflammation (egg yolk ceolomitis, pancreatitis, septic and other non-septic coelomitis) and pressure upon the kidneys (intra and extra renal masses) may also result in increased urinary GGT values.

Arterial Supply to the Kidneys

The cranial, middle, and caudal divisions of the kidney are supplied by their respective arteries that arise from the abdominal aorta. The caudal division has a branch off the aorta that bifurcates on the surface of the kidney resulting in 2 arteries that enter the parenchyma of this larger division. These arteries then branch within the substance of the kidneys to eventually form the intralobular arteries. The intralobular arteries are found in the lobule approximately half-way between the inter- and intralobular veins.

The intralobular arteries form the short afferent glomerular arteries that immediately form the glomerulus. Each of these capillary tufts is simpler and much smaller than mammalian ones, as they consist of only 2–3 capillaries that have limited interconnections compared with mammalian capillary tufts. The capillary tufts of the loopless nephrons can consist of only one capillary loop that such an afferent arteriole forms a single capillary loop and then widens to form an efferent arteriole.

The arterial supply continues as efferent glomerular arterioles. These arterioles divide to form the second capillary plexus, the peritubular capillary plexus, which surrounds the epithelium of the convoluted tubules in the cortical region. In the medullary region, these arterioles from the efferent glomerular arterioles form arteriolar recta that lie alongside the loops of Henle before forming the venulae recta that drains the area.

Venous Supply to the Kidneys

Blood from the area around the loops of Henle are drained by the venulae recta into the intralobular veins. The intralobular veins drain blood into the efferent renal veins or branches. Depending on the division of the kidney, the blood then flows into renal veins of its appropriate division. In the cranial division of the kidney, there can be several cranial renal vessels that then drain into the common iliac vein after the renal portal valve or into the abdominal vena cava directly. The caudal renal vein drains the middle and caudal divisions of the kidneys. Blood from the intralobular veins drains into the efferent renal veins or branches before draining into the caudal renal vein. The caudal renal vein, like the cranial renal vein, empties into the common iliac vein after the renal portal valve.

The renal portal system is involved in the secretion of urates in the blood so that it can be excreted by the kidneys in birds. The renal portal system supplies venous blood to the peritubular capillary plexuses that surround the proximal convoluted tubules at the periphery of the lobule. These tubules are responsible for the secretion of the urates into its lumen from these blood vessels. Urates are also filtered by the glomerulus but its rate is insufficient. It has been suggested that the renal portal system provides about two-thirds of the blood supply to the kidneys that bypasses the glomeruli. This has important ramifications for the removal of urates along with other components including drugs.

The renal portal system forms a venous ring with both kidneys. It consists of the right and left cranial and caudal renal portal veins. At the cranial end, the right and left cranial renal portal veins are connected via the internal vertebral venous sinus, which drains the vertebral column. The caudal end of the ring is completed by its anastomoses with the caudal mesenteric vein. There are numerous small venous branches that take origin from this ring to penetrate the parenchyma of the kidney. Each afferent renal branch has a muscular sphincter at its base to control the volume of blood that enters the substance of the kidney. These afferent renal venous branches are connected with the interlobular veins which are connected to the peritubular capillary plexus at the periphery of each lobule.

In addition to the valves in the afferent renal venous branches, there are renal portal valves that control blood flow as well. These valves are found within the lumen of the common iliac veins between the renal and the portal veins. Blood can enter the portal venous ring from the external iliac veins, the internal iliac veins, the caudal mesenteric vein and the ischiatic veins. Each valve is innervated by adrenergic and acetylcholine receptors. If the valves are open, then blood will flow directly into the caudal vena cava and not the substance of the kidney. Valve closure is inhibited by norepinephrine and epinephrine so that with "flight" there is available venous return to the heart. When the valve is closed, blood flows into the parenchyma of the kidney. Acetylcholine stimulates its closure.

From a clinical perspective, understanding these connections with the renal portal system helps explain the spread of neoplasia and infectious agents to other parts of the body. Blood from the renal portal system can flow through the portal valve into the vena cava, into the caudal mesenteric vein to the liver, and /or into the internal vertebral venous plexus within the vertebral canal. This shunt can result in the bypass of the kidney completely but often is only partially activated so that blood may go a variety of directions.

Retroperistalsis and the Lower GI Tract

Urine from the mammalian nephrons and urates from the reptilian nephrons are excreted from the avian kidney via the ureter to the urodeal compartment of the cloaca. The urine and urates are moved from the urodeum of the cloaca by retroperistalsis into the large intestine or colon and in species with a cecum, there as well. It appears that the urine/urates move around a central fecal core where it can come in contact with the brush boarder of the colonic epithelium and its resident bacterial population. This bacterial population when grown in culture only flourishes when uric acid is added to the medium.

The lower GI tract that includes the cloaca, colon and digestive ceca, plays an important role in avian osmoregulation. The ureteral fluid of urine and urates is moved from the urodeum into the lower GI tract and that movement is controlled by tonicity of the fluid. The control of the movement occurs in the GI tract itself and not the central osmoreceptors of the hypothalamus. When the tonicity is 200 mOsm/kh H2O higher than that of the plasma, retrograde peristalsis is stopped or significantly slowed.

Sodium transport occurs in the coprodeum and the colon. A low sodium diet increases the plasma aldosterone levels which results in increased absorption of sodium in the coprodeum and colon. With high salt levels, there is still some transport of sodium but by another mechanism using organic molecules from the bacterial breakdown of protein in ureteral urine. The continued transport of sodium then results in the recovery of urinary water. These colonic cotransporters play an important physiologic role in osmoregulation as well as nutrition of the GI tract by assimilation of protein and uric acid in its lower extent. Since the avian kidney has a limited capacity to concentrate urine, the GI tract plays a significant role to produce hyperosmolar urine in relation to plasma.

Interpreting avian urinalysis

Several significant factors complicate interpreting avian urinalysis. First, urine is mixed with feces in the cloaca of most avian species. Second, in many species ureteral urine is refluxed orad into the lower intestines to the ceca where water, and sometimes electrolyte, reabsorption takes place. In addition, gastrointestinal bleeding, inflammation, normal and abnormal organisms, reproductive status, among other things may alter urine production and composition. In short, the urine present in a dropping is not the same urine produced from the kidneys. Urinalysis results should be carefully interpreted.

Respecting Parrot Intelligence and Cognitive Abilities in Daily Practice

Irene M. Pepperberg, PhD, and
Jan Hooimeijer, DVM

Session #114

Affiliation: From Boston University (Pepperberg), Adviespraktijk voor Vogels, Meppel, Netherlands (Hooimeijer).

Abstract: In part 1 of this session, Irene Pepperberg discusses contrafreeloading. In nature, animals must expend energy foraging, competing with others to access resources while avoiding predation. Hence, most aim to forage optimally to obtain the highest caloric intake in the shortest time with the least effort. In captivity, we observe contrafreeloading, which is the occasional choice to expend excess energy by, for example, extracting food from a puzzle box rather than eating what is freely available. We compare contrafreeloading in several parrot species. Results apply to enrichment programs for captive birds. We tested grey parrots, then kea and cockatoos, on varying tasks. On one task, more kea contrafreeloaded than did grey parrots and did so slightly more often, but kea completely ignored a second task. Cockatoos were tested on the second task only; they also exceeded the grey parrots but results were affected by dominant or subordinate behavior and other individual differences. Acknowledging parrots' intelligence and playfulness can ensure their welfare. In part 2 of this session, Jan Hooimeijer discusses how to avoid undesirable behavior. Parrots are non-domesticated, highly intelligent prey animals that are kept in captivity under unnatural circumstances. If we do not understand their normal behavior and underestimate their high intelligence, a wide variety of behavioral problems are apt to occur depending upon the individual parrot, species, and circumstances surrounding the bird. To understand behavior, we must understand and learn about the principles of ethology. Looking at the world through the eyes of parrots, we must consider that our body language and attitude determine their body language and behavior.

Part 1. Contrafreeloading

Introduction

Contrafreeloading is the choice to perform a physical task to access food over simultaneously freely available food, behavior contrary to predictions of both optimal foraging and learning theories; that is, it should never occur: Why would a subject expend time and effort working for calories that could be obtained without such expenditures? Nevertheless, many instances of contrafreeloading are found in the literature, but no existent theories completely explain or provide a rationale for the behavior.[1-3] This paper examines the presence and degree of contrafreeloading in various tasks for grey parrots, kea and cockatoos, comparing their data with respect to the possible interpretation of such behavior in the context of play. The idea presented in Smith et al. 2021[1] and tested in several studies[1-3] is that if the task involved in contrafreeloading provides some value to the subject (i.e., specifically, provides a positive additional experience that can be viewed as play rather than as work), subjects are likely to engage in contrafreeloading. Kea and cockatoos are particularly good subjects for such a comparative study, purportedly being more playful species than grey parrots.[1-3] Cockatoos were tested on the second task only, and their results were affected by dominant or subordinate behavior and other individual differences.[3] Given that foraging toys used for enrichment exploit contrafreeloading, if play drives contrafreeloading, foraging tasks viewed as play will likely encourage engagement. Empahsis is on how acknowledging individual differences in parrots' intelligence and playfulness can ensure their welfare.

Methods

Subjects were five grey parrots, eight kea, four umbrella cockatoos. Experiment 1, which included grey parrots and kea, presented subjects with all permutations of pairs of more-, less-, or equally valued food rewards placed in lidded or unlidded cups (i.e., the latter providing freely available food) and birds were asked to make a choice. Types of contrafreeloading (see Table 1 in Smith et al. 2022[2]) were defined as calculated (i.e., working to access preferred food over less-preferred, free food), classic (i.e., working to access food identical to the free food), or super (i.e., working to access less-preferred food over preferred, free food). Control trials tested the extent to which the tasks were considered positive or aversive. Experiment 2, which included grey parrots, kea, and cockatoos, involved contrafreeloading for nuts as the reward (i.e., nut contrafreeloading). The task was that of choosing to crack a nutshell to access the nutmeat inside over an identical, already-shelled nut.

Results

Experiment 1

Grey parrots contrafreeloaded as follows.[1] Griffin significantly classically contrafreeloaded on 26/30 trials (87%; binomial test, chance 1/2, $P < 0.001$, one-tailed). He super contrafreeloaded in 12/30 trials, which was not significant (40%; $P = 0.181$, one-tailed), and performed calculated contrafreeloading on 27/30 trials, which was significant (90%; $P < 0.001$). Athena significantly failed to classically contrafreeload, doing so on only 8/30 trials (27%; $P < 0.008$). She super contrafreeloaded on only 2/30 trials (7%; $P < 0.001$) but did calculated contrafreeloading on 28/30 trials (93%; $P < 0.001$). Pepper significantly chose not to classically contrafreeload, doing so on only 9/30 trials (30%; $P = 0.021$). She never super contrafreeloaded, 0/30 trials (0%; $P < 0.001$), but significantly performed calculated contrafreeloading, 20/30 trials (67%; $P = 0.050$). Franco classically contrafreeloaded on only 11/30 trials (36%; $P = 0.100$). He significantly chose not to super contrafreeload, doing so on only 1/30 trials (3%; $P < 0.001$), but significantly did calculated contrafreeloading, 27/30 trials (90%; $P < 0.001$). Lucci did not participate.

Kea contrafreeloaded as follows.[2] Blofield did not significantly classically contrafreeload, 18/36 trials (50%, $P = 0.567$), but did significantly perform calculated contrafreeloading on 34/36 trials (94%, $P < 0.001$) and significantly failed to super contrafreeload, 6/36 trials (17%, $P < 0.001$). Bruce also classically contrafreeloaded on only 18/36 trials (50%, $P = 0.567$), did calculated contrafreeloading at 30/36 trials (83%, $P < 0.001$) and failed to super contrafreeload, 6/36 trials (17%, $P < 0.001$). Harley Quinn classically contrafreeloaded on only 14/36 trials (39%, $P = 0.122$), did calculated contrafreeloading on 30/36 trials (83%, $P < 0.001$) and failed to super contrafreeload on 1/36 trials (3%, $P < 0.001$). Loki classically contrafreeloaded on 14/36 trials (39%, $P = 0.122$), significantly did calculated contrafreeloading on 35/36 trials, (97%, $P < 0.001$, and super contrfreeloaded on only 2/36 trials (6%, $P < 0.001$). Moriarity classically contrafreeloaded on 14/36 trials (39%, $P = 0.122$), did calculated contrafreeloading on 30/36 trials (83%, $P < 0.001$) and super contrafreeloaded on 4/36 trials (11%, $P < 0.001$). Neo classically contrafreeloaded on 15/36 trials (42%, $P = 0.203$), did calculated contrafreeloading on 34/36 trials (94%, $P < 0.001$) and failed to super contrafreeload at all 0/36 trials (0%, $P < 0.001$). Plankton classically contrafreeloaded on 16/36 trials (44%, $P = 0.309$), did calculated contrafreeloading on 32/36 trials (89%, $P < 0.001$) and supercontrafreeloaded on 2/36 trials (6%, $P < 0.001$). Taz classically contrafreeloaded on 17/36 trials (47%, $P = 0.434$), did calculated contrafreeloading on 33/36 trials (92%, $P < 0.001$) and super contrafreeloaded on 2/36 trials (6%, $P < 0.001$).

Experiment 2

Grey parrots contrafreeloaded as follows.[1] Griffin selected the shell-less nut on 18/20 trials (90%; $P < 0.001$), Lucci on 17/20 (85%, $P = 0.001$), both of which were significantly different from chance, and Pepper chose the shell-less nut on 13/20 trials (65%, $P = 0.131$). Athena and Franco, in contrast, significantly contrafreeloaded: She selected the almond-in-shell on 19/20 trials (95%; $P < 0.001$), and he did so on 16/20 trials (80%; $P = 0.006$).

Kea contrafreeloaded as follows.[2] Blofield, Bruce, and Moriarity never contrafreeloaded, choosing the shell-less nut on 20/20 trials (100%, $P < 0.001$). Loki chose shell-less 19/20 times ($P < 0.001$), Neo and Taz 18/20 times ($P < 0.001$), Plankton 17/20 times ($P < 0.01$), and Harley Quinn 15/20 times ($P < 0.05$).

Cockatoos contrafreeloaded as follows.[3] JJ and Teddy Bear significantly preferred free food, respectively, on 23/24 trials (96%, $P < 0.001$) and 18/24 trials (75%, $P = 0.011$). Dudley contrafreeloaded at chance on 14/24 trials (58%, $P = 0.271$). Poly significantly preferred to contrafreeload on 17/24 trials (71%, $P = 0.032$). Note that difference in trial numbers reflected the slightly different conditions of testing.[1-3]

Conclusions

Given that contrafreeloading should be extremely rare or nonexistent, the fact that it occurs at all is surprising, and yet it does, often with significant frequency depending upon the task involved. Several theories have been proposed to provide a rationale for instances of contrafreeloading, but none fully explain the phenomenon. Recently, researchers have proposed the role of play,[1] suggesting that contrafreeloading is likely when the subjects view the task not (negatively) as work, but (positively) as play, where the task adds value to the situation.[1-3] Our studies are the first to examine contrafreeloading in parrots that are under no general food restrictions and that live in enriched environments (e.g., food and toys are available ad libitum), and that have tested both ethologically relevant and non-relevant tasks. We also compared behavior in ostensibly less playful species (i.e., grey parrots) and more playful species (i.e., kea, cockatoos). We found that almost all birds in all species engaged in some degree of contrafreeloading, but that their behavior was highly individualistic and task-dependent. The levels of the behavior were rarely statistically significant [e.g., Experiment 1(kea[2]: calc, 89%, classic, 44%, super, 8%; grey parrot[1]: calc, 85%, classic, 41%, super, 13%); Experiment 2 (kea[2], 7%; grey parrot[1], 47%; cockatoo[3], 40%)], but such a finding is not surprising given that the behavior is not expected to occur at all. Interestingly, the more playful species were indeed likely to contrafreeload in accordance with their group behavior slightly more than the less playful species. This was not at statistically significantly different levels and was dependent on the task and context.[3] We must also take into account that the nut-shelling task seemed to be aversive to the kea, for which it is a highly irrelevant behavior ethologically[2] and that one subordinate cockatoo in a co-housing situation seemed inhibited.[3] Given that parrots are often given foraging toys as enrichment to relieve boredom behaviors (e.g., feather plucking, screaming), our results demonstrate the importance of tailoring the type of toy to the individual bird and possibly the species. Our data demonstrate that a task that is considered extremely appealing to one bird (i.e., something with which it will consistently choose to play) may be considered aversive to another, even within the same species. Thus, parrot owners must respect the intelligence and these individual differences, and be willing to introduce multiple types of foraging toys for enrichment purposes.

Acknowledgments: The author thanks donors to The Alex Foundation.

References

1. Smith GE, Greene D, Hartsfield LA, Pepperberg IM. Initial evidence for contrafreeloading in grey parrots (*Psittacus erithacus*) via the opportunity for playful foraging. *J Comp Psychol.* 2021;135(4):516-533.

2. Smith GE, Bastos APM, Taylor AH, Pepperberg IM. Contrafreeloading in kea (*Nestor notabilis*) compared to grey parrots (*Psittacus erithacus*). *Sci Rep.* 2022;12:17415.

3. Caroll A, Pepperberg IM. *J Comp Psychol.* 2024. in prep.

Part 2. Avoiding Undesirable Behavior

Parrots are nondomesticated, highly intelligent prey animals that are kept in captivity under unnatural circumstances. If we do not understand their normal behavior and we underestimate their high intelligence, a wide variety of behavioral problems are apt to occur depending upon the individual parrot, species, and circumstances surrounding the bird.[1] To understand behavior, we have to understand and learn about the principles of ethology.[2] Looking at the world through the eyes of parrots, we must consider that our body language and attitude determine their body language and behavior.

Background

Parrots are living in captivity under unnatural circumstances as non-domesticated prey animals. If humans fail to acknowledge their high intelligence and remarkable cognitive abilities, a wide variety of behavioral problems are apt to occur depending upon the individual parrot, species, and circumstances.[1] To avoid or redirect problematic behavior, we must first understand more about these birds. For parrots, their responses to their environment, both in nature and within captivity, are based on innate behavior and founded on observational learning, experience, and intelligence/cognitive abilities. Thus, to understand their behavior we have to understand and learn about the principles of ethology.[2] Importantly, parrots are prey animals and their anatomy and behavior both reflect this central aspect of their lives. For example, their eyes are positioned at the sides of their head, and if we observe the body language between two bonded parrots, we note that they look at each other sideways, with one eye. In that way, they can still observe the whole environment; as prey animals, they are always aware of the possibility of predation. Thus, normal behavior for such prey animals, in nature as well as in captivity, is to demonstrate various levels of insecurity. For example, they may view humans with binocular vision (i.e., being watched by humans face-to-face with two eyes positioned front facing, as in the case of predators in nature) also as predators. This situation can thus be a reason for developing insecure or defensive behavior around humans. Parrots are intelligent birds with amazing cognitive abilities. As noted above, parrots learn by observing other birds, other animals, and humans as part of their environment. The Model/Rival technique of Todt and Pepperberg (see references[1]) acknowledges the observational learning skills of parrots. Parrots understand our meaning, posture and attitude; they draw their conclusions and behave accordingly. Humans can adapt their behavior towards parrots and parrots can learn appropriate responses. But before humans can adapt their behavior, they must understand the different categories of parrot behavior that birds most often exhibit in captivity.[1] These categories are defined below as part of the methodology.

Normal behavior

Normal behavior is not meant to manipulate the caretaker, not meant to get any response and not meant to create a problem or conflict. Examples of normal behavior are eating, drinking, grooming, playing, sitting in the cage, sitting on the hand. Rewarding normal behavior (e.g., giving treats) can result in the development of undesired behavior (i.e., the bird will develop behavior to manipulate the caretaker to get attention).

Undesired behavior

Undesired behavior is behavior that is meant to manipulate the behavior of the caretaker, to create a problem or conflict. Examples may include screaming, biting, and feather picking. If the human rewards the behavior, even in ways that the human does not see as a reward, by showing fear, frustration, anger, concerns and laughter, the behavior will increase. The outcome of rewarding undesired behavior is obvious.

Desired behavior

Desired behavior is behavior that is created or manipulated by the caretaker. It is behavior that the caretaker would like to see repeated in the future, such as normal inquisitiveness. It is not behavior that occurs as the result of a command or a request and occurs not to get a treat, a response, or out of fear of punishment.

Enforced behavior

Enforced behavior is the result of a command, any kind of force, or behavior motivated by a treat or food. Enforced behavior can also occur out of fear of punishment. Note that such behavior is not necessarily induced by force, but is enforced by the consequences of its not being performed.

Displacement behavior

Displacement behavior[2] can be considered as the ultimate alternative behavior in a conflict situation when neither the expected fight or flight response is the best choice or is even impossible. Displacement behavior in nature or in captivity appears to be completely out of order (i.e., unexpected). Displacement behavior involves demonstrating that an individual feels at ease in the presence of the other parrot or the human caretaker, rather than demonstrating the expected defensive or aggressive behavior. Displacement behavior involves creating a new situation that de-escalates the conflict situation. Grooming and preening can be observed as part of displacement behavior in a conflict situation.[2] It is critical to consider displacement behavior as normal behavior in a conflict situation. An example of a conflict situation is when humans, usually unintentionally, intimidate their bird by staring at it directly (i.e., with their frontal, binocular eyes, inadvertently acting like a predator). To understand the situation, humans must realize that parrots analyze and combine their perceptual experiences to understand what is happening in the surrounding world. They do not simply react mindlessly to environmental stimuli, but process the stimuli and choose to react in certain ways that are consistent with their situation as prey. What starts out as normal behavior in such a conflict situation, if not acknowledged by humans as displacement behavior, can result in undesired, repetitive behavior. Specifically, if humans do not realize that their behavior is repeatedly causing a conflict situation, the parrots' initially normative displacement actions (i.e., grooming and preening) can lead to complex behavioral problems like feather destructive behavior. As Niko Tinbergen has observed,[2] pairs of gulls that are initially in conflict with one another defuse the situation by both engaging in displacement behavior. Similarly, it is important that humans respond to a parrot's displacement behavior by also showing their own displacement behavior (i.e., by demonstrating that they feel comfortable, in a non-intimidating way, in the presence of the parrot). By doing so, they resolve the conflict situation. Notably, by not understanding the ethology of the parrot, humans can cause serious issues and fail to understand how use of ethology can instead lead to positive consequences.

Another instance of failure to understand the basics of ethology is how human hand-feeding of baby parrots can cause the birds to develop serious behavior and welfare problems. In nature, baby parrots engage in what is called imprinting on their parents. They need to learn by observing normal behavior of their parents, including normal displacement behavior in conflict situations. A bird that does not learn appropriate use of displacement activities can exhibit other serious behavioral problems.[1] To help parrot owners understand how their behavior affects that of their parrot, and how to adapt their behavior to prevent or redirect problematic behavior, Hooimeijer has developed a five-step behavioral protocol for dealing with parrots.[3]

Methods and observations

Attitude and posture are the most important aspects of dealing with parrots. A 5-Step Protocol is described below.

Step 1

You, as the owner or the veterinarian, must convince the parrot that you have no negative intentions by showing a posture and non-intimidating body language, by acknowledging the fact that parrots are prey animals. Avoid directly staring at the bird. Allow the bird to observe what is happening in each situation without allowing it to be the focus of attention. Show that you are comfortable in the situation, that you don't have a problem with anything and show that you are happy in this situation. Show that you are nonthreatening and part of an interesting environment that the bird is being encouraged to explore. You must act and play a role around the parrot. By so doing, you are creating desired behavior (e.g., calm inquisitive observation). In this way, you become a positive role model. You must act in this situation without any direct interaction with the parrot, as if the bird is not present. For example, you can play with one of the parrot's toys.

Step 2

Because the bird is not threatened, it is likely to simply sit quietly, observing what is happening. Reward this behavior, that has been created and manipulated by your attitude performed in step 1, by telling the bird, while looking at it sideways, not directly, that it is the most wonderful parrot on this planet. Do not reward by providing an extrinsic, unrelated reward such as food; that would negate what you are trying to achieve.

Step 3

Be a teacher showing respect for the bird's intelligence, and enable observational learning, by sharing, in detail, what there is to see (e.g., toys on the table, colors, shapes, sizes, objects) or even talking about the color of the bird's tail in a non-intimidating way. While you are handling a toy, tell the bird that its color is yellow and that its shape is round and that its material is wood.

Step 4

Allow the bird to touch your pen, toy, paper or towel. Allow the bird (but do not command it) to step on the hand and, when doing so, praise the bird enthusiastically. Your actions are the opposite of giving a command or using a treat to enforce behavior!

Step 5

You can then introduce some slightly novel situations. The parrot should now experience that accepting new situations is okay because you have created reasons for it to trust you under the circumstances that you have created. When the parrot shows fear, do not reassure the bird. Reassuring can easily create more fear and insecurity, because the bird (not understanding the words) likely will interpret the comment as a reward for acting fearful. It is the experience of the author that after steps 1-4 there is an established situation of mutual respect and mutual trust. When the parrot fears a place or object like a towel that is introduced in Step 5, be understanding and start over again with step 1 of the protocol. Soon the bird will realize that there is no reason for fear. Then repeat step 2, telling the bird how smart and wonderful it is. This protocol will help to stimulate the self-esteem that all parrots need and will stimulate normal parrot behavior. The outcome is to create a situation in which a parrot can accept novel situations. These situations can be an examination, taking blood, grooming, wing clipping, gently towelling, among others.[1]

Conclusions

Dealing with behavior problems means looking for a diagnosis, including medical reasons like malnutrition, but also within environmental problems stemming from the basic ethology of parrots. For most hand-reared parrots, it is most often the latter, although the former must be ruled out. In general, undesired behavior can be considered as behavior that started as normal behavior, which has improperly developed under unnatural circumstances. Preventing and redirecting undesired behavior requires understanding the background or the initial circumstances that caused this behavior and the subsequent consequences. For example, what was the body language of humans and how did the parrots react as social, intelligent prey animals? We must consider that every response to undesired behavior can be perceived as a reward. Instead of responding to the behavior of the parrot and instead of ignoring the behavior, the advice is to act in a way that shows that there is no problem. By using the 5-Step Behavioral Protocol, humans can redirect undesired behavior, demonstrate appropriate use of displacement behavior (e.g., playing with toys when the parrot is in an unfamiliar situation that could be threatening to a prey animal), and prevent and redirect potential problematic behavior. It is vital to reward desired behavior. It is vital to reward the intelligence and acknowledge the cognitive abilities in parrots. It is vital not to reward normal behavior. It is vital not to reward undesired behavior. It is important to recognize displacement behavior and to use displacement behavior ourselves when we are in a conflict situation. We must learn from parrots, what they have already known for millions of years, by looking through the eyes of parrots. Creating a relationship based on mutual respect and mutual trust prevents and solves behavioral problems and ensures the welfare of parrots.

Acknowledgments: The author thanks my partner and co-workers within the Consultancy Practice for Birds. They, and an amazing anonymous sponsor, have supported me over the years enabling me to contribute as "retired" avian veterinarian. Dr. Greg Harrison, Linda Harrison and Dr. Fern Van Sant have been sources of inspiration.

References

1. Hooimeijer J. Welfare and behavior in parrots. *Proc Annu Conf Assoc Avian Vet.* 2013;289-297.

2. Tinbergen N. Curious Naturalists. London, UK: Country Life; 1958.

3. Pepperberg I, Hooimeijer J. Studying the behavior of captive parrots: respecting their cognitive abilities creates mutual respect and trust. *Proc ExoticsCon.* 2022; 49-56.

Behavior Problems in Psittacine Birds: A Tough Nut to Crack?

Yvonne R.A. van Zeeland, DVM, MVR, PhD, DipECZM (Avian), DipECZM (Small Mammal), CPBC

Session #176

Affiliation: From the Division of Zoological Medicine, Department of Clinical Sciences of Companion Animals, Faculty of Veterinary Medicine, Utrecht University, Utrecht, The Netherlands.

Note: Previously published and presented in van Zeeland YRA. Behavior problems in psittacine birds: a tough nut to crack? *Proc Annu Conf VMX Sm Anim Exot.* 2019 and van Zeeland YRA. Behavior problems in psittacine birds: a tough nut to crack? *Proc ICARE.* 2024.

Abstract: In the hands of an inexperienced caregiver with insufficient knowledge of a parrot's behavior, needs Once established, these behaviors may be difficult to eliminate due to complicating factors (e.g., inadvertent reinforcement by the owner). To increase the chances of a successful outcome, correct diagnosis and management of the problem behavior are essential. For this purpose, so-called applied behavior analysis can be helpful, as the generated functional behavioral assessments aid in developing and testing hypotheses about the purpose or function that a (problem) behavior serves for the individual, including the environmental circumstances that contribute to maintaining it. The findings of this assessment are key to designing an effective intervention plan to help replace or modify the problem behavior, which primarily includes the use of behavior modification techniques, adaptations of the parrot's living environment, and treatment of any underlying nutritional and or medical conditions as these are generally less intrusive and more effective than punishment strategies or the use of psychoactive drugs. Throughout the implementation of a behavior modification plan, close monitoring and regular rechecks are recommended to help document the effectiveness of the intervention(s) and adjust or refine the behavior modification plan where needed.

Behavior Problems: Tough to Crack?

Many problem behaviors are intentionally or unintentionally taught through reinforcement. Reinforcement will help to strengthen a future behavior as the bird learns that this behavior is effective to either gain (i.e., positive reinforcement, such as gaining attention or access to tangible items and/or sensory stimulation) or escape or avoid something (i.e., negative reinforcement, such as removal of unwanted attention, aversive sensory stimulation, or a fear-eliciting tangible item). For example, anxiety-related behaviors and feather damaging may worsen with attempts of the caregiver to reassure or distract the bird (e.g., by petting it or telling the bird it's okay or to please stop pulling feathers). Similarly, birds may learn that biting is an effective way to keep people from approaching it. Unless these reinforcing conditions are recognized and (permanently) removed from the equation, behavior problems are likely to remain present. Recognition of the reinforcing conditions is not always easily accomplished due to lack of experience or inability of the observer (e.g., caregiver, veterinarian, behaviorist) to objectively assess and catch the stimuli associated with the problem behavior. This inability could be due to clouded judgment, incorrect information, altered bird behavior in the presence of a stranger, problem behavior that is not easily distinguished from normal behavior (e.g., preening versus feather-damaging behavior), or behavior that occurs outside of the caregiver's presence (e.g., feather-damaging behavior occurring at night). Similarly, the elimination of the reinforcers is often more complex than one may anticipate.

The process of permanent removal (i.e., withholding) of the reinforcer(s) of problem behavior is referred to as extinction. Once a behavior is no longer reinforced, the rate at which it is shown will decrease to baseline levels (i.e., before reinforcement started occurring; this thus does not necessarily mean complete suppression of the behavior). Several aspects complicate the use of extinction as a sole technique to solve a problem behavior. These aspects are described below.

- The procedure often requires more time to take effect, especially if the reinforcement occurred on an unpredictable and intermittent basis (i.e., the behavior was reinforced only part of the time or by one person, and at random moments, as is generally the case for most problem behaviors) because this will result in higher persistence of the behavior, which is comparable to the process of gambling that only requires an occasional jackpot to maintain the behavior;

- Many problem behaviors (e.g., extreme biting, screaming, or chewing on furniture) can simply not be ignored, assuming that the caregiver's reaction is the maintaining reinforcer;

- Prior to cessation or decrease in the behavior, the frequency, intensity, and duration of the problem behavior may increase temporarily, which is referred to as the extinction burst; this escalation of the behavior is rarely tolerated by caregivers, thereby resulting in inadvertent reinforcement of the behavior, but at a new, often higher level of intensity;

- Various problem behaviors, especially those associated with frustration (e.g., aggression) are actually elicited by extinction, thereby making it more likely for the behavior to occur during this procedure;

- Uncontrolled internal and/or external reinforcers may have reinforced the problem behavior, thus successful extinction of a problem behavior will rely highly on the identification and subsequent elimination of these reinforcing conditions;

- Following extinction, spontaneous recovery, reinstatement, reacquisition or resurgence of this, or other previously extinguished problem behaviors is regularly seen, thereby rendering behavioral change more difficult to sustain in the long term.

Other factors that may complicate the successful elimination of problem behaviors include the complexity and the multi-factorial nature of the problem behavior etiology; the influence of secondary reinforcers that may hinder the recognition of the original, inciting events for the problem behaviors and can result in maintenance of the problem behavior, even in absence of the original trigger; and the influence of previous life experiences, socialization and imprinting, which may have long-lasting, disrupting effects on the animal's behavior potentially as a result of abnormal brain development. In addition, some behaviors perceived as problematic by the owner may represent normal behavior. For example, many vocalizations serve as contact calls to locate other flock members and promote flock cohesion, and should be considered part of the normal behavior pattern of parrots.

Eliminating Problem Behaviors Effectively: Step 1. Diagnosis

The effective elimination of a problem behavior starts with a correct diagnosis of the underlying causes, and differentiating whether the behavior represents: 1) normal behavior thereby requiring education of the owner; 2) the result of environmental circumstances and/or learning thereby requiring environmental and/or behavior modification; or 3) a side effect of a medical problem or abnormal psyche thereby requiring medical treatment or treatment with psychoactive drugs. For this purpose, a sound understanding of the behavior and development of the species and the basic principles of learning theory are necessary. In addition, a complete diagnostic work-up should be performed to identify all behavioral and medical signs, as well as the circumstances under which the problem behavior is displayed.

The workup of a behavioral problem roughly involves the same systematic and critical thinking processes as those of a medical problem. The differences relate to what is being investigated and not to the framework of critical thought required. With medical problems, the investigation focuses primarily on internal pathologic processes that affect the animal's health, whereas the main focus during workup of behavior problems is placed on the animal's environment and how this affects the animal's behavior (i.e., external influences). In both, however, a similar sequence is followed, in which data are gathered (from the signalment, history, physical examination, or behavioral and environmental assessment), followed by compilation of a list of differential diagnoses or hypotheses as to why the behavior is occurring (i.e., what function does the behavior serve to the animal?). With no two individuals being exactly alike, each case should be considered as a "study of one," which warrants the formulation of a unique working hypothesis that is specifically tailored to the individual and its living environment.

Applied behavior analysis

Applied behavior analysis (ABA) quantifies and measures observable behavior in applied settings and uses functional behavioral assessments to develop and test hypotheses about the purpose or function that a (problem) behavior serves for the individual and which environmental circumstances are maintaining this behavior. ABA assumes that behavior is purposeful and never occurs randomly or independent of its environmental surroundings. This cause-and-effect relationship between the occurrence of a behavior and the environment implies that no behavior can be understood (or changed) without knowledge of the environmental circumstances that immediately precede or follow the behavior. Skinner described this relationship in a three-term contingency, also known as the ABCs or behavior. Dependent on the consequences (C), the frequency with which an animal displays a (problem) behavior (B) may increase (reinforcement) or decrease (punishment), whereas the antecedents (A) help identify and predict the physical and environmental conditions in which the behavior is likely to occur. These ABCs form the basis of the functional assessment, and aid in the understanding, prediction and modification of (problem) behavior.

A correct functional assessment requires the following four steps to be taken.

1. Observe and operationally define the target behavior in unambiguous, objective (i.e., observable) terms. Specifically, what does the bird do that can be observed and measured?

2. Identify the physical and environmental conditions in which the target behavior is likely to occur. Specifically, what are the immediate antecedents or predictors of the bird's behavior (i.e., where, when, and with whom is the behavior most likely to occur) and under which circumstances does the behavior not occur?

3. Identify the consequence (i.e., the immediate purpose) of the target behavior for the bird. Specifically, what does the bird gain, escape, or avoid by behaving this way, and to what extent does the bird's natural environment support this behavior?

4. Develop a summary statement that describes the relationship among the antecedents, behaviors, and consequences for each situation in which the behavior occurs.

Based on the functional assessment described above, hypotheses can be generated regarding the function(s) that the problem behavior has for the bird. These include which environmental aspects are maintaining this behavior and which of these environmental variables can be altered or eliminated, thereby also altering or eliminating the problem behavior, provided that the generated hypothesis was correct.

Eliminating Problem Behaviors Effectively: Step 2. Treatment

Following the identification of the potential risk factors, antecedent cues and reinforcing outcomes, an intervention plan can be designed to replace or change the problem behavior. To modify a parrot's behavior, various strategies may be considered, including behavior modification techniques such as rewarding right behavior and teaching the parrot new, replacement behaviors, and adaptations to the parrot's living environment. Moreover, if any underlying nutritional or medical problems are suspected or confirmed, treatment should be aimed at eliminating these underlying conditions. The strategies mentioned above are generally preferred because these are less intrusive and more effective than other methods such as the use of psychoactive drugs or punishment strategies. That is, rather than treating the symptoms or teaching the parrot what not to do (i.e., punishment strategies), they help to rearrange the antecedent and consequent environment in such a manner that environmental conditions that led to the development of the problem behavior in the first place are eliminated and the right behavior becomes easier and more reinforcing than the wrong behavior.

Behavior modification therapy

Behavior modification therapy not only aims to reduce problem behavior, but also to provide the bird with opportunities to replace the function served by the problem behavior with more acceptable, alternative behaviors, as this will render the problem behavior less likely to (re)occur in future. To increase the effectiveness and efficiency of any behavior modification strategy, behavior experts recommend that the following four conditions are taken into consideration.

1. The proposed changes should be described in as much detail as possible and describe exactly what each family member is supposed to do and when to promote and maintain appropriate behavior.

2. A clear linkage should be present between the functional assessment and the intervention plan and focus on what the bird should do to achieve its goal rather than what it should not do.

3. The behavior modification strategy should be technically sound; it should adhere to the scientific principles of learning to render the problem behavior irrelevant (i.e., an alternative behavior will provide the same or more reinforcement), ineffective (i.e., the alternative behavior will require less effort or fewer responses), or inefficient (i.e., the problem behavior results in little to no reinforcement).

4. The behavior modification strategy should correspond with the owner's routine, living conditions, values, resources, and skills.

Antecedent rearrangement with environmental adaptations

Thoughtful (re)arrangement of the living environment is often a very effective and practical method to solve behavior problems. It may also help to effectuate behavior change by providing the appropriate stimuli needed to perform natural behaviors and newly learned behaviors while reducing the occurrence of abnormal (repetitive) problems. Rearrangement can include the following actions.

Alter or remove the antecedent stimulus that signals the opportunity for reinforcement.

Altering or removing the antecedent stimulus that signals the opportunity for reinforcement makes the problem behavior less likely to occur. For example, if a bird bites when the owner reaches out with a hand inside the cage to take the bird out, the biting may simply be avoided by using a perch rather than the hand, or allowing the bird to come out of the cage on its own.

Change the value of (i.e., reinforce) consequences through antecedent arrangement.

Changing the value of consequences through antecedent arrangement decreases the chances of the problem behavior occurring. For example, when a bird vocalizes to draw attention from the owner when s/he comes home from work, the owner may consider first paying attention to the bird and interacting with it, to decrease the chances of the bird starting to scream for attention.

Managing the consequences of a behavior with commonly used behavior modification techniques

Managing the consequences of a behavior is a common strategy employed when dealing with a behavior problem. This strategy is based on the four basic elements of operant conditioning: positive reinforcement or reward training and negative reinforcement or escape training, which both result in an increase or maintenance of the target behavior; and positive and negative punishment, which result in a decrease or suppression of the behavior through applying aversive stimuli, or withholding or removal of positive stimuli, respectively. Although punishment strategies help to decrease behavior, they may be harmful to the bird, especially positive punishment strategies, and only teach it what not to do rather than what it can do to achieve its goal. Thus, these strategies can potentially lead to stress, frustration and anxiety and increase the chances for other, undesirable behaviors (e.g., aggression or fear) to occur. As such, (positive) punishment strategies should rarely, if ever, be used.

Cessation of all reinforcement of a problem behavior (i.e., extinction) may also be used to reduce problem behavior. Many factors may complicate the effective use of extinction as a behavior modification technique as mentioned above. Thus, extinction is generally only used in combination with other behavioral interventions such as differential reinforcement, in which an alternative behavior that serves a similar function for the animal as the problem behavior, but in a more appropriate way, is reinforced. Various types of differential reinforcement exist, of which the three most used schedules are described below.

Differential reinforcement of other behavior (DRO)

The animal is reinforced for any behavior other than the problem behavior.

Differential reinforcement of alternative behavior (DRA)

The animal is reinforced for performing a particular alternative behavior that is either in the animal's repertoire or is specifically taught to the animal to serve as a replacement behavior, once mastered.

Differential reinforcement of incompatible behavior (DRI)

The animal is reinforced for engaging in a behavior that cannot be performed simultaneously with the problem behavior and as a result, is the most powerful behavior reduction tool available.

Systematic desensitization is another commonly employed behavior modification technique that is used primarily to reduce or extinguish an elicited (i.e., reflexive) emotional response (e.g., fear, aggression) to a conditioned stimulus (CS). It involves a gradual and systematic exposure to a stimulus that subsequently results in an extinction of the conditioned emotional response to the stimulus. To boost its effectiveness, systematic desensitization may be combined with counter-conditioning. During this procedure, a positive (i.e., appetitive) stimulus is non-contingently paired with the aversive stimulus to change the value of the conditioned stimulus from CS− to CS+.

Eliminating Problem Behaviors Effectively: Step 3. Evaluation

Following the implementation of a behavior modification plan, close monitoring and regular rechecks of the bird are recommended. Such follow-up is not only important to help evaluate whether and to what extent the plan has been effectuated, but also because new, undesired behaviors as well as an augmentation of the initial problem behavior may occur, subsequently warranting re-evaluation of the plan. To properly evaluate the effects of a specific intervention, the behavior modification plan is preferably implemented in a stepwise fashion. Moreover, owners may be asked to keep a record of the bird's behavior to allow tracking of the frequency, severity, and moments at which the problem behavior occurs, including the environmental conditions that may influence the onset of this behavior. This record not only helps document the effectiveness of the treatment but also facilitates additional definition of the problem and adjustment or refinement of the behavior modification plan, if needed.

Recommended Reading

1. Bergman L, Reinisch US. Parrot vocalization. In: Luescher AU, ed. *Manual of Parrot Behavior*. Ames, IA: Blackwell Publishing; 2006:219-223.

2. Evans M. Environmental enrichment for pet parrots. *In Pract.* 2001;23(10):596-605.

3. Friedman SG. A framework for solving behavior problems: functional assessment and intervention planning. *J Exot Pet Med.* 2007;16(1):6-10.

4. Orosz SE. Diagnostic work-up of suspected behavioral problems. In: Luescher AU, ed. *Manual of Parrot Behavior*. Ames, IA: Blackwell Publishing; 2006:195-210.

5. Speer B. Normal and abnormal parrot behavior. *J Exot Pet Med.* 2014;23(3):230-233.

6. Van Sant F. Problem sexual behaviors of companion parrots. In: Luescher AU, ed. *Manual of Parrot Behavior*. Ames, IA: Blackwell Publishing; 2006:233-245.

7. Van Zeeland YRA, Friedman SG, Bergman L. Behavior. In: Speer BL, ed. *Current Therapy in Avian Medicine and Surgery.* St. Louis, MO: Elsevier; 2016:177-251.

8. Van Zeeland YRA. Medication for behavior modification in birds. *Vet Clin North Am Exot Anim Pract.* 2018; 21(1):115-149.

9. Welle KR. Aggressive behavior in pet birds. In: Luescher AU, ed. *Manual of Parrot Behavior*. Ames, IA: Blackwell Publishing; 2006:211-217.

10. Wilson L, Luescher AU. Parrots and fear. In: Luescher AU, ed. *Manual of Parrot Behavior*. Ames, IA: Blackwell Publishing; 2006:225-231.

Advances in Sinusitis and Upper Respiratory Disease in Birds

Lauren Thielen, DVM, DABVP (Avian Practice)

Session #178

Affiliation: From Cornell University Hospital for Animals, Ithaca, NY 14853, USA.

Abstract: Upper respiratory disease and sinusitis are common conditions of patients presenting to the avian practitioner. Since patients often have subtle clinical signs associated with acute upper respiratory infection, persistent infections and consequent sinusitis are not uncommon. The unique anatomy of the avian upper respiratory system makes prevention, treatment, and resolution of avian sinusitis challenging. Resolution of sinusitis in birds is accomplished by both identifying the offending agent and addressing any underlying predispositions. This lecture discusses pertinent sinus anatomy, common etiologic agents, clinical signs associated with sinusitis, proper methods in obtaining diagnostics, effective treatment strategies, and novel techniques. Species covered primarily focus on psittacine birds and poultry.

Avian Sinusitis

Upper respiratory infections are common in avian patients. If left untreated, upper respiratory disease can progress over time to sinusitis or rhinitis. Sinusitis is defined as inflammation of the sinuses, and it may be purulent or non-purulent, acute or chronic.

Predispositions

The underlying cause of respiratory disease can be complicated by secondary infections that take advantage of damaged respiratory epithelium. All primary and secondary pathologic entities must be addressed for a successful treatment of respiratory disease. Hypovitaminosis A, low humidity, poor ventilation, and environmental inhalant irritations can all predispose the avian patient to respiratory disease.[1,2] The skin and mucosal linings are the bird's primary immune defense against pathogen colonization. Nutritional deficiencies, such as hypovitaminosis A, can predispose a bird to respiratory infections by creating an unhealthy respiratory epithelium from the development of squamous metaplasia and hyperkeratosis in the sinus and nasal passages.

Avian sinus anatomy

The unique anatomy of the avian upper respiratory system makes prevention, treatment, and resolution of avian sinusitis challenging. A septum divides the nasal cavity into two halves. In the majority of avian species, each half of the nasal cavity contains the rostral, caudal, and middle nasal concha.[3] The caudal nasal concha does not connect to the nasal cavity directly but communicates instead with the infraorbital sinus by a dorsal opening. The infraorbital sinus is a triangular cavity located lateral to the nasal cavity and ventrally surrounding the eye. Once air passes into the infraorbital sinus, it can flow caudally to the communicating cervicocephalic air sacs.[3]

The avian sinuses are found within the restricted space of the lateral skin and the subcutaneous tissues of the face and are not completely enclosed within the bones of the skull as in mammals. The infraorbital sinus is a frequent site of accumulation of mucopurulent debris that occurs with upper respiratory infections due to the sinus' anatomic complexity and limited access. There are only two openings to the infraorbital sinus: the dorsal opening to the medial and caudal nasal concha and the caudo-dorsal opening into the cervicocephalic air sac.[3] Therefore, the only passageway for drainage of mucopurulent material that ends up in the infraorbital sinus is the caudal nasal concha up through the dorsal opening. This difficult anatomic passageway is the reason pus and cellular debris become trapped in the ventral portion of the infraorbital sinus, and why infections can go subclinical for extended periods of time. The trapping of mucopurulent material leads to periorbital swelling.

Clinical signs

Often birds have mild or even absent nasal discharge and instead are scratching their face, rubbing their beak frequently and sometimes violently on perches, or yawning often to attempt and relieve sinus pressure. Other clinical signs commonly associated with infection of the upper respiratory tract are open-mouth breathing, sneezing, ocular discharge, periorbital sinus swelling, conjunctivitis, nasal discharge, development of nasal granulomas, feather loss on top of the head, voice change, head-shaking, dyspnea, exercise intolerance, plugged nares, neck stretching, and/or swollen cere.[1,2,4] On physical exam an audible upper respiratory congestion can often be heard along with open beak breathing in cases of bilateral sinus involvement. Breathing is usually altered by an increased rate, but not necessarily an increase in effort.

Causative agents of sinusitis

Normal upper respiratory flora found in birds is considered to consist of predominately gram-positive bacilli bacteria such as *Lactobacillus* sp., *Streptococcus* spp., *Staphylococcus* sp., *Bacillus* sp. and *Micrococcus* sp.[5-7] Occasional non-budding yeast may also be seen. Abnormal flora would include a high percentage of gram-negative organisms, or more than ten budding yeast per high power field;[5] however, in a study culturing the choana of 36 healthy parrots, 26% of the bacteria were gram-positive bacteria and 62% were gram negative.[6] Some of these bacteria are known potential pathogens, indicating that resident bacteria of the upper respiratory tract in parrots can cause infection if given the right opportunities. It should also be noted that birds with malnutrition are expected to have a higher percent or normal gram-negative respiratory flora due to chronic malnutrition. In these situations, the presence of gram-negative bacteria requires clinical signs of an infection to warrant antibiotics.[7]

Common etiologic agents causing upper respiratory disease leading to sinusitis in birds include bacteria, fungi, viruses, and parasites. Anatomic issues arise with congenital or developed choanal atresia. Gram-positive bacteria that are typically considered normal flora, such as *Staphylococcus* spp. and *Streptococcus* spp. can become opportunistic pathogens. Mycobacterial infections can present as periorbital nodules and should be placed on the initial differential diagnosis list of birds presenting with periorbital swelling.[5] Common gram-negative pathogenic bacterial isolates that may cause sinusitis in birds include *Bordetella* spp., *Pasteurella multocida*, *Avibacterium paragallinarum*, *Enterobacter* sp., *Pseudomonas* spp., *Helicobacter sp* in cockatiels, *Salmonella* spp., and Coliforms such as *Escherichia* spp., *Klebsiella* spp., and *Enterococcus* spp.[1,5,6,7] Among the most common primary bacterial pathogens causing sinusitis in birds are *Chlamydia psittici* and *Mycoplasma* spp. Avian fungal sinusitis can be due to infection with *Cryptococcus* sp., *Aspergillosis, Mucor* sp. and *Candida* spp.[1,8] Parasites including *Trichomonas* sp., *Knemidokoptes pilae* in budgies or *Cryptosporidia* spp. can affect the avian upper respiratory system as well, contributing to sinusitis.[8] Infections secondary to foreign bodies such as grass awns, feathers, or other foreign bodies must be ruled out as underlying causes of sinusitis. Viral etiologies (although rarely affecting psittacine birds) include paramyxovirus, Newcastle disease, influenza virus, poxvirus, and herpesvirus.[8] Neoplasia of the avian sinuses include nasal adenoma or adenocarcinoma, lymphoma, fibrosarcoma, squamous cell carcinoma, fibroma, basal cell carcinoma, hemangioma and melanoma.[5,8] Rhinoliths, granulomas or foreign bodies consisting of food particles, wood shavings, or bedding particles have also been reported causes of sinusitis as well as the possibility of environmental allergies.[1]

Mycoplasma

Mycoplasma spp. are a common cause of sinusitis in chickens and other poultry, as well as conjunctivitis in wild passerines, notably in house finches (*Haemorhous mexicanus*), and can also occur in psittacine birds. The most common organism in poultry is *M. gallisepticum*, which causes a chronic respiratory infection in chickens, resulting in sinusitis. Its pathogenicity is enhanced with concurrent infections, such as *E. coli.* In turkeys and peafowl *Mycoplasma* infections commonly lead to a dilated infraorbital sinus with a caseous or stringy mucus abscess within.[4] Individuals infected can have subclinical infections chronically, possibly noted by a mild cough.[9] Due to mild clinical signs in immunocompetent individuals, infections can be easily overlooked. Suspicions for *Mycoplasma* infections should increase with recurrent upper respiratory infections or in birds with chronic conjunctivitis, nasal exudates, or sneezing.[9,10] Transmission is both horizontal and vertical via inhalation. Fomites can play a role in the spread and persistence. The gold standard diagnosis is a specialized culture utilizing Hayflick's media, but culture can be difficult.[10] Diagnosis is often based on polymerase chain reaction (PCR) samples taken from swabs of the conjunctiva or fluid from flushes of the nasal cavity in symptomatic birds.[9] There are several serologic test available, but due to variability in specificity and sensitivity, they are recommended only for flock screening.[10] Treatment is by use of tetracyclines, macrolides, or aminoglycosides.[10] In poultry, tylosin is regarded as the best approved treatment.[4] One must keep in mind in regarding to management that treatment alleviates clinical signs but does not eradicate the organism. Once infection is confirmed, birds are thought to be infected for life. Vaccination (killed) for poultry prevents and reduces some clinical signs but does not prevent birds from transmitting.[10]

Bordetella avium

Bordetella avium is a gram-negative aerobic bacterium that is highly contagious and causes an upper respiratory tract infection in a variety of bird species, but most notably leads to death in cockatiel chicks. *B. avium* affects the epithelium of the respiratory tract by producing toxins that propagate tissue damage. *B. avium* does not appear to clinically affect adult cockatiels, but it does cause severe disease in young cockatiels (e.g., 3 days to 4 wk old). Clinical signs include sneezing, constant begging, swelling of the suborbital chamber of the infraorbital sinus, inflamed nasal cavity and infraorbital sinus. The most severe of clinical signs cause a temporomandibular rigidity or lockjaw syndrome, that occurs typically 5 days after initial clinical signs. Due to the swelling of the temporomandibular muscles, birds cannot open their mouths to eat and death can occur. *Bordetella avium* is transmitted through close contact with an infected bird or through exposure to contaminated litter or water. Infected parent birds will pass it to their young post hatch. Clinical signs in chicks begin typically 3 days of exposure. There is no clear treatment for cockatiel chicks that contract *B. avium* and develop lockjaw. Once the lock-jaw syndrome sets in, death within a couple of days is typically inevitable. If caught early, antibiotics such as doxycycline or tylosin and anti-inflammatories may be beneficial. Nebulization with gentamycin can also be beneficial. It is important to perform passive range of motion every few hours to help maintain mobility of the jaw. Many birds require frequent gavage feedings until the infection is under control.

Diagnostics

Cytologic and culture samples of the respiratory tract are collected with a deep nasal swab, choana swab, or sinus aspirate/flush.[2] Nasal discharge may not be the best sample collection site due to contamination. Cytologic samples can be Gram stained to help guide treatment selection while waiting for the results of aerobic and fungal culture and sensitivities. Samples can also be evaluated for neoplasia or possible parasites.

The rostral choanal slit communicates with the nasal passageways and the infraorbital sinus. It should be examined carefully and in some cases via the use of endoscopy. Careful use of a metal speculum can be used to inspect and obtain samples from the choana, but in larger conscious psittacine birds this technique can cause damage to the beak as the bird tries to bite down and therefore the author prefers instead the use of gauze strips or tape to hold the beak open. Sample collection for cytology and culture from the rostral choanal slit may help with determining etiologic agents associated with infection of the infraorbital sinus.[5] In contrast, sample collection from the caudal choanal slit is of little diagnostic value.[5] Keep in mind that not all bacteria will culture from routine aerobic methods, examples include as *Chlamydia psittaci*, *Mycoplasma* spp., *Bordetella avium*, and anaerobic bacteria.

A nasal flush or sinus flush with a buffered sterile saline may provide both diagnostic and therapeutic benefit to the patient. The therapeutic benefits of flushing include dislocation of mucus and material trapped within the affected sinus and consequently improved breathing and overall patient comfort. Keep in mind, organisms cultured from a nasal flush sample may have oral flora contamination. For this reason, it may be helpful to culture the rostral choanal cleft first prior to flushing. Centrifugation of sample fluid can also aid in cytologic analysis.

The flush is performed in a well-restrained patient as they are held upside down over a sterile collection cup. The patient's nare is fitted with a tight seal using either the end of a lure lock syringe, sterile metal feeding tube, or cut red-rubber tubing. Then using 1-12 ml of solution, depending on patient size, the wash solution is flushed slowly through both nares, into the periorbital sinus and exits the bird's choana, clearing debris.[11] With obstruction, fluid can come out through the lacrimal gland and be seen around the eye. The procedure can be repeated until excess debris or thick mucus is no longer accumulating with the flush. Take care to examine the patient's oral cavity and remove excess fluid post flush. It has been suggested to use a buffered flushing solution with an optimal pH of 7.2 because a low pH solution can cause additional irritation and edema to the tissues.[12] An example of a buffered solution is eye wash. The author recommends keeping the solution in a fluid warmer to enhance patient comfort for the flushing procedure.

Cytology from a sinus aspirate of the infraorbital diverticulum may also be beneficial. This is performed in one of two ways in a sedated bird. The author prefers passing the needle into the sinus at a perpendicular angle through the skin at a point midway between the imaginary line traced between the naris and the medial canthus of the eye.[8] Also described is passing a needle through the skin located at the commissure of the mouth. The needle is held parallel to and along the side of the head, and as it slides under the zygomatic bone, it is directed towards a point midway between the eye and the external nares. Once the needle is inserted in the sinus, the sample can be aspirated.[13] With either technique, a sinus aspirate should be performed cautiously because a caudally misdirected needle could result in the penetration of the ocular orbit.[13] A sinus flush utilizes the same technique described for a sinus aspirate, with the difference that the bird is held with its head parallel to the floor and the affected sinus down, while a buffered sterile solution is infused into the sinus.[13]

Biopsy of the affected sinus should be considered in some cases to investigate both infectious and noninfectious causes. Careful evaluation of the choana and trachea are important for understanding the extent of clinical disease and ruling out or facilitating the removal of a potential foreign body.

Imaging such as radiographs, preferred CT scan or MRI, is important in obtaining a definitive diagnosis, underlying etiology (infections vs foreign body vs neoplasia), and for surgical planning.

Therapeutics

Therapy for sinusitis is dependent on properly addressing the underlying predisposing factors as well as the underlying etiologic agent. Antimicrobials should be chosen accordingly. In the author's experience, the use of systemic anti-inflammatories, such as meloxicam, is paramount to control of the initial stages of sinusitis in both poultry and psittacine birds.

In cases where vitamin A deficiency is suspected, the author initially utilizes a one-time vitamin A injection followed by oral supplementation with the vitamin A precursor, β-carotene, in the form of red palm oil; however, current sourcing of injectable vitamin A has been a challenge in recent years. Ideally, water-soluble formulations would be utilized to avoid post-injection irritation. Conversion to an appropriate diet is also paramount.

Medicated nasal flushes are utilized by some practitioners in cases of sinusitis. Intranasal antibiotics or antifungals, typically in the form of ophthalmic preparations, may be applied to the nares as a way to allow for topical therapy even at home.

Nebulization in birds with respiratory disease, especially upper, is helpful as it helps maintain proper hydration to epithelium, breaks up necrotic debris, and can deliver antimicrobials to the source of infection. Nebulization is indicated for sinusitis, pharyngitis, and bronchitis. It is recommended to have a treatment regimen of 3-4 times a day for 10-15 minutes at a time, and continued until all clinical signs are resolved for 3 days. Aerosolized particles can be delivered to lungs and some air sacs, but only if particle size is 1-5 microns, thus requiring a specialized nebulizer. The new generation mesh nebulizers are preferred as they have a higher respiratory tract deposition compared to the standard jet or ultrasonic nebulizers;[14] however, in cases of localized infection to just the upper respiratory tract, a standard nebulizer (jet or ultrasonic) can be effective.[11] Nebulization of various compounds can be considered based on etiology. Some practitioners prefer saline nebulization alone. In cases with bacterial sensitivity to it, a gentamicin solution of 3-6 mg/ml has been suggested.[8] One must keep in mind the possibility of systemic absorption of nebulized drugs. Amphotericin B is commonly used in nebulization with cases of fungal infections, such as Aspergillosis.[14]

In chronic or refractory cases, frequent, but no more than one to two times per week, nasal flushes can be performed to help manage excessively thick mucus or abundant debris. Flushing too often, with low pH solution, or flushing too aggressively can cause the development of inflammation of the turbinates and subsequent worsening of clinical signs.

In advanced cases of sinusitis, surgical intervention via a sinusotomy is commonly necessary to remove infectious or noninfectious debris trapped in avian sinuses. CT imaging can help guide surgical intervention by means of sinus trephination, although incisions are often best made directly into the skin over the affected sinus. Incisions can be made easily into the skin of the infraorbital diverticulum and the site flushed and left open for continued flushing. [15] In poultry, frequently a solid caseous mass is removed *en bloc* and has the appearance of an impression mold of the sinus space. Once the debris is removed, the site can be left open to drain and for further wound management, such as daily flushing. Acceptable flushing solution examples include dilute chlorhexidine, dilute iodine, or sterile saline solutions.[15] The sinusotomy wound typically heals rapidly and flushing usually can only be performed for three to five days.[15]

Trephination involves drilling a small hole through the frontal bone of the skull into the infraorbital sinus to allow retrieval of material that a nasal or sinus flush cannot access. Trephination site location is aided by advanced imaging (CT or MRI), but the hole is typically made one-half to two-fifths of the way between the naris and the medial canthus of the eye.[16] Once entered into the sinus the site can be curetted out, flushed, or biopsied. In psittacine birds, the left and right infraorbital sinuses communicate so it is possible to make only one trephination site. The bone defect is left open and skin heals around the trephination hole relatively quickly. Drains using IV tubing can be placed to delay closure and allow for prolonged flushing or medication administration.[16]

References

1. Crosta L. Respiratory diseases of parrots: anatomy, physiology, diagnosis and treatment. *Vet Clin North Am Exot Anim Pract.* 2021;24(2):397-418.

2. Orosz SE, Lichtenberger M. Avian respiratory distress: etiology, diagnosis, and treatment. *Vet Clin North Am Exot Anim Pract.* 2011;14(2):241-255.

3. King AS, McLelland J. Respiratory system. In; *Birds: Their Structure and Function.* Sussex, England: Bailliere Tindall, 1984;110-114.

4. Greenacre CB, Morishita TY. Respiratory disease. In: Fulton RM, eds. *Backyard Poultry Medicine and Surgery: A Guide for Veterinary Practitioners.* Ames, IA: John Wiley & Sons; 2015:137-143.

5. Ritchie RW, Harrison GJ, Harrison LR. Pneumonology. In: Tully TN, Harrison GJ, eds. *Avian Medicine: Principles and Application.* Lake Worth, FL: Wingers Publishing; 1994:556-579.

6. Jesus SO, Correia JHD. Potential pathogens recovered from the upper respiratory tract of psittacine birds. *Proc Int Virtual Conf Vet Med.* 1998.

7. Doneley RJ. Bacterial and parasitic diseases of parrots. *Vet Clin North Am Exot Anim Pract.* 2009;12(3): 417-432.

8. Samour J. Disorders of the respiratory system. In: Bárbara Arca-Ruiba, eds. *Avian Medicine.* 3rd Ed. St. Louis, MO: Elsevier; 2016;385-394.

9. Levisohn S, Kleven SH. Avian mycoplasmosis (*Mycoplasma gallisepticum*). *Revue Scientifique et Technique.* 2000;19(2):425-442.

10. World Organization for Animal Health. Manual of Diagnostic Tests and Vaccines for Terrestrial Animals 2017. Avian Mycoplasmosis. Available at: www.oie.int. Accessed Dec 2017.

11. Powers LV. Common procedures in psittacines. *Vet Clin North Am Exot Anim Pract.* 2006;9(2):287-302.

12. Orosz SE. Avian anatomy and physiology of the respiratory system: a review. *Proc Annu Conf Assoc Avian Vet.* 1997;3-7.

13. Ritchie RW, Harrison GJ, Harrison LR. Cytology. In: Campbell TW, eds. *Avian Medicine: Principles and Application.* Lake Worth, FL: Wingers Publishing; 1994:199-222.

14. Antonissen G, Martel A. Antifungal therapy in birds: old drugs in a new jacket. *Vet Clin North Am Exot Anim Pract.* 2018;21(2):355-377.

15. Greenacre CB, Morishita TY. Soft tissue surgery. In: Echols SM, eds. *Backyard Poultry Medicine and Surgery: A Guide for Veterinary Practitioners.* Ames, IA: John Wiley & Sons; 2015;222-225.

16. Pye GW. Surgery of the avian respiratory system. *Vet Clin North Am Exot Anim Pract.* 2000;3(3):693-713.

Nowhere To Go: Bird Sanctuaries and Rescue Organizations

Anthony A. Pilny, DVM, DABVP (Avian Practice)

Session #149

Affiliation: From Arizona Exotic Animal Hospital, 2340 E Beardsley Rd. Ste 100 Phoenix, AZ 85024, USA.

Abstract: Pet birds, especially the larger parrot species such as cockatoos and macaws, are frequently surrendered or rehomed and typically only after a few years of ownership. Most people are attracted to the lore surrounding bird ownership and are not prepared for all aspects of care. Behavioral issues such as biting and vocalizations, expenses, mess, time, and physical trauma to owners are some of the leading reasons owners will give up on the ownership commitment. The issue, like with most other surrendered pets, is where do they go when most shelters are unable and ill-equipped to take them, and rescue organizations and sanctuaries are full? Around the world, many entities describe themselves as a sanctuary or rescue organization for unwanted birds and animals; however, the quality of animal care and sustainability of the organization varies widely. Currently, most rescue organizations and sanctuaries are full and unwanted birds have nowhere to go!

Sanctuaries and Rescue Organizations

The Global Federation of Animal Sanctuaries defines a sanctuary as "any facility providing temporary or permanent safe haven to animals in need while meeting the principles of true sanctuaries: providing excellent and humane care for their animals in a non-exploitative environment and having ethical policies in place, regarding tours, commercial trade, exhibition, acquisition and disposition, and breeding."[1]

A growing number of bird rescue and sanctuary organizations are facing the challenge of caring for large numbers of unwanted parrots. Many shelters and animal control organizations are also being forced to gain the skills needed to provide suitable care and placement for the rising numbers of unwanted exotic birds. It is important to distinguish between legitimate organizations and substandard operations that are exploiting the situation for their personal benefit at the public and bird's expense.[2]

A true sanctuary is one where there is:

- no commercial buying or selling;
- no invasive research;
- no public visitation or contact with wild animals;
- no removal of wild animals for exhibition, education, or research; and
- no breeding (Exceptions include rehabilitation and release centers engaged in a breeding-for-release program with available release sites within the state/province, conducted with specific conservation goals).

Pseudo Sanctuaries, Hoarders, and Fake Rescue Groups

Some groups claim to be rescue or sanctuary facilities but are animal collectors or hobby breeders that formed a nonprofit organization and promote themselves as a rescue organization able to take unwanted birds. Many unwanted birds also fall victim to hoarders who warehouse them in grossly substandard conditions while others acquire them to use for entertainment or to re-sell at bird marts, auctions, or over the internet.

What to Look for in a Legitimate Avian Rescue or Sanctuary Organization

Avian rescue and shelter organizations range from small rescue groups that provide temporary care and placement for birds to large sanctuaries that provide lifetime care. The quality of animal care, ethics, organizational management, education and advocacy are among the factors to be considered.[3,4] Regardless of their differences in size, location, or available resources, certain qualities are universal among ethical avian rescue organizations.

Superior care

A safe, healthy, and clean environment where birds are provided with ample space, a nutritious diet, qualified medical care, and a variety of enrichment activities that allow them to enjoy their full range of natural behaviors. Each bird's individual physical and behavioral needs are always put before the desires and expectations of their human caretakers.

Professional management

It takes more than good intentions and taking in large numbers of birds, especially long-lived species like macaws and cockatoos, without sufficient resources can easily overwhelm a facility.

Advocacy and collaboration

Legitimate rescue and sanctuary organizations also stand against practices that contribute to the suffering and exploitation of animals and work in partnership with other avian and animal protection groups to help ensure that birds everywhere are afforded quality care and humane treatment.

Education

An effective program that focuses on the facts, promotes a better understanding of the true needs of birds, and offers solutions to the complex issues involved in addressing their welfare. Most importantly, the organization's stance should be clear; all parrots belong in their rightful place in the wild.

Public outreach

Parrots' highly emotional nature and repertoire of behaviors are often entertaining to humans, but reputable groups refrain from exploiting them for profit or self-aggrandizement.

Integrity

Reputable groups do not promote the commercial trade in birds, nor do they breed them or place them in breeding situations.

References

1. Global Federation of Animal Sanctuaries. Available at: https://sanctuaryfederation.org/about-gfas/what-is-a-sanctuary/ Accessed February 29, 2024.

2. Avian Welfare Coalition. Available at: https://www.avianwelfare.org/issues/sheltering.htm. Accessed October 2023.

3. International Alliance for Protection of Parrots. Available at: https://www.allianceforparrots.org/. Accessed February 29, 2024.

4. Great Nonprofits. Find Bird Sanctuaries Nonprofits and Charities. Available at: https://greatnonprofits.org/categories/view/bird-sanctuaries?gad=1&gclid=Cj0KCQjwxYOiBhC9ARIsANiEIfbUF8N1ZB_EEVzA-9TWG84Weeysj1fgpmIPVEFaEH0RZq0jiOajbAjIaAv5vEALw_wcB. Accessed February 29, 2024.

What Avian Clinicians Need to Know about the U.S. Department of Agriculture Standards for Birds under the Animal Welfare Act

Anthony Pilny, DVM, DABVP (Avian Practice)

Session #150

Affiliation: From Arizona Exotic Animal Hospital, 2340 E Beardsley Rd. Ste 100 Phoenix, AZ 85024, USA.

Abstract: New regulations have been issued under the Animal Welfare Act that affect individuals selling, exhibiting, and transporting birds as well as those using birds in research. The new regulations define pet bird species and categorize them by size to determine thresholds for licensure as a breeder, broker/ dealer, or exhibitor.

As of February 21, 2024, individuals and businesses that sell, exhibit to the public, transport commercially, or use birds in research may need to be licensed or registered with the U.S. Department of Agriculture (USDA) to comply with new regulations under the Animal Welfare Act (AWA).[1-4] The regulations define specifically what avian species are considered pet bird species and categorize what pet bird species are small or large size. The designations of pet bird species and size are used in determining thresholds for licensure.

- A Class A license (Breeder) is required when selling birds for use as pets or for exhibition that were hatched on your premises, if you sell more than 200 pet bird species having an average adult body weight of 250 grams or less, annually, or if you sell more than eight pet bird species having an average adult body weight of more than 250 grams, annually. This is breeding and selling birds hatched on your property.
    - Covers all species as listed by the Species List of Pet Birds as Defined by Animal Welfare Regulations[5]
    - Any bird not on the above list requires a license regardless of weight or number unless already listed as exempt

- A Class B license (Broker/Dealer) is required for those who purchase and resell birds not hatched on their property. This includes auctions, random source dealers, and some rescues that receive any remuneration.

- A Class C license (Exhibitor) is required if you are exhibiting to the public more than eight pet bird species, any non-pet bird, or greater than four raptor species. This includes carnivals, circuses, animal acts, zoos and petting zoos, walk-in aviaries, photo shoots, and birds given away as prizes.

Birds protected under the Migratory Bird Treaty Act are excluded from the definition of a pet bird species as defined by the AWA and regulations. Face-to-face sales of pet bird species in which the seller, buyer, and pet birds are physically present are exempted from licensure (the Retail Pet Store Exemption).

References

1. U.S. Department of Agriculture. Animal and Plant Health Inspection Service. Animal Welfare Act Standards for Birds. Available at: https://www.aphis.usda.gov/aphis/ourfocus/animalwelfare/new-bird-rule/awa-standards-for-birds. Accessed March 1, 2024.

2. 7 U.S.C. 2131–2159; 9 CFR Part 3 Subpart G. Specifications for the Humane Handling, Care, Treatment, and Transportation of Birds. Available at: https://www.ecfr.gov/current/title-9/chapter-I/subchapter-A/part-3/subpart-G. Accessed March 1, 2024.

3. U.S. Department of Agriculture. Animal and Plant Health Inspection Service. Animal Care Licensing Registration Assistant. Available at: https://efile.aphis.usda.gov/LRAssistant/s/. Accessed March 1, 2024.

4. U.S. Department of Agriculture. Animal and Plant Health Inspection Service. USDA Animal Care Welfare Inspection Procedures for Birds. Available at: https://www.aphis.usda.gov/aphis/ourfocus/animalwelfare/new-bird-rule/awa-standards-for-birds-faq. Accessed March 1, 2024.

5. U.S. Department of Agriculture. Animal and Plant Health Inspection Service. List of Pet Birds as Defined by the Animal Welfare Regulations. Available at: https://www.aphis.usda.gov/aphis/ourfocus/animalwelfare/new-bird-rule/pet-bird-species. Accessed March 1, 2024.

Section 9

Wildlife and Conservation Masterclasses

Ritchie Burdeaux, Jr, DVM, DABVP (Avian Practice)
Michael Lutz, DVM
Session Chairs

Care, Rehabilitation and Release of Confiscated Wild Parrots

Patricia J. Latas, DVM, MS, BS

Session #104

Affiliation: From Wild Parrot Coalition, 7223 E. Camino Valle Verde, Tucson, AZ 85715, USA, SoCal Parrot, Jamul, CA 91935, USA.

Abstract: Veterinarians and wildlife rehabilitators are becoming increasingly important members of an expanding network caring for and determining the disposition of wild parrots seized by authorities. As veterinarians, we have unique skills and insights but also gaps in knowledge that need to be addressed to ensure a high degree of success and collaboration. The fate of victims of illegal wildlife trade is often forgotten in the rush to justice, but our community can collect data, recommend high standards of care, offer best practices, inform decision-making, participate in high-level networking, assist policymakers, and facilitate safe and appropriate return to the wild. The first hour of the masterclass includes an introduction to wildlife trafficking and how we can contribute to successful outcomes. The second half includes participatory tabletop exercises and lively discussions based on current and past cases. Supplemental materials may be found at: https://www.wildparrotcoalition.world/wild-parrot-confiscation-network/supplemental-material.

Introduction

By the time a veterinarian examines a confiscated parrot, the bird has experienced unimaginable suffering. Minimizing pain and suffering is integral to the veterinary oath, best practices and highest standards of care. The wild psittacine veterinarian must balance the ultimate goal of successful release to the wild with the amount of suffering induced by confiscation and subsequent care through rigorous triage, consideration of euthanasia as a first option, and minimum invasiveness for each case and an entire facility's operations.

Working with wild psittacine birds is first and foremost working with wildlife, and veterinary care of wild psittacine birds is wildlife medicine. Where a psittacine ends up will likely be determined by people and agencies not related to rehabilitation or veterinary care; a typical decision tree for the deposition of confiscated parrots is shown in the Decision-making Tree for Confiscated Psittacines in the Supplemental Materials. The goal is to return the wild parrots to the wild: as a functional member of family, flock, and ecosystem; and an active contributor to system stability and biodiversity. This approach differs greatly from companion animal, aviary, and captive breeding situations, in which captive parrots will live their entire lives with marginal contribution to ecosystem function and biodiversity.

Introduction to wildlife trafficking

Legalities

In most nations, the government owns wildlife, and specific agencies within a region, nation or locality determine who can do what with these animals, often dictating care, licensing and control measures. Permits or licenses may be required to handle wildlife. For endangered or threatened species the controls may be even tighter and permission more difficult to obtain. Most psittacine birds fall under the Convention on International Trade in Endangered Species of Wild Fauna and Flora (CITES), an international agreement regarding trade in wildlife;

however, not all countries are signatories. Almost all psittacine species are listed as CITES Appendix categories I (i.e., threatened with extinction) or II (i.e., in which trade must be controlled in order to avoid utilization incompatible with their survival).

Wildlife trafficking is organized crime.[1-3] Care and treatment of confiscated birds will require legal involvement and strict constraints of confidentiality and evidence management and associated reports and testimony.

Depending on locale, various agencies and laws may influence both ability and scope when working with wild psittacine birds, and may include requirements for correct credentials or licensure for the locale; and in addition there may be municipal, state/province/district, national and international permits required from health, agriculture and wildlife agencies. Special permits may be required for the use of pharmaceuticals, antibiotics, and other veterinary products. Since there is little uniformity on a global or national scale, it will be up to the individual facility to determine in advance exactly what paperwork needs to be filed and maintained.

In some countries and local regions, humane euthanasia is tightly controlled or illegal. For example, Indonesia and several other nations and legal entities do not allow euthanasia under any circumstances. A variety of issues may arise from efforts to care for psittacine birds. Treatment may be regulated by laws or other regulations affecting who can treat psittacine birds; for example, veterinarians versus support staff. Facilities accepting impacted psittacine birds must comply with any regulations for the hospitalization of wildlife.

Personnel involved in caring for psittacine birds must be aware of, and follow, any regulations regarding the use of pharmaceuticals. Personnel should be aware of who is allowed to purchase and prescribe pharmaceuticals. There may also be regulations regarding the disposal of unused or outdated medications. Regulations stating who can perform humane euthanasia and whether there are any restrictions on the methods used need to be known and followed.

Working as a veterinarian for a group of confiscated birds can be complicated and confusing. Getting to know people at the relevant agencies and cultivating a relationship with them can be key to being able to initiate and continue rescue operations. These relationships may also help when dealing with the regulations discussed below, because the agency in charge of wildlife may have the authority to overrule local regulations in a crisis.

The realities

Authorities and regulatory agencies may be confusing and bureaucratic. Their actions may inhibit, delay or prohibit appropriate care. Rehabilitators may be unprepared for governmental interference. Veterinarians may not be cognizant of regulations or wildlife care. All parties must combat these issues through open and honest relationships, patience, proactive collaboration and above all, pre-planning preparation where possible.

The Wild Psittacine Rehabilitation Team

Wild psittacine bird rehabilitation can be daunting at many levels. People are accustomed to pet psittacine birds as companions and family members, and thus have prior prejudice and assumptions that must be considered. Their rescue, care, rehabilitation and release can become highly emotional.

Proper liberation of psittacine patients into a safe, secure, and appropriate environment and maintaining scientific and emotional integrity along the way will require a team of participants. To maintain professional quality, mental health and emotional intelligence during the rehabilitation process, team members should support each other and assist with difficult decisions. Diversity of opinion can be harnessed to improve the eventual successful release of birds into the wild.

The team

A wild psittacine rehabilitation team comprises people with diverse skills and specialties who have a common goal: liberating wild psittacine birds back where they belong.

Rehabilitation of parrots for release to the wild is a challenging undertaking. Parrots are intelligent, sociable, curious, and strong problem-solvers requiring the rehabilitator to re-think and prepare for changes in the most routine planning. Enclosures, carriers, structures, equipment, tools, techniques and procedures have to be modified to accommodate psittacine behavior and constraints. Care of young birds, especially, requires giving them the greatest psychological support and ensuring their mental health while avoiding serious imprinting; from both the human and the psittacine point of view.

Psittacine birds may recover from serious trauma yet be unfit for release to the wild. Unless an appropriate and high-quality sanctuary is available, birds must be triaged.

The ultimate goals of rehabilitation are to prepare the individual birds for release as quickly as practicable and provide the best chance at successful re-introduction to the wild while addressing health and mobility, minimizing stress and the negative effects of confinement.

Who works with confiscated wild psittacine birds?

Wildlife rescuers and rehabilitators, animal shelters, zoos, government facilities, and sometimes private individuals or NGO facilities are potential destinations for confiscated parrots. Skills, training, education, awareness of conservation reintroduction principles, competence and financial support will be variable.

Who pays for their care?

The sad truth is that no one entity or agency may be financially responsible for the care of wild psittacine birds, even if they are IUCN red-listed endangered species. Wildlife care centers are often financially constrained and unable to pay for extensive testing and treatment. Government agencies rarely contribute to psittacine care. The biggest obstacle to highest veterinary standards and best practices is lack of funding. The most serious impingement on a practical and serious approach to producing a viable bird for release or re-introduction is money.

The role of the veterinarian in management of confiscated birds

The veterinarian for confiscations must take responsibility in planning: the strategic, tactical, and logistical aspects of screening, rehabilitation and release of psittacine birds. Veterinarians are uniquely equipped to understand epidemiology and plan accordingly to protect the birds, rehabilitation facilities, and the environment.[4] Please refer to the *Flowchart: Assessing Risk for the Release of Psittacines* in Supplemental Materials link at the end of the manuscript.

Veterinarians can ensure that captive husbandry is based on best practices and the highest standards with successful release to the wild, and plan protocols and procedures guiding health screening and evaluation for rehabilitation, release, translocation and repatriation. Pre-planning by veterinary staff should include proper and appropriate clinical evaluation before, during and after captivity; clinical recommendations to improve fitness for release; and appropriate protocols and training for staff, ancillary personnel, and the public. Veterinarians are also in the unique position to bridge multiple disciplines and network with various organizations and agencies in confiscation planning at many levels.

Veterinary planning and risk assessment are essential for successful release to the wild; overall evaluation of the facilities, husbandry, care and health protocols for wild parrots in rehabilitation have a single goal: successful release of wild psittacine birds. Evaluation of the wild population may also need to be considered before release, and advanced planning for disease outbreaks in wild or rehabilitation populations needs to be devised and administered by the veterinary staff.

Pathogen screening

Confiscated birds may be the single most vulnerable population to infectious disease and stress-related immunodeficiency. The victims of illegal trade have been subjected to the worst of all captive conditions which include exposure to a wide diversity of taxa (primates, reptiles, domestic poultry, invertebrates just to name a few) and their non-avian pathogens; human sewage; human pathogens of poverty; contaminated and inadequate water, feed, shelter; hypo- and hyperthermia; sleep deprivation; rigors of transport; and extreme physiological and immunological stress. Pathogen detection, treatment and prevention are essential if confiscated parrots are to meet standards for rehabilitation and release and minimize disease risk. Veterinarians are trained in clinical practice in the art of relevant and economical pathogen analysis and can contribute expertise on many levels.

The overall and highly subjective conclusions from publications,[5,6] (personal observations; PJL, oral communication, January 2024) and communications from colleagues, is that trafficked birds are extremely ill, critically and fatally so in many cases (PJL, written communication and literature review in process; Supplemental Materials link at the end of the manuscript.). Multiple infections are common, diverse and resistant to treatment; repeated testing is essential; survival decreases with co-morbidity; chronic, long-term changes are possible.

Data suggest that immediate and aggressive treatment coupled with available testing is essential for the individuals and the future populations in the wild, for public health, and to protect agricultural and pet trade interests (Latas, et al., unpublished data, January 2023).

Fitness for release

Fitness-for-release evaluations are team efforts, but a veterinarian needs to plan the clinical and screening aspects, as well as determine physical, physiologic, and mental fitness for release to the wild.

The role of the veterinarian in policymaking

The psittacine veterinarian can contribute to evidence-based confiscation decision-making by bringing valuable information, data, and advice to those who formulate such. Policymaking involves politicians, councils, legislation, and many levels of engagement. Veterinary communication with decision-makers is imperative to protecting the health, welfare, and well-being of confiscated wild psittacine birds.

The Basics

Evaluation and examination

Physical evaluation of the wild psittacine bird differs from that of a patient in the clinical setting. Pre-planning and a well-rehearsed team are essential to rapid and efficient examination of the wild psittacine patient. Biometrics, marking/banding and other non-veterinary procedures may add to the workload. Completing a thorough physical examination, taking numerous samples for clinical pathology, parasitology, microbiology and pathogen/gender PCRs and other testing has to be done quickly and efficiently. Sedation and anesthesia are not recommended unless painful or prolonged procedures are planned, such as imaging and microchip implants. The average restraint time should not exceed 5 min.

The non-invasive visual examination may be the only option in some cases and is well worth being the first step. Using "stand back and observe" to evaluate external appearance and general behavior, "sick bird syndrome," eating, drinking, perching, ambulation and flight, and droppings can gather important information without undue stress. Thermal imaging or UV lighting are additional useful hands-off techniques.[7-16]

Quarantine and isolation of patients

Quarantine is absolutely necessary. An important and often overlooked consequence of quarantine is isolation, from not only contagion, but from companionship and social structure. Confinement of the individual bird suspected of, or diagnosed with, treatable zoonotic disease in isolation is necessary yet can be a severe psychologic injury to a tightly pair- or family-bonded parrot. Many psittacine birds actively care for injured and ill members of the flock. In most confiscations, more than one individual will be involved, and an effort can be made to house the birds together or at least in close visual proximity. This fundamental need for companionship must not be ignored and requires careful planning.

Treatment

Early and aggressive intervention and protocols emplaced by experienced veterinarians can save lives and aid successful release. Appropriate treatment regimens should use the shortest and most efficient timeline for effective treatment and recovery. Long-acting or time-released antibiotics, pain management and other treatments are options to reduce restraint, handling and distress in wild birds.

Common Physical Injuries and Illnesses

Confiscated birds represent a unique population. The stress associated with capture, confinement, transport, and various smuggling techniques (including hiding birds and drugging them) puts them at imminent risk of severe immunodeficiency. Filthy, toxic, unhygienic conditions predispose them to infections from pathogens of human and domestic animal origin. Combined with poor, non-existent or inadequate nutrition, severe overcrowding and human harassment, the majority of birds die before they can be confiscated. The survivors are usually severely ill and show multiple and repeated traumatic injury. They are the sickest birds a veterinarian is likely to see in their career.

An overlooked but serious injury is the psychological damage they suffer from the entire poaching process, which can ultimately lead to fearful, aggressive, timid or other maladapted states impairing social development and eventually may affect the success of integration to a flock and successful release.

Birds that have been confiscated go through many stages prior to the authorities removing them, and sometimes during the authorities' holding. Physical trauma can result from almost any aspect of capture, trafficking, exhibit, sale, or recovery operations and all transitions between. Psittacine birds are wild, intelligent and powerful and their efforts to escape or avoid trauma add to the mix.

How injuries happen

Capture injury

Traumatic injuries result from the physical action of snares, traps, bird lime and resin "sticky glue" traps, nets, projectiles, removal of young and adults from nests, felling of trees in which nests are found, and many more. These capture methods often result in fractures of wing and leg bones, luxations, soft tissue avulsion/crushing/laceration/maceration, myopathy, hypoglycemia and exhaustion. Psittacine birds fight back and often are not handled with compassion at capture.

Smuggling and trafficking

Smuggling techniques are myriad and creative; but birds usually show the signs of severe confinement (in tubes, bottles, bags, etc.) including fractures, luxations, bruising, suffocation or respiratory injury. In addition, they may have been dosed with inappropriate or toxic sedatives (acepromazine, diazepam, diphenhydramine, chloral hydrate, and others). Typical are compression injuries to the body; internal organ damage; fractures of wing and leg bones; luxations; soft tissue avulsion, crushing, laceration, or maceration; myopathy; hypoglycemia; and exhaustion.[17,18]

Cage trauma

Confiscated birds very likely were housed inappropriately, in overcrowded conditions and without regard to cage safety issues. There may be chew marks on the beak that indicate escape attempts and severe lesions of the feet and nails. Typical injuries include those resulting from entanglement of the head or feet; hanging or strangulation; crushing by cagemates; intense soiling; fractures of wing and leg bones; luxations; soft tissue avulsion, crushing, laceration, or maceration; myopathy; hypoglycemia; and exhaustion.

Transport injury

Captured birds must be transported; and trafficking, especially smuggling activity, involves many dangers associated with confinement and transport. Psittacine birds may be moved on foot, horseback, moto, vehicle, boat, or aircraft. Little regard will be paid to their welfare or viability (there are recorded instances of smuggling confined birds within tires to cross inspection stations). Poachers apprehended during water transport may throw birds overboard. The veterinarian can expect hyper- and hypothermia, respiratory distress and damage from fumes, carbon monoxide and/or dioxide and dust inhalation, lesions of near-suffocation, fractures of wing and leg bones, luxations, soft tissue avulsion/crushing/laceration/maceration, myopathy, hypoglycemia and exhaustion.

Confiscation injury

Birds may be harmed during apprehension and confiscation activities, and during holding periods prior to release to proper facilities. Most large-scale counter-trafficking operations have few psittacine-trained personnel, thus accidental or inadvertent damage to the birds may result. Recapture during confiscation operations may happen if birds need to be collected from holding pens or areas, and the same injuries as seen in the original capture from the wild often result from inexpert techniques during such operations. Expedient or emergency transport to appropriate facilities may lead to transportation injuries.

A preventable problem is the need for publicity or publicity-directed "selfie syndrome" that goes beyond collecting data, evidence and curation. Prolonging stress and delaying triage and treatment can lead to unnecessary morbidity and mortality. The level of these practices need to be clearly delineated and managed at the planning stage of anti-trafficking operations and during collaborative planning with wildlife rehabilitators. The veterinarian should be prepared to have a voice in this.

General traumatic injury

Fractures: Multiple long bone and facial/skull fractures are common in confiscated birds, generally accompanied by massive soft tissue injury and internal organ damage. Euthanasia at intake is the most humane action. Parrots with single fractures or external and superficial injury can recover, with the caveat that they MUST be able to fly and have normal prehension of food. Some parrots with head injury can recover to normal functionality and should be given the benefit of the doubt (for more information and suggestions see the Supplemental Materials link at the end of the manuscript for Flowchart: Trauma).

Extensive or complex traditional orthopedic repair should not be considered, as the eventual return to full flight and successful release is unlikely in the psittacine species. It is inhumane to hold a wild bird for invasive and painful procedures, over an extended time period, only to have loss of flight capability at the end and euthanasia.

Entanglement: Entanglement is a broad term for situations that entrap and immobilize the victim. Scenarios include inappropriate caging; strings/wire/rope/fiber/line deliberately tethering the bird or wrapping it in bondage or acquired accidentally; netting entrapment; glue traps; bird lime; snares and traps, and many other situations. Entanglements are often much worse than they look on superficial examination, as soft tissues, bones and joints can sustain devastating injury. Some damage, for example to the esophagus and crop, and from strangulated tissues, may not appear for days. Pulmonary and cardiac issues are likely. Edema may result in circulatory compromise with secondary skin slough or necrosis of skin and deeper tissues. Extensive loss of flight feathers, tail feathers and contour feathers can occur and may be associated with deeper damage, not immediately apparent to follicles and other structures.

Projectiles: Projectiles include any object, water or air, that can be aimed with force and cause physical and psychological damage. Confiscated birds may have been collected in the wild using pellets, bb's, bullets, shotgun (firing into a flock and collecting all the survivors, similar to dynamiting a reef), darts, arrows, spears, rocks, slingshot ammunition, and so on. Forcible water streams from garden or high-pressure hoses can inflict serious injury. Compressed air or other gases are also used to collect birds. Projectile injuries are often much worse than they look and are also commonly associated with secondary fall injuries. Pressure waves and rebound effects can be serious, even with "low velocity" weapons such as bb and pellet guns. Introduction of foreign material and detritus into wounds is likely, even from tiny-appearing injuries.

Birds that survive projectile attack will need extensive after care, antibiotics, and multiple procedures. Mortality is formidable even under the best of circumstances. Projectile injury to the wing bones rarely indicates recovery of flight. Euthanasia must be considered as a first option.

Burns: Thermal burns may result from inexperienced people feeding nestlings and juveniles formula or food that is too hot or wildfires that have been deliberately set for poaching. Tissue damage may not be apparent for days or weeks. Crop burns are usually extensive but not discovered for days to weeks and often require humane euthanasia. Superficial burns to the feet may be treatable and compatible with release. Ocular and respiratory damage may require euthanasia as a first option. Singed feathers may require an entire molt cycle (possibly 2 years in some species) if imping is not available, and removal by "pulling", months after complete healing of the integument, may be the only option compatible with potential release (please refer to the Supplemental Materials link at the end of the manuscript: Imping). Externally local-appearing burns can be given a week to show further deterioration or no worsening.

Chemical burns may be associated with capture techniques, such as resins or other bird-lime type trapping, or may be the result of inappropriate transport or smuggling and are a challenge to diagnose, treat and prognosticate especially as what substances are involved may not be known. Chemical burns can involve the pedal surface of the feet or extensive whole-body exposure. Vapor or fume burns to respiratory tissues from chlorine, diesel exhaust, petroleum exhaust and so on are usually fatal; if the bird survives, euthanasia must be considered as a first option as recovery to flight ability with respiratory scarring is unlikely.

Intentional bleaching: To obtain higher prices for trafficked psittacine birds, dealers often engage in the practice of bleaching and dyeing feathers to mimic more valuable, rare, or desired species. The cruel techniques include use of hydrogen peroxide-based hair bleaches, chlorine-based bleaches, or ammonia to remove pigment. Food coloring, hair coloring or other pigments are then applied. In the process, the birds suffer severe stress; chemical burns over the face, head, neck, wings and body, often where the bleaching products drip; and irreversible respiratory damage (please refer to examples of Bleaching in the Supplemental Materials link at the end of the manuscript).

Falls: Nestlings and fledglings are the most common victims of fall injuries. Adults fall secondary to other capture impact or injury. The severity of injury depends on distance fallen, substrate, and plumage development. External bruising, internal injury (usually liver contusion or fracture), cardiac and pulmonary contusions, ruptured crop, head and facial injury, and fractures are common. Young birds should be very carefully examined as imaging may be inconclusive and bones may sustain greenstick-type fractures that are not palpable. Hemorrhage and bruising may take 48 hours to become apparent. Young birds are likely to recover completely from fall injury if they survive the first 48 hours in care. If a fall is associated with capture technique in an adult bird, survival or fitness for release are unlikely.

Captivity and inappropriate care

Due to deplorable conditions prior to confiscation, most birds will be suffering the effects of captivity and inappropriate (if any) care. The conditions range from the acute and life-threatening, to the chronic and fitness-reducing. All birds from a confiscation will be exhibiting these conditions to some extent, and many may require humane euthanasia because of the resulting inhumane levels of suffering.

Dehydration

Over-crowding and inappropriate caging during captivity will virtually guarantee that water availability to all birds will be curtailed or absent. Consequently, many birds in a seizure will be clinically dehydrated and some will die as a result. Every individual should be carefully evaluated and triaged accordingly. An additive problem is that water that is available may be heavily contaminated with pathogens; toxins; and illicit antibiotics, sedatives and other treatments which render it unpalatable. The veterinarian must be prepared to treat dehydration immediately with parenteral fluids, tube-feeding, and other means.

Malnutrition

Due to struggle and exhaustion, many birds will be hypoglycemic and initial fluid therapy for any debilitated birds should be accompanied by 50% dextrose orally. Confiscated birds may not have been fed at all, or may have food withheld during smuggling operations, or not have access to food because of competition, inappropriate feeding bowls or presentation of food (they may not recognize food bowls). For example, psittacine birds are routinely offered wild bird seed or local available grains, which they do not recognize as food or which are nutritionally incomplete or hazardous (as in offering nectivorous lorikeets a bird seed mix). Nestlings and juveniles may be hand- or syringe-fed inappropriate formulas, and in some cases human masticated food is offered to these young birds. Food may be contaminated by pathogens, feces, contaminated water, rotten fruit, mold. Gastrointestinal infections, parasites, and stasis, diarrhea or obstruction should be expected. Birds may be presented with profound emaciation. The veterinarian must assume all birds are suffering from malnutrition of some sort and must correct nutritional issues immediately, as well as maintain proper husbandry.

Metabolic bone disease

Physical or radiographic evidence of bone deformity is justification for euthanasia. Birds that were held in captivity for long periods from the neonate stage should be considered potential victims of metabolic bone disease and an effort to rule the syndrome out should be attempted.

Trimmed feathers

Control of birds during smuggling and at bird markets may require that wing and tail feathers be trimmed. Long feathers may be damaged from soiling and overcrowding, disease, and human interactions. Ex-pet birds may have trimmed flight feathers. Damaged feathers impair fitness for release for up to 2 years and will either need to be pulled or allowed a full molt cycle to regenerate. The process of imping has the potential to speed the process and maintain flight fitness during rehabilitation. Please refer to Imping in the Supplemental Materials link at the end of the manuscript for recommendations.

Pinioning and intentional wing fractures or amputation

Psittacine birds may suffer deliberate fracture of wings to curtail escape from captivity. Amputation of the distal wing and pinioning (amputation of the distal wing to remove the primaries) can result in infection, pain, decreased mobility and severe debilitation. None of these procedures are compatible with release.

Tethering

Psittacine birds are usually tethered by a foot, but collar, wing, or body harness may also be used. Tethering may result in irreversible damage to a limb, but the bird may still be a candidate for rehabilitation if flight is intact and general mobility and prehension are functional.

Infectious disease

Confiscated birds are exposed to infectious pathogens at every step of trafficking. They are the victims of and the source of many potential zoonotic diseases. The pathogens of concern may not be confined to typical avian diseases but can also originate from and affect humans, domestic animals, and other wildlife. Pet and aviary birds can be a source of infection. Wild psittacine birds have been shown to be remarkably free of pathogens,[19] but birds which have spent time in captivity or which have originated from captive situations can harbor infectious pathogens.[20,21] The pet trade and animal trafficking are often the nidus of epizootics,[22] thus veterinarians must identify potential threats and reduce risk to wild populations. Please see *Flowchart: Infectious Disease* in the Supplemental Materials link at the end of the manuscript for more details and suggestions.

Some non-psittacine pathogens may be present in confiscated parrots, since they are often housed and transported with, or in proximity to, primates, reptiles, domestic poultry, invertebrates (just to name a few). Veterinarians should not be hesitant to test for organisms rare or unknown in psittacine birds; for example, toxoplasmosis, *Giardia* species, *Cryptosporidium* species, *Tritrichomonas*-like protozoans, water-borne parasites (including various protozoans, flukes, and nematodes), enteric mammalian bacteria, and rodent- and arthropod-borne pathogens.

Chlamydia psittaci may be the most feared zoonotic disease, as it has a wide range of host susceptibility and pathogenicity including humans, domestic poultry and livestock, and avian and wildlife taxa. Confiscated psittacine birds should be considered suspect and need to be tested immediately upon arrival. Other potential zoonotic organisms include *Mycobacterium* spp., dermatophytes, avian influenza, and certain protozoans, all of which are favored by overcrowding and close proximity to diverse species, as well as human poverty, poor husbandry and unhygienic conditions, and sub-minimal nutrition.

Pathogens which may be acquired from or infect poultry and aviculture include avian influenza and Newcastle's disease (paramyxovirus PMV-1), avian bornavirus, psittacine feather and beak (circovirus) disease, Pacheco's disease (avian herpesvirus), polyomavirus, and numerous parasites are a risk to other psittacine birds.

Salmonella sp., *E. coli*, *Pasteurella* sp., *Klebsiella* sp., food- and water-borne pathogens are acquired in captivity and are associated with sub-optimal or filthy conditions. Psittacine birds are highly susceptible to these infections, and microbiological culture with antimicrobial sensitivity testing should be an obligatory screening procedure and a condition of quarantine.

Non-hygienic conditions in close quarters with humans and domestic animals can expose the parrots to myriad pathogens to which they have no prior immunological knowledge. Stress-related immunosuppression will guarantee that a large percentage will succumb or be sickened. The veterinarian needs to be prepared for every bird to be in critical condition and septic on presentation. Microbial screening is essential, but ill birds need immediate emergency treatment with parenteral antibiotics such as piperacillin, amikacin, ceftazidime, cefotaxime, and/or piperacillin/tazobactam. Implicit is the caution that the birds and their fecal material, cages, etc. may be contaminated with infectious human pathogens and numerous parasites are a risk to other psittacine birds.

Confiscated birds should be placed in quarantine on admission, and aggressive treatment should start if birds exhibit typical clinical signs and/or when test results indicate.

Environment, weather and climate crisis

Forage disruption

Anthropogenic impact (i.e., via habitat destruction or development) or natural disasters can disrupt reliable and dependable forage sites. Birds may be faced with starvation or minimal nutrition when completing long journeys between sites and have diminished reproductive capability. Water resources and roosting sites can also be devastated.[23]

Loss of food sources

Weather or climate events may impact the usual food sources, and the climate crisis may impact agricultural, urban and wild landscapes. Psittacine species are known to seek urban areas as refugia and for food opportunities; changing or variable climate will put further pressure and competition on urban species.[24]

Urban development and human incursion

Climate crisis will ultimately impact urban areas and will cause expansion of human habitat to rural and wild areas. Anthropogenic habitat disruption will greatly alter the scope of urban psittacine interactions.[25]

Human incursion can include agriculture, housing, tourism, and other activities that negatively impact the environment. This includes the removal of native food sources and proliferation of exotic and cultivated species of potentially lesser nutritional value.

The egg

If eggs cannot be saved or incubated, follow humane options for embryo euthanasia. If eggs are viable and there is appropriate planning and equipment to incubate, hatch and release to the wild. Authorities and rearing facilities must be prepared well in advance to institute agreed procedures.

Prognostication

Prognostication for confiscated psittacine birds entering a rehabilitation facility can be challenging because the prognosis for these patients is based on the likelihood of successful release. Sadly, most birds are fixable, but the question to be answered should be: Will this bird not only be healed of this injury and illness, and not just survive in the wild, but will it also thrive? The prognosis for return to the wild includes raising a family, finding food and water, keeping up with the flock, and not suffering chronic pain and disability.

Guidelines

Level of suffering

How much is the parrot suffering? Will the pain be minimized with reasonable treatment and care? Chronic conditions have a very poor prognosis for a successful return to flock functionality.

Flight

Will the bird fly? Prognosis of condition and response to treatment is null if the bird cannot recover flight.

Functionality

Will the bird return to normal functionality? Normal functionality is a spectrum, and prognosis can be subjective. If there is flight, appropriate feeding behavior, and ability to groom self and others, prognosis can be favorable even in the less-than-perfect patient. If the bird cannot possibly obtain release qualifications, prognosis is null. If the bird improves in rehabilitation, it may well reach those goals; these birds have a guarded but positive prognosis.

Age

How old is the bird? Healthy young birds usually have a good prognosis for release. Healthy adults likewise have a good prognosis. Tiny neonates have a poor-to-guarded prognosis since most came into human hands under dire conditions.

Injury

What is the severity of injury? The severity of injury and response to treatment guide prognosis for release. It is more humane to make decisions early in the course of diagnosis and treatment.

Viable alternatives and options to euthanasia

Appropriate sanctuaries

There are almost no appropriate sanctuaries for wild psittacine birds anywhere in the world. Where such sanctuaries exist, they are unlikely to have space for additional admissions.

Endangered species

Endangered and protected species may be held to different euthanasia guidelines, depending on local, country, and international law. Alternatives for long-term care may exist and be mandated.

Conclusions, Suggestions, and Resources

The plight of parrots in the wild is dire. A majority of the 400+ species are under direct assault from anthropogenic impact, and wildlife trafficking is the greatest threat to the majority of species.[26-29] Wildlife crime law enforcement has an impossible task to curtail the trade, and confiscations of smuggled birds are ever more numerous, complex, and urgent. Dedicated and well-trained veterinarians are desperately needed for collaborative efforts to render appropriate care, guide decisions, inform and maintain highest welfare standards, and assist with placement of the unfortunate victims.

The first step is awareness of the problem. The next steps on the journey are acknowledgment, education, training, networking, collaboration, and eventually release to the wild. You, as an informed and competent veterinarian, can make a huge difference in the life of even a simple individual parrot, and you can influence the future of the entire species.

The author has made online handbooks available for veterinarians and rehabilitators working with parrots.[30,31] Supplemental materials and helpful links may be found at the following:

Supplemental Material for AAV Confiscated Parrots Masterclass 2024: Wild Parrot Coalition Bookshelf

Acknowledgments: I am deeply indebted to the many caregivers, rehabbers and veterinarians across the globe who dedicate themselves to this horrendous work. I gather strength from their determination, knowledge and wisdom and deep compassion. Thank you to my dear collaborators, especially Brooke Durham at SoCal Parrot and Nikki Buxton at Belize Bird Recue, for teaching me so much and allowing access to your mountains of data. And lastly and most importantly, I acknowledge all those parrots who have given their lives to human greed and folly, and the brave survivors who have made it back to the wild.

References

1. CITES Secretariat. *World Wildlife Trade Report 2022*. Geneva, Switzerland: Secretariat of the Convention on International Trade in Endangered Species of Wild Fauna and Flora (CITES); 2022.

2. Mitsilegas V, Fasoli E, Giuffrida F, Fitzmaurice M. Environmental Crime at the International Level: Criminalisation of Illegal Trade in Wildlife under the Convention on International Trade in Endangered Species of Wild Fauna and Flora (CITES). *The Legal Regulation of Environmental Crime.* Leiden, NL: Brill; 2022:7-54.

3. van Uhm D, South N, Wyatt T. Connections between trades and trafficking in wildlife and drugs. *Trends Organ Crim*. 2021;24:425-446. doi: 10.1007/s12117-021-09416-z.

4. Jakob-Hoff R, MacDiarmid S, Lees C, et al. In: *Manual of Procedures for Wildlife Disease Risk Analysis. World Organisation for Animal Health.* Paris, FR: International Union for Conservation of Nature and the Species Survival Commission; 2014.

5. Hidasi HW, Neto JH, Moraes DM, , et al. Enterobacterial detection and *Escherichia coli* antimicrobial resistance in parrots seized from the illegal wildlife trade. *J Zoo Wildl Med*. 2013;44:1-7.

6. Lopes ES, Cardoso WM, Nishi DM, et al. Serogroup identification, phenotypic detection of hemolysis and extended spectrum beta-lactamases of *Escherichia coli* isolated from psittacine of illegal wildlife trade in Fortaleza, Brazil. *Arquivo Brasileiro de Medicina Veterinária e Zootecnia*. 2018; 70:823-829.

7. Denny G, Khanna R, Hornstra I, et al. Rapid detection of fungal elements using calcofluor white and hand-held ultraviolet illumination. *J Am Acad Dermatol.* 2020;82:1000-1.

8. Jerem P, Herborn K, McCafferty D, et al. 2015. Thermal imaging to study stress non-invasively in unrestrained birds. *J Vis Exp*. 2015;6:e53184.

9. Jerem P, Jenni-Eiermann S, Herborn K, et al. Eye region surface temperature reflects both energy reserves and circulating glucocorticoids in a wild bird. *Sci Rep*. 2018;8:1907.

10. McCafferty DJ. Applications of thermal imaging in avian science. *Ibis*. 2013;155:4-15.

11. Mota-Rojas D, Pereira AM, Martínez-Burnes J, et al. Thermal imaging to assess the health status in wildlife animals under human care: limitations and perspectives. *Animals*. 2022;12:3558.

12. Noh JY, Kim KJ, Lee SH, et al. Thermal image scanning for the early detection of fever induced by highly pathogenic avian influenza virus infection in chickens and ducks and its application in farms. *Front Vet Sci.* 2021;8:616755.

13. Prosser DJ, Collier T, Sullivan JD, et al. Using thermal infrared cameras to detect avian chicks at various distances and vegetative coverages. *J Fish Wildl Manag*. 2020;11:245-257.

14. Rzucidlo CL, Curry E, Shero MR. Non-invasive measurements of respiration and heart rate across wildlife species using Eulerian video magnification of infrared thermal imagery. *BMC Biol*. 2023;21:61.

15. Barreira AS, Lagorio MG, Lijtmaer DA, et al. Fluorescent and ultraviolet sexual dichromatism in the blue-winged parrotlet. *J Zool*. 2012;288:135-142.

16. Joyner L, Ramcharan S, Lingaard M. Guide to Understanding and Reducing Harm to Parrots in Suriname. Available at: https://www.oneearthconservation.org/_files/ugd/d204d4_309fef0f70d4496893889975bc30b257.pdf. Accessed March 1, 2024.

17. Martin RO, Senni C, D'cruze N, Bruschi N. Tricks of the trade—legal trade used to conceal Endangered African grey parrots on commercial flights. *Oryx*. 2019;53:213-217.

18. Blanco K, Niehaus C, Salinas A, et al. Apparent lack of evidence on selected infectious agents in wild Yellow-naped Amazon parrots: implications for releasing attempts. *Rev Ciencias Veterinarias* 2013:31:7-17.

19. Burton O, Brightsmith D. Testing for *Salmonella* spp. in released parrots, wild parrots, and domestic fowl in lowland Peru. *J Wildl Dis* 2010;46:718-723. doi: 10.7589/0090-3558-46.

20. Xenoulis P, Gray P, Brightsmith D, et al. 2010. Molecular characterization of the cloacal microbiota of wild and captive parrots. *Vet Microbiol* 2010; 146:320–325. doi: 10.1016/j.vetmic.2010.05.024.

21. Marković I. Wildlife trafficking as catalyst of (COVID 19) pandemic. *Archibald Reiss Days*. 2021;26:11.

22. Intergovernmental Panel on Climate Change 2018. Summary for Policymakers of IPCC Special Report on Global Warming of 1.5 C° approved by governments. Available at: https://www.ipcc.ch/site/assets/uploads/2018/11/pr_181008_P48_spm_en.pdf. Accessed January 10, 2024.

23. Paravisini-Gebert L. The parrots of the Caribbean facing the uncertainties of climate change. *Revista*. 2018;13:24-27.

24. Díaz S, Settele J, Brondízio E, et al. 2019. Summary for policymakers of the global assessment report on biodiversity and ecosystem services of the Intergovernmental Science-Policy Platform on Biodiversity and Ecosystem Services 44. Available at: http://www.mari-odu.org/academics/2018su_Leadership/commons/library/Summary%20for%20Policymakers%20IPBES%20Global%20Assessment.pdf. Accessed January 10, 2024.

25. Aloysius SL, Yong DL, Lee JG, Jain A. Flying into extinction: Understanding the role of Singapore's international parrot trade in growing domestic demand. *Bird Conserv Int*. 2020;30:139-155.

26. Berkunsky I, Quillfeldt P, Brightsmith DJ, et al. Current threats faced by Neotropical parrot populations. *Biol Conserv*. 2017;214:278-87.

27. Chan DT, Poon ES, Wong AT, Sin SY. Global trade in parrots–Influential factors of trade and implications for conservation. *Glob Ecol Conserv*. 2021;30:e01784.

28. Nandika D, Agustina D, Heinsohn R, Olah G. Wildlife trade influencing natural parrot populations on a biodiverse Indonesian Island. *Diversity*. 2021;13(10):483.

29. Wild Parrot Coalition. Wildlife Rehabilitation for Confiscated Psittacines. Available at: https://www.flipsnack.com/wildparrotcoalition/wildlife-rehabilitation-for-confiscated-psittacines-xtkfhzoayr/full-view.html. Accessed January 5, 2024.

30. Wild Parrot Coalition. Wild Psittacine Rehabilitation and Appendix. Available at: https://www.flipsnack.com/wildparrotcoalition/appendix-wild-psittacine-rehabilitation/full-view.html. Accessed January 5, 2024.

31. Wild Parrot Coalition. Wild Urban Psittacines Veterinary Handbook. Available at: https://www.flipsnack.com/wildparrotcoalition/wild-urban-psittacines-veterinary-handbook/full-view.html. Accessed January 5, 2024.

Raptor Management: Admission to Discharge

Michael P. Jones, DVM, DABVP (Avian Practice),
John Chitty, BVetMed, CertZooMed, CBiol, MRSB, MRCVS,
Scott Ford, DVM, DABVP (Avian Practice)

Session #169

Affiliation: From American Eagle Foundation, 510 Kodak Rd., Kodak, TN 37764, USA (Jones), Wombourne, Allington Track, Allington, Salisbury, SP4 0DD, UK (Chitty), Avian Specialty Veterinary Services, 519 S 68th St., Milwaukee, WI 53214, USA (Ford).

Abstract: This class provides an overview of the medical management of birds of prey, both free-ranging and those in managed care, from admission to discharge or release. Topics include regulatory and legal concerns, admission and triage, sample collection and diagnostic testing, nutrition, selected infectious and non-infectious diseases, husbandry, rehabilitation, release criteria and conditioning, and final disposition options for raptors (i.e., discharge, euthanasia, release, or permanent placement in managed care).

Regulatory and Legal Concerns

Free-ranging raptors face significant threats including habitat alteration or destruction through agricultural practices, intentional take, intentional or unintentional poisonings and even climate change.[1] In many countries and regions around the world, raptors are protected species with regulatory and legislative activities directed at protection from various threats they face around the world. In many instances, international treaties or memorandums of understanding (MoU) are established between countries to mitigate threats to the raptors and their ecosystems.[1] Examples include the following:

1. United States Fish and Wildlife Services (USFWS) Migratory Bird Treaty Act (MBTA) of 1918 (16 U.S.C. 703-712).[2]

2. Memorandum of Understanding (MoU) on the Conservation of Migratory Birds of Prey in Africa and Eurasia (the "Raptors MoU") concluded, signed and came into effect in 2008.[3]

3. Convention on International Trade in Endangered Species (CITES) of Wild Fauna and Flora is a global agreement among governments to regulate or ban international trade in species under threat.[4]

Raptor legislation in the United States

In addition to the Migratory Bird Treaty Act, other key regulations that are important to know about include:

1. Endangered Species Act (ESA) in 1973 (16 U.S.C. § 1531 et seq.).[5]

2. 50 Code of Federal Regulations (50 CFR, especially Parts 17, 21 and 22).[6]

3. Bald and Golden Eagle Protection Act of 1940 subsequently amended in 1962 to include golden eagles (16 U.S.C. 668-668d).[7]

4. Lacey Act of 1900 (18 U.S.C. 42).[8]

5. Animal Welfare Act: United States Department of Agriculture (USDA) and the Animal and Plant Health Inspection Service (APHIS) developed new welfare standards for birds that were published in the Federal Register on February 21, 2023.[9,10] Currently, raptors used in falconry are exempt.

6. States, tribes or territories may have regulations that are as strict or more stringent than Federal regulations, and each State, tribe or territory's regulations must receive Federal approval.

Raptor legislation in the United Kingdom

Free-ranging raptors

All United Kingdon (UK) wild raptors are protected and may not be killed, persecuted or taken from the wild as eggs or birds.[11] In terms of rehabilitation in the UK,[12] a useful summary is found at http://www.raptorrescue.org.uk/wp-content/uploads/2014/06/part8-Law-v2.2.pdf.

Raptors in managed care

There is minimal registration of raptors required; only nine species require registration. One of these, the peregrine falcon (*Falco peregrinus*), does not require registration if captive bred, and they are rung or microchipped with the keeper holding an Article 10 Certificate. The Convention on International Trade in Endangered Species of Wild Fauna and Flora (CITES) species require an Article 10 certificate if traded, exchanged or used for public display. Raptors used for public display require licensing under the Zoo Licensing Act or Licensing of Activities Involving Animals legislation (LAIA).

Captive animal welfare is regulated under the Animal Welfare Act and there are no formal welfare guidelines (other than for zoos), although there are several voluntary codes. Hunting with birds of prey is unregulated other than requiring permission from a landowner to hunt on their land, and the prohibition to hunt protected species.

Admission, Handling and Restraint, and Triage

While trauma is the most common reason that free-ranging raptors are presented to veterinarians or rehabilitators for care, other conditions such as toxicosis, starvation, infectious diseases and even orphaned raptors may present for veterinary care. Often the trauma to free-ranging raptors is so significant that important and challenging decisions need to be made during initial triage. As veterinarians, we must advocate for the health and well-being of our patients; however, the severity of injury and regulations may dictate what should or should not be done when receiving and endangered or threatened species (50 CFR § 21.76 (e)(5)(i-iii)) and when deciding treatment or euthanasia options (50 CFR § 21.76 (e)(4)(iii-v).[6] If the veterinarian does not have a rehabilitation permit, most states require the veterinarian to transfer the raptor to a rehabilitator within 24 hours after the raptor is considered stable.

Handling and restraint

Care should be taken to minimize the amount of stress and discomfort the raptor experiences during restraint while allowing procedures to be completed. Gloves, towels or blankets, and even protective eye wear or thick clothing are essential. Nonessential but extremely useful equipment includes hoods, stockinettes, self-adherent bandaging, casting jackets, or abbas.

When capturing a free-ranging raptor an appropriately sized towel or blanket is quickly placed over the raptor, and the handler will then move to secure the body and legs followed by the head. Manual restraint for most raptors normally requires one handler; however, for larger species (i.e., eagles, Old World vultures) two handlers may be required. To protect personnel from being footed, self-adherent dressing may be wrapped around each clenched foot during prolonged restraint. The foot is clenched around several gauze sponges during wrapping to prevent self-induced talon trauma.

Good communication between the handler(s) and others performing the procedures (e.g., physical examination, radiographs, and treatments) is essential to minimize the potential for serious injury to all involved. Never take the raptor, no matter how ill it appears, for granted.

Trained birds (i.e., raptors used for falconry and educational ambassadors) may be cast directly from the falconer's glove with a soft towel or blanket. When handling these birds, the veterinarians must make every effort to do no harm to the feather or feet. Falcons are notorious for grabbing their own feet if they are not properly separated during restraint. In some cases, the majority of a physical exam can be performed without casting or restraint. Many of these birds will generally perch on a scale for weighing and your expectation of this from the falconer or handler will raise their estimation of your knowledge and professionalism. Untrained birds will need to be weighed in a container or can be gently wrapped in a towel.

Clinical examination

The clinical examination of a bird of prey should consist of three parts: the observation of the patient, examination of its environment where possible and the actual physical examination. The major objective of raptor rehabilitation is to restore an animal to its former healthy state, prepare and condition it physically and mentally, and then release it at a suitable time and place. Therefore, the goal of the clinical examination is to allow the veterinary team to thoroughly evaluate if the raptor is an acceptable candidate for release. Raptors suitable for release should have fully functional appendages, good visual capability, good athletic ability and appropriate social conditioning. Every raptor should receive a complete ophthalmologic examination on admission, to include a fundic examination. Any raptor that is unable to mentally or physically meet the demands placed upon it in order to survive should not be considered for release. Regarding euthanasia, veterinarians are allowed to euthanize raptors to end suffering without any additional permits; however, veterinarians should make every attempt to obtain permission to euthanize any threatened, endangered eagle species (bald and golden eagles).

Hospitalization

Proper hospitalization of raptors in a busy practice can be difficult; especially if there are no rehabilitators in the near vicinity to assume care of a convalescing raptor. Generally, raptors should be housed in an appropriately sized cage in such a way as to minimize stress. Consideration should also be given to appropriate perch size and shape, ventilation and protection from other predators, including other birds of prey. In this situation isolation is acceptable. The cages should be easily cleaned and disinfected, and preferably not stacked above one another. Regulations governing permanently captive raptors in captivity require that housing minimizes the risk of injury, disturbances, protection from extreme weather and chance of injury.

Raptor Clinical Pathology

Minimum database

Depending on resources allocated for care of injured, sick and debilitated raptors, a minimum database may include a physical examination and body condition score, diagnostic testing (i.e., complete blood count or packed cell volume and estimated white blood cell count/differential, and a biochemistry panel), radiographs, a fecal examination, and ancillary diagnostic tests as indicated. Jones and Chitty (2020)[13] provide a detailed description of clinical pathology and diagnostic testing in birds of prey, including serologic testing and laboratory resources information is available elsewhere.[13]

Sample Collection

Restraint

Chemical restraint is rarely necessary for sample collection in raptors. If the animal is unstable, in a critical state, or the clinician is inexperienced in sampling smaller species, then sedation or mask induction of anesthesia using isoflurane or sevoflurane may be suitable for short periods for sample taking. The authors refer you to your favorite formulary for dosing and frequency. Anecdotally, it may be necessary to reverse some raptors that may be overly sensitive to medications such as midazolam.

Venipuncture sites

Talon clip

Talon clip is considered inappropriate in all species as it is only likely to obtain a small, contaminated sample, especially with respect to uric acid levels, and likely to cause excessive pain and damage.

Jugular vein

Positioning: The right jugular is preferred as the left is much smaller in most birds including raptors. The raptor should be restrained, sedated or anesthetized, placed on its left side to allow access to the right jugular vein, and the neck extended. Parting of the feathers reveals the vein running under a large apterium. Placement of a digit at the base of the neck raises the vein.

Advantages: Large volumes can be taken easily. It is generally the first-choice venipuncture site, except in large vultures and eagles. Hemostasis is relatively simple, although large subcutaneous hematoma may form if not careful. This may preclude further use of this vein until the hematoma has resolved.

Disadvantages: Restraint may be stressful to sensitive or dyspneic birds. This site may be difficult for left-handed operators. Handling may be difficult in larger raptors (e.g., eagles or vultures) or in large falcons (e.g., female gyrfalcons) where the muscular neck can be hard to extend. Struggling birds may cause laceration of the vein and fatal subcutaneous hemorrhage. This is very rare and may only occur where there is pre-existing coagulopathy.

Superficial ulnar (basilic) vein

Positioning: The bird is restrained or anesthetized and placed on its back. One wing is extended, and the vein is visualized. The operator raises the vein with his/her free hand.

Advantages: This site might be easier for left-handed operators. Restraint is simpler than for the jugular vein, especially in larger species. It is easier to place and maintain an intravenous catheter for repeated sampling.

Disadvantages: This is a more fragile vein so it may be hard to draw large volumes, especially in smaller birds. Hemostasis can be hard to achieve, and the bird should be replaced in the carrier or bay to allow it to calm down and its blood pressure to drop. This is preferable to prolonged handling. Unlike jugular venipuncture, this hemorrhage is unlikely to be fatal and has the advantage of being easily visible; however, small losses may cause a lot of mess and distress to falconers. A drop of tissue glue may be used to facilitate hemostasis. Occasionally a hematoma may form along with a slight wing droop that may last a day or two.

Superficial plantar metatarsal (caudal tibial) vein

Positioning: The bird is restrained and the leg extended. The operator can raise the vein with their free hand. Chicks can be left in a sitting position with feet in front of them. Restraint is normally minimal; however, in adults the leg must be fixed to avoid operator injury.

Advantages: This site is superficial and simple to visualize and raise. Restraint is easy, especially in juveniles or chicks. In chicks, it is an excellent site for catheter placement.

Disadvantages: In large raptors, this site requires working uncomfortably close to the talons and there is risk of injury if inadequately restrained. It is an excellent site for single venipuncture or catheter placement in chicks and vultures that do not have such powerful feet. In vultures, the venipuncture site must be thoroughly cleaned prior to sample collection to reduce the chances of uric acid contamination. Hemostasis difficulties are similar to those described when using the superficial ulnar (basilic) vein. Tissue glue may be applied, or a light dressing applied after sampling. This should certainly be considered when sampling birds kept in collections on view to the public.

Nutrition and Nutritional Disorders

As in other species, proper balanced nutrition is fundamental to the correct husbandry and performance of the captive raptor. They therefore require a diet balanced in terms of quality, quantity and affordability that should reflect their natural diet and meet their requirements in terms of water, carbohydrates, protein, fat, as well as micronutrients (i.e., vitamins and minerals). In simplistic form, raptors are whole carcass feeders, and the diet should reflect this.

Failure to provide the whole carcass of prey of an appropriate size for the raptor may result in selective feeding and nutritional imbalance. Nutrient requirements vary according to life stage with extra and specific needs during periods of breeding, rearing of young, growth, and performances.

All birds of prey produce castings. These are the regurgitated indigestible parts of the carcass (feather, fur, bones). The volume, appearance and timing of casting varies according to diet fed and, to a lesser extent, the individual bird. Some raptors may cast every day while others may accumulate casting material over several days before egesting the cast. Experienced falconers or trainers should monitor castings very closely as variance from the normal often indicates a digestive upset and may even represent an urgent medical problem.

Nutrients

Energy

Raptors eat, primarily, to fulfil their daily calorie requirements by consuming a diet that is typically low in carbohydrates (2%) and high in fats (2-28%) and protein (17-20%).[14] Of these, fat represents the major source of energy in terms of gross energy (GE) being approximately twice as energy dense (kJ/g) as protein or carbohydrates.

Any shortfall in energy intake will be made up from body stores. In general, this will come from fat stores laid down in times of dietary energy excess. Inactive birds will utilize protein stores (i.e., muscle) in cold weather representing a need to not only increase dietary fat, but also dietary protein in winter. As stated, birds eat to fulfil their daily energy requirements. For raptors this means they hunt when hungry. This is the basis of falconry training or management of raptors in managed care as educational ambassadors: strict weight management to maintain the raptor within an optimal weight range (i.e., flight and hunting weight) to encourage hunting behavior. To achieve this, falconers monitor weight and body condition very closely during the flying season. An experienced falconer will have an excellent knowledge of the individual bird's hunting performance in relation to its body weight and feeding regime. Any variance from this may represent disease and should be taken seriously, meriting in-depth investigation in spite of a bird that often overtly seems well. Overall, the bird is judged by weight gain or loss as expected in proportion to the amount of food it is given, body condition in relation to body weight, behavior and performance at a given weight or body condition, and appetite in relation to body weight or condition.

Weight, condition, and performance: At the start of each flying season the bird's weight may be reduced to an expected (i.e., based on previous seasons) or estimated (i.e., based on other birds of similar size) bodyweight during training. This is a part of the manning procedure. Kirwkood (1981)[15] studied rates of weight loss in starved raptors. Rates of energy loss did not vary significantly between species indicating the same type of tissue being broken down in each case. Rate of loss (g/day) was shown to be $28.27 \times \text{bodyweight (kg)}^{0.723}$ with time to achieve 25% bodyweight loss being $9.7 \times \text{bodyweight (kg)}^{0.276}$ The presence of bodyweight in these equations indicates that larger birds are much more capable of withstanding fasting than smaller birds

Clinically, two syndromes may be seen concerning energy. They are related to excess energy and energy deficiency.

Excess: This is very unlikely in the trained bird during the flying season. However, this tightly regulated feeding contrasts with the feeding of aviary birds that are often fed to excess in spite of doing less exercise! Hepatic lipidosis is relatively common in aviary birds, especially those in permanent breeding aviaries especially in zoological collections. While prolonged calorie excess is a major factor in the development of this syndrome, other factors may also play a part. In particular, Fatty Liver-Kidney Syndrome of merlins is well-described[16] where it is proposed that the feeding of day-old-cockerel to birds that feed predominantly in the wild on insects may result in a higher proportion of fat: protein in the diet and a high proportion of avidin resulting in binding of dietary biotin and reduced hepatic gluconeogenesis. Atherosclerosis in older birds may also be a consequence of long-term excessive dietary energy combined with reduced activity, although genetic factors may also be involved. In both cases, signs may be minimal or absent with birds dying spontaneously when stressed. Certainly, aviary birds should be handled on several occasions throughout the year allowing assessment of bodyweight and condition and, therefore, modification of diet. Serum cholesterol measurements may also be of assistance in screening for potential hepatic lipidosis or atherosclerosis cases.

Deficiency: Energy deficiency is seen commonly in trained birds. The most frequent presentation is that of a thin bird that collapses during or immediately after flight. Diagnosis is by the typical history and blood samples may reveal hypoglycemia. Response to intravenous glucose should be rapid. Many falconers will carry glucose or gluconeogenic compounds (e.g., Liquid Life-Aid, Norbrook) for immediate oral use should this happen in the field. In most cases, history reveals a starved bird. Increased feeding and a short break from hunting will enable weight gain. The bird should then be flown at a slightly higher weight, given suitable breaks during hunting, or fed during hunting trips. In some cases hypoglycemia may be the result of an underlying disease process (e.g., parasitism, malabsorption, hepatic disease). A thorough history and clinical investigation should be recommended for any case of hypoglycemia, even those that seem simple, particularly for those that fail to gain weight with increased feeding.

Protein

Given a good quality whole carcass diet, sufficient essential amino acids should be present and, indeed, deficiencies are rarely seen. Failures to absorb and process protein due to gastrointestinal disease, parasites, or metabolic disease may result in temporary deficiency that are evidenced by stress or fret marks in growing feathers.

Micronutrients

Calcium, phosphorus, and vitamin D: It is impossible to discuss one aspect of these without addressing the others. Vitamin D is required as D_3 and is ingested in both active and inactive forms from prey items. Activation of vitamin D_3 in the raptor requires access to unfiltered sunlight for at least 45 min/day. Inactive vitamin is excreted from the preen gland. It is spread over the plumage and re-ingested during preening. Calcium and phosphorus are readily available from the whole carcass, provided bones can be ingested (i.e., prey is small enough that selective picking does not occur). The ratio of these minerals is vital and should ideally be between 1:1 and 2:1.

Deficiency of calcium or vitamin D_3 (or relative excess of phosphorus) can result in:

- Reproductive failure including soft-shelled eggs, egg binding, egg-related peritonitis
- Nutritional secondary hyperparathyroidism including rickets in young birds. This will most commonly be seen in second or third clutches during a breeding season. Females lay down stores of calcium before the first clutch in the form of increased bone density and intra-medullary hyperostosis. These stores may become depleted with successive clutches and bone deformities may develop in chicks. Otherwise, disease is seen in birds reared in the total absence of unfiltered sunlight (or artificial UV-B light) and those fed on deficient diets during rearing.
- Neurologic signs. Usually in growing birds or breeding females, signs include tremors, weakness or fits. Typical signalment should arouse clinical suspicion and assessment of plasma ionized calcium levels.

Excess calcium in the diet is unlikely to cause a clinical problem as excess will not be absorbed from the diet. Overdosage of activated vitamin D_3 supplements should be avoided as it will result in hypercalcemia and mineralization of soft tissues, especially the kidneys.

Other minerals and electrolytes: Given a good quality whole carcass diet, deficiency or excess of these components is extremely unlikely and clinical syndromes are correspondingly rare. Iron supplementation may, however, be required after blood loss or hemolytic episodes.

Other fat-soluble vitamins: Vitamin A is stored in the liver and, to a lesser extent, fat and should be available in required amounts from almost all whole prey items. Deficiency is seen occasionally especially if only eviscerated carcass or muscle meat are given, or poor-quality food is provided. Signs include the following:

- Weight loss and failure to grow properly
- Poor hatch rates
- Xerophthalmia
- Opacification of the third eyelids and enlargement of the lacrimal glands. This may result in infection and/or keratoconjunctivitis sicca
- Hyperkeratosis of the eyelids
- Squamous metaplasia resulting in lesions or infections of the mucous membranes, especially the mouth and conjunctiva
- Squamous metaplasia of the ureters will result in renal disease
- Vitamin A deficiency has also been implicated in bumblefoot pathogenesis

Diagnosis is by suggestive signs, biopsy, and nutritional analysis.

Therapy involves diversification of the diet and oral supplementation of vitamin A. Injectable vitamin A is rarely indicated in the author's opinion. Pathogenesis of the disease is slow and recovery will be accordingly slow. There is also the risk of overdosage with both oral and injectable preparations although it is less common when given by mouth. Signs of excess are very similar to those of deficiency.

Vitamin E is an antioxidant. It is stored in body fat so, while deficiency should be rare, any spoilage of the carcass will deplete vitamin E stores. Opinions vary on the likelihood of deficiency. Vitamin E concentrations in the tissues of chicks, mice rats and quail are low and deficiency is likely in raptors fed these;[17] however, Forbes and Flint (2000)[18] state that levels in rats and chicks should be adequate. Both sources agree that levels in quail are low unless these birds are fed vitamin-E-supplemented diets. Requirements vary between species with piscivorous birds having a higher requirement. It is essential for fertility red blood cell formation and stabilization of cell membranes. Deficiency will result in white muscle disease (a muscular dystrophy in young birds), reduced libido in males, reduced fertility, reduced hatchability and chick survival and thrift, splayed legs, encephalomalacia, and edema around the neck, wings, and ventrum possibly linked to cardiac lesions of white muscle disease. Vitamin E excess is unlikely to cause clinical signs; however, all fat-soluble vitamins compete for absorption from the gut. Therefore, excess vitamin E may result in deficiency of vitamins A, E or K.

Water-soluble vitamins: There are many B vitamins and dietary deficiency is rare. Good-quality whole carcass feeding should provide these; however, some deficiencies may be seen. Thiamine (B1) deficiency may occur especially in fish-eaters fed thawed frozen fish, where activation of thiaminase will destroy the available vitamin. To avoid this, frozen fish should be thawed in boiling water or (preferably) fresh fish only should be used. In both cases, a B1 supplement should be given. Clinical signs are dramatic, with incoordination progressing to seizure. Classically, affected birds will stargaze. The response to injectable thiamine is rapid. Riboflavin (B2) deficiency has been reported by Forbes and Rees-Davies (2000)[19] who report curled toes in very young chicks fed diets deficient in riboflavin.

Wild-caught prey

Wild-caught prey seems the obvious alternative to free hunting by the bird; however, there are a number of problems associated with this type of feeding whether the prey is shot, caught by the bird itself, or picked up as roadkill. Problems include the following.

- Toxins. Wild-caught prey may contain various potential toxins including alphachlorulose, coumarol rodenticides, and, especially, lead. Certainly, any quarry shot with lead shot should be regarded as a possible source of lead as metal may be scattered through the carcass.

- Infectious agents. Various infectious agents can be contracted from wild-caught quarry. Of particular importance are trichomoniasis (frounce), salmonellosis, *E. coli*, mycobacteriosis, campylobacteriosis, chlamydiosis, paramyxovirus, falcon herpesvirus, avian influenza, as well as various enteric parasites.

Many of these may be contracted from wild pigeon or doves, many of which are symptomless carriers of trichomoniasis, salmonellosis, chlamydiosis, paramyxovirus and herpesvirus. Some advocate feeding pigeon meat after it has been frozen for several weeks then defrosted. While this will remove trichmonads it will not prevent viral infection. Therefore, feeding wild pigeon is not recommended. Most birds are allowed to feed briefly on any quarry they have caught. This needs to be allowed, but the bird should be removed from the quarry at the earliest opportunity and only allowed to continue feeding after a thorough inspection of the carcass. While the quarry may have appeared fit and well in flight, it may be that the hunting raptor selected a weaker individual in hunting.

Nutritional support options for critically ill raptors

There is a significant relationship between poor nutrition and immunosuppression. Therefore, nutritional support is paramount to successfully managing and supporting critically ill or starving patients and those that are convalescing from illness or injury. The aim of nutritional support for malnourished or starving patients is to rehydrate then provide a highly nutritious diet that is easily digestible, quickly absorbed and readily utilized. Emaciated raptors may be offered food on their own, often in the form of small frequent meals, or supported by assisted feeding or gavage of appropriate nutritional supplements. Commercial formulated products are commonly used due to ease of administration (i.e., gavage, esophagostomy tube) and include Lafeber Company's EmerAid Intensive Care Carnivore diet and Oxbow Animal Health's Critical Care Carnivore Diet. At an appropriate time, when gastrointestinal function is normal, the raptor should be offered whole prey sans casting material (i.e., beak, feathers, fur, feet or even bone). Variably sized pieces of meat may also be briefly soaked in water or electrolyte solutions and offered to convalescing raptors. Casting material may be judiciously added to the diet as the bird's condition improves.

Calculating basal metabolic rate (BMR):

$BMR = K \times \text{weight (kg)}^{0.75}$ where $K = 78$ for non-passerine species.

Example: The BMR of a female 800-g red-tailed hawk with severe trauma would be $BMR = 78\ (0.800^{0.75})$ $= 65.9$ kcal/day. Estimating maintenance energy requirement (MER) as $1.5 \times$ BMR, approximate MER $= 1.5 \times 65.9$ kcal/day $= 99$ kcal/day. Severe trauma adjustment is 1.1×99 kcal/day $= 108.8$ kcal/day. If the caloric density of the enteral product is known, the exact amount can be determined. The stomach capacity of raptors is approximately 40 ml/kg body weight.[20] It is also important to remember that the actual metabolizable energy obtained from vertebrates is approximately 75% since approximately 25% of the energy in prey is lost through egestion of the casting or pellet (10%) depending on the species and excreta (15%).

Refeeding syndrome

Refeeding syndrome is associated with metabolic and electrolyte disturbances that occur as a result of feeding a patient that has been starved or severely malnourished. This syndrome is associated with increased insulin release that may occur within a few days of renourishment, especially carbohydrates. In contrast to granivorous species, insulin may be important for maintaining normal blood glucose levels in raptors[21-23] and refeeding syndrome, while controversial, may be a significant consequence of starvation and renourishment in birds of prey. Succinctly, as patients are renourished increased insulin secretions results in increased intracellular uptake of glucose along with phosphorus and other electrolytes, namely potassium, magnesium as well as vitamin B1 (thiamine). This combined with already depleted electrolytes may lead to dangerously low levels of plasma phosphorous, potassium and magnesium. To combat this phenomenon plasma electrolytes should be monitored in critically ill patients and managed during fluid therapy.

Selected Disease Conditions

Chlamydia spp.

Until recently, chlamydial infections associated with *Chlamydia psittaci* and the newly identified *C. buteonis*, were rarely reported in raptors.[24,25] Clinically, raptors may be asymptomatic or may exhibit non-specific clinical signs such as lethargy, depression, anorexia, inability to fly and biliverdinuria. Clinical signs more specific to the respiratory tract may include oculonasal discharge, sinusitis, conjunctivitis, or acute death. Diagnosis of *C. psittaci* in raptors is like that in other avian species in that a combination of hematology, biochemical analysis, radiography, and appropriate antigen and antibody detections methods are utilized.[19] Certainly, the zoonotic potential of this organism must be considered when evaluating management, and treatment and prevention, especially when managing collections of raptors. Treatment of chlamydial infections in raptors is no different than treatment in other avian species, except that raptors are usually more difficult to handle and potentially more easily stressed.

Herpesvirus

Herpesvirus infections reported in raptors include inclusion body hepatitis of falcons (FHV, falcon herpesvirus), hepatosplenitis infectosa strigorum (OHV, owl herpesvirus), and eagle herpesvirus.[8,9] FHV preferentially affects reticuloendothelial cells and hepatocytes causing marked depression, anorexia and high mortality. OHV usually affects epithelial and mesenchymal cells, clinical signs and histopathologic lesions may resemble FHV. Necrotic foci are seen within the liver, spleen, intestine and jugular veins.[8] Diagnosis of herpesviruses in raptors is based upon clinical signs, histologic lesions, serologic identification and virus isolation. Treatment consists of antiviral medications (i.e., acyclovir) along with supportive care. Adenovirus infections in falcons may resemble herpesvirus infections clinically.

West Nile virus

West Nile virus (WNV) infection is caused by a flavivirus and may lead to high morbidity and mortality in corvids, raptors, and many other avian species. Transmission occurs primarily through *C. pipiens* in Europe and North America, although louse flies and even ingestion of infected prey species (house sparrows and pigeons) may also serve as a means of acquiring the WNV. Clinical signs of WNV infection are highly variable and infected birds may show signs ranging from slight depression and weight loss to marked depression, anorexia, blindness, ataxia, head tremors, seizures and sudden death.[14] Ante mortem diagnosis of WNV infection can be difficult; however, consistent clinical signs coupled with a rise in antibody levels and highly suggestive of the diagnosis. Unfortunately, most cases are confirmed through post-mortem examination. Kidney and brain are appropriate samples to submit for histopathologic evaluation.[26] Nemeth et al, (2006)[27] reported the most common histopathologic lesions in naturally and experimentally infected raptors included subacute myocarditis and encephalitis. Wünschmann (2004) described a triad of inflammatory lesions, including chronic lymphoplasmacytic and histiocytic encephalitis, endophthalmitis, and myocarditis, was common in red tailed hawks and Cooper's hawks.[27] Treatment for WNV infections centers around supportive care and anti-inflammatory medications (i.e., meloxicam). Preventive measures (vaccinations) involve the use of a vaccines available for horses. Angenvoort et al (2014)[28] Evaluated the efficacy of commercially available vaccines (an adjuvanted inactivated WNV lineage 1 immunogens vaccine and a canarypox recombinant live virus vector vaccine) and vaccination regimes in falcons both serologically and by challenge with WNV strain of homologous lineage 1. The report concluded that both vaccines provide good partial protection when administered three times.[28] This may hold true for other raptor species as well. Moving birds indoors, covering outdoor enclosures with mosquito netting, isolating infected birds and removal of standing water are effective means of control.

Avian influenza (influenza A virus)

Avian influenza is an extremely contagious and potentially zoonotic disease that targets the respiratory, gastrointestinal and nervous system of many different avian species. Wild birds, especially waterfowl and shorebirds, act as reservoirs and shed the virus in feces as well as ocular or nasal secretions. Raptors become infected by consuming infected prey. The winter of 2021 marked the start of an unprecedented and historic outbreak of highly pathogenic avian influenza (HPAI). This epidemic involving the Eurasian H5N1 clade 2.3.4.4b differs from previous outbreaks in its persistent and wide geographic distribution, the number of avian deaths including free-ranging raptors and raptors in managed care, and its significant spill over into mammalian species.[29,30] Detections of HPAI in wild birds is available at: https://www.aphis.usda.gov/aphis/ourfocus/animalhealth/animal-disease-information/avian/avian-influenza/hpai-2022/2022-hpai-wild-birds.

If clinical signs are present, they may include mild to severe respiratory disease, lethargy, depression, anorexia, neurologic signs, diarrhea and acute death with or without clearly identifiable clinical signs, which is common in raptors.[17,20-23]

Diagnosis of influenza A infections in raptors requires due diligence and thorough evaluation of the patient as other infectious agents may cause similar clinical signs. Raptors suspected of suffering from HPAI must be isolated and managed using careful isolation and control measures including personal protective equipment. When evaluating live birds virus isolation from cloacal or tracheal swabs (preferably) are considered the methods of choice.[17,20,21] Polymerase chain reaction (PCR) testing is the method of choice for confirming a diagnosis. Point of care testing (i.e., Zoetis FluDETECT®) is available; however, the FluDETECT® has a high rate of false negative test results compared to PCR, suggesting that this diagnostic tool has limited value in veterinary or rehabilitation centers.[31] Serology is also of limited value as patients may die before seroconversion. HPAI is a reportable disease, and veterinarians should contact their state veterinarian if there is a high suspicion of this disease. Kits are available for point of care testing in the UK; however, the UK Animal and Plant Health Agency (APHA) strongly recommends against their use and advises veterinary surgeons to report and test suspected HPAI cases via Government laboratories. This has been advised to reduce the chances of delays in reporting suspected disease (a statutory obligation) by false negative results from non-validated test kits. Veterinary surgeons have been advised by APHA that prosecution may result from such delays.

Management of infected birds appears to center around supportive care, however, because of its high transmissibility to other avian species (i.e., commercial flocks) and zoonotic potential, veterinarians should follow policies and procedures as determined by the appropriate state veterinary offices, USFWS, and USDA APHIS. As a precaution, falconers should attempt to avoid pursuing waterfowl and shorebirds or feeding multiple birds from caught game. Additionally, adherence to sound biosecurity principles including facilities, equipment, and protocols is required to appropriately manage these cases. The Association of Avian Veterinarians provides additional information and resources for veterinarians and clients that may come in contact with HPAI,[32] and these resources should also prove useful for raptor facilities.

Presently, vaccination is not allowed in the United States; however, a vaccine trial was initiated in black vultures (*Coragyps atratus*) in late summer of 2023 in response to multiple deaths of California condors (*Gymnogyps californianus*) due to HPAI in 2023.[33] Preliminary details of the vaccine trials in black vultures and subsequent plans for trials in California condor are available at: https://www.fws.gov/story/2023-08/california-condor-hpai-response-update-august-18-2023.[33]

Ectoparasites

Birds of prey are commonly infected with ectoparasites including feather lice (*Malophaga* sp) that feed on skin, feathers and blood, and louse flies (Hippoboscidae) that feed on blood. Usually, these ectoparasites cause few clinical signs; however, marked feather damage, anemia, poor performance, and even death may be seen with excessive numbers. Interestingly, while louse flies are often reported on raptors, little is known of their life cycle, prevalence, or geographic distribution.[34] Levesque-Beaudin et al (2021)[34] reported the three most common louse flies found on raptors in southern Canada including the most prevalent *Icosta americana*, *Ornithomya avicularia*, *O. anchineuria*, and *O. bequaerti.* A fun fact is the female louse fly will rear one larva in its uterus, fed on milk that it produces.[34] Once developed, the female releases a prepuparium into the surrounding substrate, pupation occurs almost immediately, and the emerging adult can easily find a host.[34] Treatment of ectoparasites such as louse flies is often managed with topical permethrin powders or sprays. Ivermectin is also effective for the treatment of lice.

Endoparasites

Nematodes

Clinical signs of nematode infestations depend on the location and life history of the parasite. *Capillaria* spp. and ascarids cause little disease in low numbers but may cause diarrhea, anorexia, emaciation, poor performance and ability to capture prey, and death; especially when comorbidities are present. *Syngamus* sp. and *Serratospiculum* sp. often cause respiratory signs including head shaking, open-mouthed breathing, dyspnea, exercise intolerance, and poor performance. Serial fecal examinations are recommended, and demonstration of the ova, eggs, oocysts or adult worms within the feces, or in other areas of the gastrointestinal tract is diagnostic. Most nematode infestation can be treated with an appropriately administered anthelmintic (i.e., ivermectin, pyrantel pamoate, fenbendazole). Fenbendazole toxicosis has been reported in a number of avian species.[35,36]

Trichomoniasis

Trichomoniasis, also called frounce, is caused by *Trichomonas gallinae* and is of major significance in captive and wild birds of prey.[2,5] Healthy pigeons (*Columba livia*) and doves (*Zenaidura macroura*) often harbor the organism. Infections are seen in captive raptors fed a diet of freshly killed pigeons and wild raptors that feed on pigeons (goshawks, falcons, and some species of owls).[2,5] Clinical signs of infection often reveals variably sized caseous plaques in and around the oropharynx, crop, and esophagus. Samour (2003)[37] also reported lesions in the infraorbital sinus, nasal cavity, and syrinx in falcons. Affected raptors often have difficulty swallowing and may flick pieces of food. Diagnosis is based upon history, clinical signs and demonstration of the organisms from swabs of the exudate expressed into saline (wet mount), direct smears, various stains, and molecular diagnostic tests. Metronidazole or carnidazole are recommended as treatment options.

Coccidia

Coccidian parasites that affect raptors include *Caryospora* spp., *Cryptosporidium* spp., *Eimeria* spp., *Fenkelia* spp., *Sarcocystis* spp., and *Toxoplasma gondii*. Most coccidia are considered nonpathogenic in raptors; however, coccidia seem to affect young raptors more commonly or those with a compromised immune system. Clinical signs may include lethargy, depression, diarrhea (with or without blood in the stool), poor body condition, weight loss, poor performance, or even death. Diagnosis is usually accomplished by demonstrating oocysts in the feces or the organisms in tissue. While a number of drugs are recommended, use of triazinone antiprotozoal agents (e.g., toltrazuril or ponazuril) provides the best opportunity for successful treatment.

Hemoparasites

Leukocytozoon sp. and *Hemoproteus* sp. are seen in both captive and wild birds of prey and may be found in large numbers within the bloodstream with no apparent clinical signs.[2,24] *Plasmodium* sp. is of clinical significance, and may show a range of clinical signs including depression, weight loss, dyspnea, decreased appetite, poor performance, biliverdinuria, anemia, and death. Mosquitoes serve as a means of transmission from host to host. It is thought that passerines may serve as reservoirs.[2] Diagnosis is based upon clinical signs and demonstration of the organisms in red blood cells. Note that intraerythocytic gametocytes of *Plasmodium* sp. and *Hemoproteus* sp. may appear very similar and can be difficult to distinguish from one another. Treatment of *Plasmodium* sp. infestations may involve supportive care, blood transfusions, use of antihistamines, and antimalarial medications (i.e., mefloquine, chloroquine, and primaquine).

Lead intoxication

Lead intoxication is a relatively common and underappreciated problem in free-ranging raptors worldwide. Acute and chronic lead intoxication occurs primarily through ingestion of anthropogenic sources of lead found in prey items or carrion that contains lead fragments. Slabe et al (2022)[38] evaluated lead levels in samples (liver, blood, and feathers) collected from 1210 bald and golden eagles, including 620 live eagles, from 38 states across North America. across the United States and found that nearly 50 percent of eagles sampled had chronic, toxic levels of lead.[38] Once ingested, lead is degraded in the stomach, unless egested, then slowly released into the gastrointestinal tract and blood stream. Lead that is absorbed through the digestive tract is sequestered in soft tissues and bone and slowly excreted by the kidneys over a period of days, weeks, and months.

The pathologic effects of lead intoxication are well described as lead has the ability to disrupt many biochemical processes within the body. Succinctly, lead interferes with sulfhydryl-dependent enzyme function, mimics calcium and so interferes with neurologic function and mitochondrial function and lead also adversely affects DNA and RNA synthesis.[39]

Clinical signs of lead intoxication may vary depending on the amount of lead ingested and include anorexia, weight loss, emaciation, regurgitation, vomiting, diarrhea, ileus, depression, ataxia and weakness, seizures, blindness, hematuria, hemoglobinuria, polyuria and polydipsia and death.

Diagnosis of lead toxicosis is based primarily on clinical signs, hematology, biochemical analysis, radiology, toxicologic analysis and postmortem evaluation. Blood lead levels greater than 20µg/dl (0.2 ppm) with consistent clinical signs is highly suggestive of a diagnosis. Blood levels greater than 50 µg/dl in most species are considered to be diagnostic.

Therapeutic goals for managing heavy metal intoxication should concentrate on stabilizing the patient's condition, removing the source of intoxication from the GI tract, removing lead from the tissues of the patient, prevention of further exposure to the toxicant, seizure management, and supportive care. Fallon et al (2017)[40] provide guidelines for evaluation and treatment of lead poisoning in wild raptors. As per their report, best practices suggest that birds whose blood-lead levels are < 40 µg/dl be released back to the wild as soon as possible after capture.[40] The decision to release or treat birds with blood-lead levels between 40 µg/dl and 60 µg/dl should be made based on the presence of clinical signs of poisoning and relevant biologic characteristics (e.g., breeding status). Finally, birds with blood-lead levels > 60 µg/dl are potentially lethally poisoned and should be removed from the wild for appropriate treatment at a licensed rehabilitation facility and later released.[40]

Chelation therapy is the mainstay of treatment. Calcium ethylenediaminetetraacetate (CaEDTA) and D-penicillamine (cuprimine) are recommended.[40]

Pre-Release Conditioning

Basic principles

After treatment and surgery, patients enter a phase of physiotherapy and flight conditioning interspersed with regular evaluations to determine progress and readiness for release. Both physical health and behavior must be normal so that the bird can support itself and interact with other animals, people, and the environment in an appropriate manner. Holistic conditioning minimizes pain and stress, encourages repair and rebuilding of damaged tissues, and increases strength, mobility, and cardiopulmonary stamina.

Stress and pain management

Every step of this process must be conducted with utmost respect for the individual bird. Wild birds are not accustomed to captivity, human conversation, or people in close proximity to them. Be calm, quiet, careful, and kind. During restraint, at the very least, utilize hoods to limit vision, which is a raptor's primary sense. Analgesics, sedation, and anesthesia are also appropriate when an evaluation does not include assessing behavior or neurologic function.

Pain will increase secretion of corticosterone which can delay healing, so it must be minimized. However, some drugs require more frequent dosing than others, or require handling for administration versus delivery via food. Frequent handling will also increase stress for wild birds and can induce further trauma during capture efforts. These factors must be weighed in choosing the appropriate regimen.

Nonsteroidal anti-inflammatory drugs (NSAIDs), such as meloxicam, are appropriate for most species of birds of prey although recent research demonstrates that pharmacokinetics are diverse between taxa with generally high rates of elimination. Therefore, once-daily administration may not be sufficient. Fortunately, meloxicam appears to have a broad safety margin, even among Old World vultures, who have shown susceptibility to toxic damage from other NSAIDs, including diclofenac and carprofen.[41] It is wise for clinicians to research current papers regarding drugs in their taxon of concern. In general, dosages of 1.5-2 mg/kg q12-24h appear safe in birds of prey for a course of 1-2 wk but use caution with birds that have compromised GI tracts, are dehydrated, or have renal compromise. If inappetence, increased water intake, lethargy, or other symptoms occur, dosage should be reduced or discontinued.

Butorphanol (1-4 mg/kg q4-12h) has a wide margin of safety but requires frequent dosing and must be given IM/IV. Its best use is for perioperative pain management or situations where frequent handling and treatment are already required. Tramadol has the advantage of demonstrated oral bioavailability in at least members of Accipitriformes and Falconiformes.[42,43] The recommended dosage for kestrels and bald eagles, based on pharmacokinetics, is 5 mg/kg PO q12h.

Diurnal birds of prey generally calm significantly when hooded. Owls do not have a head shape that allows hooding and blocking their vision seems, at least anecdotally, less effective. Where hooding is not possible or found to be ineffective, sedation with midazolam (0.5-2 mg/kg IM) is recommended. Full effect requires 10-15 min, in a quiet, darkened environment, prior to manipulation of the patient. Midazolam is not recommended in cases where medical behaviors are to be taught as it is known to have an amnestic effect in humans and, per one study, in chickens.[44] On the flip side, forgetting the experience of capture and manipulation may be beneficial to reducing the stress of repeated capture episodes.

Raptors are highly food-motivated so immediate feeding after return to an enclosure may have benefit in reducing the stress response, or at least hastening a return to normal physiologic function after handling episodes.

Tracking progress

Evaluation starts with observing the bird in its enclosure and noting responses to captors and conspecifics, ability to evade capture, and ability to walk and run. In a large chamber, both vertical and horizontal flight modes should be observed. Look for asymmetry in posture, ease of flight and landing, and cardiopulmonary stamina. Digital video at 60 frames per second aids in this review. Pay special attention to wingtip symmetry and direction of the legs and tail during straight flight. Shifts to one side may indicate compensation for pain, limitations in range of motion, or aerodynamic deficiencies (e.g., the legs/tail will generally point away from the affected wing).[45,46]

Before starting a flight conditioning program, all injuries must be resolved, the bird must have excellent vision, and an initial set of goniometry measurements of the major wing joints should be obtained. If more than 3 flight feathers from any wing or 5 from the tail are broken or less than half grown, they should be imped or allowed to grow in prior to beginning forced exercise regimens.[45] Imping is the process of cutting off a damaged grown feather and splicing in a replacement.[47]

During manipulation of a limb, it should first be palpated for swelling or obvious injury before being extended. Note hyperesthetic responses which may indicate pain and use caution during further manipulation. Second, the limb is extended gently to assess passive range of motion. In the case of the wing, the quality of the feathers is also assessed and the symmetry of how the feathers spread. Note any crepitus, pain, or extension deficits. If a deficit is unclear a comparison with the contralateral limb is recommended. Next, each joint is individually manipulated. It is critical, particularly with the wings, that extension is performed in a consistent plane. For example, the wing and each joint should be extended straight outward from the body maintaining it in the same plane as the shoulders. Goniometer measurements are taken at the maximum angle where resistance is felt but discomfort is not expressed. All evaluators must use the same reference points and technique to ensure comparable measurements. Finally, the function of tendons and joints is assessed by manipulation of the reciprocal apparatus of the wings and legs. In the case of the wings, the carpus is pressed caudally while the elbow is pulled cranially. These combined forces should cause the elbow and carpus to pivot open and spread the wing utilizing the tendons primarily.

Diagnostics are covered elsewhere, but at a minimum should include a recheck of any previously abnormal results or radiographic lesions. A minimum laboratory database of packed cell volume and total serum solids is advisable as a simple, albeit incomplete, assessment of oxygen-carrying capacity. Serum lactate, determined from blood samples collected pre- and post-exercise, has been reported in the literature as a means for measuring improvements in cardiopulmonary fitness.[45,48,49] Blood is sampled before and after exercise sessions, and again 20 min later. Samples can be rapidly and inexpensively checked utilizing a point-of-care analyzer (e.g., Lactate Plus, Nova Biomedical, Waltham, MA). The testing regimen is performed weekly during the exercise phase of a patient's conditioning. The expectation is a downward trend at all points from session to session, but most especially between the 2 post-exercise samples which should presumably indicate increased ability to clear lactate and deliver oxygen to tissues. Rather than focusing on specific ranges, trends are the emphasis. With repeated measurement, a point should be reached where the decrease in lactate or improvement in lactate recovery from session-to-session plateaus, which would prompt final evaluation for release.

Physiotherapy

Following trauma or surgery, contracture of tendons and muscles occurs accompanied by formation of adhesions between damaged tissue layers resulting in decreased limb mobility.[52] Early application of physiotherapy diminishes the untoward effects of these processes and has many short- and long-term benefits.[50] The inactivity accompanying injury and convalescence also results in atrophy of muscles and blood vessels, so their capacity must be rebuilt prior to release.

There are three primary physiotherapy modalities: Physical agent modalities (PAMs), limb manipulation therapy, and aerobic exercise.[51] For best results, use least one component from each of these categories. Early in rehabilitation, PAMs are preferred for their low impact and tendency to encourage anatomically correct healing. PAMs include massage, therapeutic ultrasound, low-level laser therapy, and application of cold or heat. At this early stage, exercise may be limited to providing a small enclosure for stretching limbs and walking short distances. Manual therapy involves passive stretching under sedation or light general anesthesia. As injuries resolve, typically after the first 2 wk, ROM improves, and the risk of adhesions diminishes, a transition is made away from PAMs and manual therapy to self-exercise in a large enclosure and eventually to forced exercise regimens that build muscle and cardiopulmonary performance.

Physical Agent Modalities

Cold therapy: Cold therapy is most indicated during acute stages (48-72 hours) of injury or surgery.[51] In summary, it reduces pain, swelling, and inflammation. Cold compresses are applied to surgical sites for up to 10 min, one to two times daily. Birds in particular are prone to hypothermia. Monitor core temperature if the bird is under general anesthesia or observe for signs of hypothermia such as shivering if they are awake. Cold gel packs covered in fabric are recommended versus direct application of ice.

Heat: Heat application, after the first 72 hours post-surgery or injury, produces vasodilation, improves tissue elasticity, relaxes muscles, and increases the pain threshold.[51] As with cold packs, heat is applied for up 10 min, one to two times daily or at the start of each physical therapy session. Heat can be applied with the use of a compress soaked in hot water, chemically activated hand-warmers, or small bags of raw rice heated briefly in a microwave. Care must be taken to avoid local thermal injury or hyperthermia. A compress should be no warmer than 43°C (110°F) at its surface (hot to the touch but not painful).

Massage: Gentle manipulation of the soft tissues diminishes adhesions, stimulates endorphin release, and increases blood flow, improving oxygen delivery, reduction of edema, and easing of muscle tension.[45] Massage should start with effleurage, a light superficial stroking. Pressure is gradually increased to gentle kneading or circular massage. Massage is contraindicated around freshly closed surgical wounds or friable tissues.

Therapeutic ultrasound: Therapeutic ultrasound uses high-frequency (> 20 kHz) acoustic waves to produce deep physiologic heat. (Hoppes 2015) It has been widely applied to the treatment of soft tissue injuries in humans, domestic animals, and birds. Its mode of action includes increasing temperature in specifically targeted tissue and induction of acoustic cell streaming.[52] The thermal effects encourage blood flow and analgesia.[45,52,53] Acoustic streaming is presumed to enhance healing through increased cell permeability to calcium ions, increased protein synthesis, and release of growth factors. The result is stimulation of collagen deposition, increased formation of new blood vessels, and fibroblast proliferation.[45] Most commonly, 1-3 MHz is used in birds and heats to depths of 5-30 mm. As frequency is increased, penetration decreases, such that 3 MHz is ideal for superficial lesions. Ultrasound waves striking bone may cause periosteal pain and damage so use sparingly over areas of decreased soft tissue coverage. Intensity, measured in watts/cm^2, is the rate of energy delivered. Increasing tissue temperature to 40-45 °C can be accomplished with 1-2 W/cm^2 with a continuous wave duty cycle for 5-10 min. The normal recommendation for transducer movement is 4 cm/sec or slower. Treatments are daily or every other day, depending upon the needs and relative fractiousness of the patient.[53] Do not use around freshly closed wounds, friable tissues, or infected joints.[53] It can be initiated sooner than passive/active extension (PAEPT) since it does not produce strain on bones and soft tissues. There are benefits seen to its use even in chronic cases of contracture of the propatagium.[52]

Phototherapy or low-level laser therapy (LLT): LLLT involves the use of focused, coherent light of various intensities and wavelengths to penetrate tissues without thermal injury. It is thought to stimulate oxygen penetration, cellular metabolism, and cell division.[45] There also appear to be pain management benefits, probably from endorphin release. There are few documented uses of LLLT in birds, but it has been applied successfully to raptors.[54] There appear to be no side effects to LLLT but clear benefits in avian species have not yet been documented.

Limb manipulation therapy

The terms active and passive limb manipulation therapy (LMT), in terms of physical therapy, refer to whether the motion is performed by the animal (active) or by an external force (passive). Range of motion (ROM) is the total amount of freedom of movement of a joint.[50] Limb manipulation increases blood flow, improves synovial fluid production, decreases pain, and reduces potential for adhesions and bony entrapment of tendons at fracture sites. The goal of limb manipulation therapy is, through repetition, to improve ROM.[55] Massage, heat application, therapeutic ultrasound, or LLLT are good preambles to LMT sessions as they will relax muscles and increase circulation.

For passive LMT, the clinician applies gentle traction to limb segments to manipulate individual joints.[45] Forces should only be sufficient to flex or extend the joint to the point of gentle resistance. After massage, the limb is first stretched and held for 30-60 seconds. This process is repeated three to five times. Then each joint is flexed and extended gently for 10-30 cycles. If discomfort is expressed then the manipulations are slowed, decreased in intensity, or discontinued. Diurnal raptors generally tolerate light non-painful ROM exercises while hooded. In other species or in birds that express severe panic or pain during manipulation, light sedation or anesthesia are preferred. ROM measurements should be taken before and after each session to track progress.

For active LMT, the bird is restrained by the legs in one hand and held upright so that its weight is supported at the hips. The hand is then tilted slowly forward or back to encourage the bird to spread its wings. In the case of raptors, a hood can be applied to diminish stress and encourage more gentle use of the wings or removed to encourage more active use of the wings. With large birds, such as bald eagles, the opposite hand can support the breast and the bird may flap softly in place for several minutes. This should be performed in a quiet and distraction-free environment, with lights dimmed for particularly fractious individual birds. An outdoor breeze or fan can be used to further encourage spreading of the wings. This method is not recommended for use in the first 4-5 days after major orthopedic or tendon repair. The degree of stress response should dictate whether this approach is appropriate. Its primary benefit tends to be close evaluation of the bird's ability to manipulate its wings if there are deficits that are difficult to localize. In cooperative individuals, it may also be an excellent introductory phase before graduation to a large flight enclosure, particularly if the bird has been housed in a small recovery enclosure for more than a couple of weeks.

Exercise and flight conditioning

Birds of prey have strong legs that lend themselves well to attachment of jesses and creances, allowing controlled flights in open spaces. Flight enclosures are still preferred for daily exercise, but in cases where a flight enclosure is not immediately available, creance flight is a suitable option. The particulars of jess application and strategies of creance use can be found in *Medical Management of Wildlife Species*.[56] Sessions should start with a fresh body weight and brief limb manipulation. Several flights on the creance are conducted until flights become significantly shorter or recovery periods between flights become noticeably longer. The bird should be allowed to rest between flights until they close their beak. Performance should be recorded as numbers and approximate distances of flights, plus a total flight distance and duration of the session. If body mass decreases more than 10% per week, except in the case where weight loss is prescribed, the session should not be conducted and a veterinary medical evaluation should occur. Sessions usually start at two times weekly and increase to three to five times per week as performance improves.

Conversely, birds can also be exercised in large flight enclosures. A person calmly walks around the walls of the flight chamber to gently encourage flight from one end to the other. Use as little visual stimulus to encourage flight as necessary. This approach is less stressful than creance since it does not require capture and handling, but it has disadvantages of not allowing weight monitoring or performance of pre-exercise physical therapy and evaluation. For these reasons, the bird should still be caught weekly for evaluation. The criteria for cessation and parameters for tracking progress are the same as with creance flight.

Falconry is also an excellent option as the final step to the rehabilitation process for many species as it builds strength and endurance, and thoroughly evaluates the flight and hunting capabilities of the bird. Performed on fledged birds, it does not permanently alter the behavior of birds of prey or create a dependency on humans.

Release

Before release, a raptor must have all its injuries resolved, be able to maintain its health without assistance, and have a demonstrated ability to hunt, kill, and manipulate prey for consumption. Cardiovascular stamina and muscle strength must be excellent for survival. They must be released at a point in their natural range that is appropriate for the time of year and is free and clear of potential conflicts with breeding pairs. They must also exhibit normal avoidance of humans and appropriate conspecific social behaviors.[45,46,57]

Soft releases provide a bird time to observe the new location, observe and interact with conspecifics, and receive food while they gradually acclimate to their wild home. Soft release is a particularly attractive option for captive-raised raptors (known as hacking) and is performed using an elevated box with slatted front that can eventually be opened. Food is delivered through blind openings in the back of the box to diminish association with humans.

Hard release is the practice letting a bird go directly into the wild with minimal opportunity for ongoing observation or assistance. This is most appropriate for birds admitted as adults and being returned to the same locale as their capture, especially when very little time has elapsed.

Releasing animals, wild or domestic, on National Park land in the United States is prohibited. There can be similar restrictions on other forms of public land so this should always be investigated. Release is accomplished with the least amount of stress. The bird should be transported in a darkened carrier, set down on the release spot, the crowd calmed while one person quietly opens the door and steps back. Allow the bird to venture out on its own accord and watch the miracle unfold.

References

1. McClure CJW, Westrip JRS, Johnson JA, et al. State of the world's raptors: distributions, threats, and conservation recommendations. *Biol Conserv*. 2018;227:390-402.

2. United States Fish & Wildlife Service. Migratory Bird Treaty Act of 1918. Available at: https://www.fws.gov/law/migratory-bird-treaty-act-1918. Accessed January 22, 2024.

3. Convention on Migratory Species (CMS) Memorandum of Understanding on the Conservation of Migratory Birds of Prey in Africa and Eurasia (Raptors MOU). Available at: https://www.cms.int/raptors/en. Accessed January 22, 2024.

4. Convention on International Trade in Endangered Species (CITES) of Wild Fauna and Flora. Available at: https://cites.org/eng. Accessed January 22, 2024.

5. United States Fish & Wildlife Service. Endangered Species Act. Available at: https://www.fws.gov/law/endangered-species-act. Accessed January 22, 2024.

6. Title 50 Code of Federal Regulations-Wildlife and Fisheries (eCFR). Available at: https://www.ecfr.gov/current/title-50. Accessed January 22, 2024.

7. United States Fish & Wildlife Service. Bald and Golden Eagle Protection Act. Available at: https://www.fws.gov/law/bald-and-golden-eagle-protection-act. Accessed January 22, 2024.

8. United States Fish & Wildlife Service. Lacey Act. Available at: https://www.fws.gov/law/lacey-act. Accessed January 22, 2024.

9. United States Department of Agriculture, Animal and Plant Health Inspection Service, Animal Welfare Act: New Bird Rule. Available at: https://www.aphis.usda.gov/aphis/ourfocus/animalwelfare/new-bird-rule/awa-standards-for-birds. Accessed January 22, 2024.

10. Association of Avian Veterinarians. USDA Resources. Available at: https://www.aav.org/general/custom.asp?page=usda. Accessed January 29, 2024.

11. UK Legislation, Wildlife and Countryside Act 1981. Available at: https://www.legislation.gov.uk/uk-pga/1981/69/contents. Accessed January 29, 2024.

12. Raptor Rescue: Raptor Rescue Rehabilitation Handbook-A Guide to Relevant Laws. Available at: http://www.raptorrescue.org.uk/wp-content/uploads/2014/06/part8-Law-v2.2.pdf. Accessed January 29, 2024.

13. Jones MP, Chitty J. Raptors. In: Russell KE, Heatley JJ, eds. *Exotic Animal Laboratory Diagnosis*. Hoboken, NJ:John Wiley & Sons, Inc.; 2020;437-482.

14. Joseph V. Emergency care of raptors. *Vet Clin North Am Exot Anim Pract.* 1998;1(1):77-98.

15. Kirkwood JK. Maintenance energy requirements and rate of weight loss during starvation in

birds of prey. In: Cooper JE, Greenwood AG, eds. *Recent Advances in the Study of Raptor Diseases*. Keighley, UK;Chiron Publications Ltd.;1980;153-158.

16. Forbes NA, Cooper JE. Fatty liver-kidney syndrome of merlins. In: Redig PT, Cooper JE, Remple JD, Hunter DB, eds. *Raptor Biomedicine.* Minneapolis, MN; University of Minnesota Press; 1993;45-48.

17. Klasing, KC. *Comparative Avian Nutrition.* Wallingford, UK: CAB International; 1998.

18. Forbes NA, Flint CG. Raptor Nutrition. Evesham, UK: Honeybrook Farm Animal Foods. Available at: https://www.honeybrookfarm.com/resources/downloads/raptornutritionEN.pdf. Accessed February 20, 2024.

19. Forbes NA, Rees-Davies R. Practical raptor nutrition. *Proc Annu Conf Assoc Avian Vet.* 2000.

20. Redig P. Nursing avian patients. In: Beynon P, Forbes N, Harcourt-Brown N, eds. *Manual of Raptors, Pigeons, and Waterfowl.* Cheltenham, UK: British Small Animal Vet Assoc Ltd; 1996:42-46.

21. Hazelwood RL. Pancreas. In: Whittow GC, ed. *Sturkie's Avian Physiology*. San Diego, CA: Academic Press; 2000:539-550.

22. Minick MC. Correlation of simultaneously determined plasma insulin and glucose levels in fasting and mildly or severely stressed raptors. *Comp Biochem Physiol.* 1986;83A(2):301-305.

23. Minick MC, Duge GE, Fishman SW, et al. Effects of intragastric nutrients upon simultaneously determined plasma insulin, glucagon, insulin/glucagon rations and glucose in red-tailed hawks, *Buteo jamaicensis* Gmelin. *Comp Biochem Physiol.* 1996;113A(2):173-179.

24. Luján-Vega C, Hawkins M, Johnson CK, et al. Atypical Chlamydiaceae in wild populations of hawks (*Buteo* spp) in California. *J Zoo Wildl Med.* 2018;49:108-115.

25. Seibert JBA, Keel MK, Kelly TR, et al. *Chlamydia buteonis* in birds of prey presented to California wildlife rehabilitation facilities. *PLOS ONE.* 2021;16(10):e0258500. doi: 10.1371/journal.pone.0258500.

26. Nemeth N, Gould D, Bowen R, et al. Natural and experimental West Nile virus infection in five raptor species. *J Wildl Dis.* 2006;41(1):1-13.

27. Wünschmann A, Shivers J, Bender J, et al. Pathologic findings in red-tailed hawks (*Buteo jamaicensis*) and Cooper's hawks (*Accipiter cooperi*) naturally infected with West Nile virus. *Avian Dis.* 2004;48(3):570-580.

28. Angenvoort J, Fischer D, Fast C, Ziegler U, et al. Limited efficacy of West Nile virus vaccines in large falcons (*Falco* spp). *Vet Res.* 2014;45(1):41.

29. Wünschmann A, Franzen-Klein D, Torchetti M, et al. Lesions and viral antigen distribution in bald eagles, red-tailed hawks, and great horned owls naturally infected with H5N1 clade 2.3.4.4b highly pathogenic avian influenza virus. *Vet Pathol.* 2024. doi: 10.1177/03009858231222227.

30. Elsmo EJ, Wünschmann A, Beckmen KB, et al. Highly pathogenic avian influenza A (H5N1) virus clade 2.3.4.4b infections in wild terrestrial mammals, United States, 2022. *Emerg Infect Dis.* 2023;29(12): 2451-2460.

31. Franzen-Klein D, Ahlmann A, Hall V. Highly pathogenic avian influenza in raptors. *Proc ExoticsCon.* 2023;242-243.

32. Association of Avian Veterinarians. Avian Influenza Resources. Available at: https://www.aav.org/page/avian_influenza. Accessed February 1, 2024.

33. California Condor HPAI Response Update. August 18, 2023. Available at: California Condor HPAI Response Update - August 18, 2023 | U.S. Fish & Wildlife Service (fws.gov). Accessed February 1, 2023.

34. Levesque-Beaudin V, Sinclair BJ. Louse fly (Diptera, Hippoboscidae) association with raptors in southern Canada, with new North American and European records. *Int J Parasitol Parasites Wildl.* 2021;16:168-174.

35. Bonar C, Lewandowski AH, Schaul J. Suspected fenbendazole toxicosis in 2 vulture species (*Gyps africanus, Torgos tracheliotus*) and marabou storks (*Leptoptilos crumeniferus*). *J Avian Med Surg.* 2003;17(1):16-19.

36. Howard LL, Papendick R, Stalis IH, et al. Fenbendazole and albendazole toxicity in pigeons and doves. *J Avian Med Surg.* 2002;16(3):203-210.

37. Samour JH, Naldo JL. Diagnosis and therapeutic management of trichomoniasis in falcons in Saudi Arabia. *J Avian Med Surg.* 2003;17(3);136-142.

38. Slabe VA, Anderson JT, Millsap BA, et al. Demographic implications of lead poisoning for eagles across North America. *Science.* 2022;375:779-782.

39. Virgolini MB, Aschner M. Molecular mechanisms of lead neurotoxicity. *Adv Neurotoxicol.* 2021;5:159-213.

40. Fallon JA, Redig P, Miller TA, et al. Guidelines for evaluation and treatment of lead poisoning of wild raptors. *Wildl Soc Bull.* 2017;41(2):205-211.

41. Jimenez-Lopez O, Ponder J, Nault A, et al. Non-steroidal anti-inflammatory drugs (NSAIDS) and their effect on Old World vultures: a scoping review. *J Raptor Res.* 2021;55(3):297-310.

42. Souza MJ, Martin-Jimenez T, Jones, MP, et al. Pharmacokinetics of intravenous and oral tramadol in the bald eagle (*Haliaeetus leucocephalus*). *J Avian Med Surg.* 2009;23(4):247-252.

43. Guzman DSM, Drazenovich TL, Olsen GH, et al. Evaluation of thermal antinociceptive effects after oral administration of tramadol hydrochloride to American kestrels (*Falco sparverius*). *Am J Vet Res.* 2014;75(2):117-123.

44. Gilbert DB, Patterson TA, Rose SP, et al. Midazolam induces amnesia in a simple, one-trial, maze-learning task in young chicks. *Pharmacol Biochem Behav.* 1989;34(2):439-442.

45. Ponder JB. Avian physiotherapy and reconditioning in a rehabilitation program. *Proc Annu Conf Assoc Avian Vet.* 2011.

46. Scott D. Internal Protocols, Carolina Raptor Center. 2014.

47. Lierz M, Fischer D. Clinical technique: imping in birds. *J Exot Pet Med.* 2011;20(2):131-137.

48. Pollard-Wright HM, Wright MT, Warren JM. Use of a portable tower and remote-controlled launcher to improve physical conditioning in a rehabilitating wild mallard (*Anas platyrhynchos*). *J Avian Med Surg.* 2010;24(4):308-315.

49. Holz PH, Naisbitt R, Mansell P. Fitness level as a determining factor in the survival of rehabilitated peregrine falcons (*Falco peregrinus*) and brown goshawks (*Accipiter fasciatus*) released back into the wild. *J Avian Med Surg.* 2006;20(1):15-20.

50. Shumway R. Rehabilitation in the first 48 hours after surgery. *Clin Tech Small Anim Pract.* 2007;22(4): 166-170.

51. Hanks J, Levine D, Bockstahler B. Physical agent modalities in physical therapy and rehabilitation of small animals. *Vet Clin North Am Small Anim Pract.* 2015;45(1):29-44.

52. Wimsatt J, Dressen P, Dennison C, et al. Ultrasound therapy for the prevention and correction of contractures and bone mineral loss associated with wing bandaging in the domestic pigeon (*Columba livia). J Zoo Wildl Med.* 2000;31(2):190-195.

53. Hoppes S, Sessum D. Therapeutic ultrasound in a blue and gold macaw. *Proc Annu Conf Assoc Avian Vet.* 2008.

54. Shaver SL, Robinson NG, Wright BD, et al. A multimodal approach to management of suspected neuropathic pain in a prairie falcon (*Falco mexicanus*). *J Avian Med Surg.* 2009;23(3):209-213.

55. Crook T, McGowan C, Pead M. Effect of passive stretching on the range of motion of osteoarthritic joints in 10 labrador retrievers. *Vet Rec.* 2007;160(16):545-547.

56. Ford S. Pre-release conditioning (avian species). In: Hernandez S, Barron HW, Miller EA, eds. *Medical Management of Wildlife Species*. Hoboken, NJ: Wiley & Sons Inc.; 2020:105-115.

57. Miller EA. Minimum Standards for Wildlife Rehabilitation. 4th Ed. St. Cloud, MN: National Wildlife Rehabilitators Association; 2012. Available at: http://theiwrc.org/wp-content/uploads/2011/05/Standards-4th-Ed-2012-final.pdf. Accessed February 10, 2024.

Stork and Heron Medicine: An Overview of Management, Medicine and Surgery of Select Shallow Wading Birds

J. Jill Heatley, DVM, MS, DABVP (Avian Practice), DABVP (Reptile and Amphibian Practice), Dipl ACZM, and
John Chitty, BVetMed, CertZooMed, MRCVS

Session #130

Affiliation: From the Department of Small Animal Clinical Sciences, College of Veterinary Medicine and Biomedical Sciences, Texas A&M University, College Station, TX 77843, USA (Heatley), School House, Allington, Salisbury, UK (Chitty).

Abstract: Ardeidae and Ciconiiformes species are commonly presented for rehabilitation and maintained or tolerated in captive collections; however, minimal veterinary literature is dedicated to the medical management of these species. As a cosmopolitan and highly mobile group, herons are affected by human encroachment, agricultural practices, toxins and infectious diseases. Despite being commonly encountered in rehabilitation and in the zoo and aviculture community, herons and storks have received relatively little specific attention to common injuries, maladies and appropriate treatments. This master class reviews handling, anesthesia, diseases, treatment and care of these species in captivity, based on a review of rehabilitation and zoo records.

Herons

Symbolism

In Buddhism, the heron symbolizes purity, transformation and the wisdom of the Buddha and the expansion of awareness and the ubiquity of consciousness. In Native American culture, the heron symbolizes renewal, rejuvenation and rebirth and serves as a reminder that we are all a part of a larger cycle of life and death.

Distribution and classification

Herons are long-legged, long-necked, long-beaked, freshwater and coastal birds of the family Ardeidae, with 72 species, some of which are called egrets or bitterns rather than herons. Ardeidae are cosmopolitan birds that occur on all continents except Antarctica, with more species found in the tropics. Herons are a very mobile group of birds, with most species migrating, and with dispersal occurring after colonial breeding.

Diet and hunting

These carnivorous species feed on aquatic prey of the wetlands they inhabit. Fish, reptiles, amphibians, crustaceans, mollusks, and aquatic insects, and even birds and bird eggs, rodents, and rarely carrion for the larger species may be included in the diet of more generalist species. How some specialist species like the yellow crowned night heron prefer crustaceans such as crabs. Acorn, peas, grain and other vegetable matter ingestion reported is considered incidental. Hunting is mainly sit, wait, and spear, but use of bait, foot stirring, probing, and use of the wings to shade and reduce water reflection are methods use to hunt prey. Three species, the black-headed heron, whistling heron, and the cattle egret, often feed far away from water.[1,2] Cattle egrets improve foraging success by following large grazing animals, catching insects flushed by their movement. One study found that the success rate of prey capture increased 3.6 times over solitary foraging.[3] In captivity, these species readily accept chicks, fish and mice and some species may occasionally egest pellets, depending on the diet.[4]

Anatomy and physiology

Herons exhibit minimal sexual dimorphism and range in size from 25 cm (10 inches) (i.e., the dwarf bittern) to 152 cm (60 inches) (i.e., the goliath heron). The neck's kink (i.e., S-shape) is based on the modified shape of the cervical vertebrae. Unlike most other long-necked birds, the neck is retracted during flight, while the legs are extended backwards. The feet of herons are anisodactyl, similar to perching birds. The bill is generally long and harpoon-like with the exception of the boat-billed heron, which has a broad, thick bill. Species may have 10, 11 or 9 (boat-billed heron) primary feathers and 12 rectrices (10 in the bitterns). The feathers of the herons are generally soft, with minimal waterproofing and can be complex. Although sexual dimorphism of plumage is uncommon for day heron, except pond herons, it more is common in night herons and smaller bitterns. Many species also have different color morphs. *Aredea cinareus* has a glottis situated relatively more aborally than many other avian species.[4]

The arteries of the pelvic limb of the great egret (*Ardea alba*) are without sex-based differences, with the ischiatic artery as the primary vessel and overall vasculature was similar to that of the ostrich and chicken. Vascularity of the legs of the egret was significantly less than that of the caracara based on perfusion and radiographic imaging.[5]

Handling

Heron handling is based on recognition and nullification of the heron's natural defenses (bill, bill stab, feet and wings) while protecting the relatively delicate head, eyes, neck, wings and legs.[4] Generally, the birds are held by the bill with their legs and wings folded against their body. Gloves and goggles or full face shields are recommended when handling these animals because they have impeccable aim and they like to stab globes of the eyes. Based on the elongate wings and neck, creance line flying for rehabilitation is likely unsuitable for these species.[4]

Anesthesia and analgesia

Sedation with buprenorphine (0.05 mg/kg) and midazolam (0.5 mg/kg) and induction of anesthesia with isoflurane have worked well for these species in the author's experience.[6] Reversal with flumazenil in these species was also rapid effective and safe. Tall plastic bottles or plastic drinking cups can be used to fashion gas induction masks for herons. Transparent plastic allows for visualization of the bird's eye during anesthetic induction.[4] A dorsal papillary protrusion of the glottis, which is set relatively far back within the oral cavity may seduce the practitioner into a choice of endotracheal tube that is of larger diameter than the trachea, which is only about 66% of the tracheal opening in the gray heron (*Ardea ardea).*[7] Endotracheal tube sizes successfully in used in some species of herons have been reported, based on body weight (Table 1).[8]

Table 1. Endotracheal tube sizes and body weight of select heron species.

Species	Weight (g)	Endotracheal tube internal diameter (mm)	Bird number
Schrenck's bittern (*Ixobrychus eurhythmus*)	89 ± 21	2	2
Green-backed heron (*Butorides striatus*)	157 ± 21	2	2
Cattle egret (*Bubulcus ibis*)	326	2	1
Little egret (*Egretta garzetta*)	380	2.5	1
Purple heron (*Ardea purpurea*)	704	3.5	1
Large egret (*Egretta alba*)	798	3.5	1
Great egret (*Ardea alba*)	1174	3.5	1

Source: Jang JH, Park YS, Kim HJ, et al. Selection of appropriate endotracheal tube in wild birds. *J Vet Clin.* 2018;35(5):184-189.

Non-infectious diseases

Trauma is undoubtedly the most common disease affecting these species, most often based on human refuse entanglement or human origin structure and vehicle interactions.[9] One author advocates that herons and storks are severely handicapped in their methods of hunting with one leg and should be euthanized.[4] As wading birds, these species are somewhat less likely to be mortally affected by oil or other environmental chemicals, nonetheless, multiple reports suggest they are commonly affected.[6,10] For biologists and toxicologists, herons commonly serve as excellent bioindicators of environmental toxicity for methylmercury and many other environmental toxins.[11-13]

Nutritional disease in these species appears most commonly reported during the nestling or juvenile phases of life. Naturally occurring secondary nutritional hyperparathyroidism has been reported from juvenile cattle egrets (*Bubulcus ibis*) in central Texas.[14] Suspected sodium toxicity was reported in and-reared great blue heron (*Ardea herodias*) chicks, vitamin E deficiency and pansteatitis in juvenile boat-billed herons (*Cochlearius cochlearius*), and vitamin D_3 responsive metabolic bone disease was reported in heron and egret chicks fed capelin (*Mallotus villosus*).[15-17]

Between 1979 and 2001, wildlife rehabilitation centers in San Diego, California, United States reported over 700 cases of steatitis among free-living herons along the Pacific coast.[18] A similar occurrence was reported in Japan in the fall of 2008 with over 70 egrets and herons found sick or dead at an agricultural water reservoir in Kanagawa Prefecture.[19] Clinical signs included weakness, lethargy, and inability to fly before death. Necropsy findings included large amounts of firm subcutaneous and cavitary fat comprised of necrotic adipose tissues with infiltrates of heterophils and macrophages, consistent with a diagnosis of steatitis or adipose tissue inflammation.[20] Egrets with steatitis had low blood levels of vitamin E and high counts of *Cyanobacteria* (*Microcystis aeruginosa*) were found in the reservoir concurrent with the steatitis outbreak.[19]

Neoplasms reported in the heron appear to most commonly affect the musculoskeletal system based on reports of osteosarcoma in a Panama boat-billed heron (*Cochledrius cochlearius*), chondrosarcoma in a great egret, and chondroma in rufescent tiger-heron (*Tigrisoma lineatum*) reported in the United States, Japan and Brazil respectively.[21-23] Other neoplastic diseases reported in Ardeidae include avian lymphoid leukosis (lymphoid) lymphoma in *Egretta alba*, a retrobulbar teratoma in a great blue heron (*Ardea herodias*) and carcinoma with concurrent hemosiderosis in an Australian bittern (*Botaurus poidloptilus*).[24-26]

Infectious diseases[27]

The Ardeadae are often associated with amplification and transmission of viral encephalitides. Japanese encephalitis virus (JEV) is the most common cause of encephalitis globally with about 70,000 human cases per year, 10,000-15,000 deaths and psychiatric or neurologic sequelae in about 30-50% of survivors. The main wild bird reservoirs of JEV are large ardeid water birds like pond herons (*Ardeola grayii*) and cattle egret (*B. ibis*). Wild Ardeidae (i.e., egrets, herons, black-crowned night herons, and bitterns) serve as both reservoir and amplifying hosts for JEV and can transmit JEV over long distances during migration. Egrets have also been associated with transmission of West Nile virus and the night heron (*Nycticorax caledonicus*) is believed to be the main vertebrate host for Murray Valley encephalitis. Highly pathogenic avian influenza has been isolated from egrets, storks and herons. To our knowledge clinical signs of encephalitis in *Ardeidae* have not been reported. Use of mosquitocidal ectoparasiticides should be considered for all herons upon entry to wildlife rehabilitation or zoological facilities.

Bacteria

Avian botulism has been associated with the deaths of grey herons and avian mycobacteriosis has been documented in captive little egret, buff-backed heron, great white egret, and bittern.[27] Great egrets can carry *Aeromonas hydrophila* among channel catfish (*Ictalurus punctatus*) culture ponds and *Salmonella* spp. and diarrheagenic *Escherichia coli* in the urban environment.[28,29] Great blue herons (*Ardea Herodias*) fed *Edwardsiella ictalurid*-infected channel catfish can be viably recovered, suggesting that this species can also transmit *E. ictaluri* among ponds.[30]

Salmonellosis has been reported in a captive heron colony.[31] Cattle egrets can carry multiple *Salmonella enterica* subsp *enterica* serotypes. While egrets, horses, feedlot cattle, and dairy cattle are likely exposed to similar sources of *Salmonella*, no evidence of transmission between these two species was found. The cattle egret[4,5,12]: i-monophasic *Salmonella* serotype was highly invasive and other isolates of this serotype are associated with food poisoning. Thus, cattle egret colonies could pose a health risk to exposed humans.[32]

As near shore, carnivorous wading species, parasitism is amazingly common in these species. *Giardia ardeae* infection as cultured from the great blue heron (*A. herodias*) is commonly reported as concerning for zoonosis; however, transmission to humans has yet to be proven.[33] *Cyathostoma* spp., liver flukes and tracheal flukes or oral flukes also appear common in these species as well as acanthocephalans, and of course *Eustrogyloides ignotus* as a cause of verminous peritonitis.[34]

Some biochemistry and complete blood count data are published for these species. Recently reference values for plasma T3 (triiodothyronine) and T4 (thyroxine) hormones have been reported for rehabilitated healthy grey herons (*Ardea cinerea*).[35] Vitamin E plasma values are likely also diagnostically important for these species, with low values possibly associated with steatitis.[36,37]

Medicine and surgical reports for Ardeidae are relatively few and are generally restricted to case reports. Use of maropitant citrate has been used successfully (injectable q24h for 5 days) to cease the regurgitation of *A. alba* in rehabilitative captivity.[9] A liposomal formulation of amphotericin B was also successfully used to treat wound aspergillosis in a Goliath heron (*Ardea goliath*).[38]

Storks

Symbolism

The great affinity between man and stork dates, in Europe, from pagan times with European white storks being linked with fertility and childbirth, and their nesting on houses being encouraged as good luck. More recently the strutting walk of storks may explain their species names as veiled insults for strutting humans (e.g., the adjutant storks of India).

Distribution and classification[39,40]

Like herons, storks are members of the Order Ciconiiformes. This classification was first proposed by T.H. Huxley in 1867. Since then, the Order has broken up into different suborders and families, with herons and the unique hamerkop now being classified apart from the Ciconiae (storks). The Ciconiae are divided into three families: Threskionithidae (ibises and spoonbills); Balaenicipitidae (Shoebill) and the Ciconiidae (storks). Other areas of controversy in this very fluid Order include the relationship to New World vultures, which have previously been classified alongside storks, but are currently considered to be members of a separate Order. This paper covers only the Ciconiidae which, like the herons, are commonly kept in zoological collections but rarely covered in veterinary texts. Storks are found on all continents, barring Antarctica and in a wide range of habitats.

Diet and hunting

Storks are carnivorous although most are non-specialized generalist feeders taking a wide range of fish, amphibia, mammals and birds. Most species will readily consume carrion with numbers of certain species (e.g., European white storks and Marabou storks) boosted by association with human refuse areas although all are active hunters as well. The general hunting technique involves walking through water with the bill in the water. The bill is held open and snaps shut when prey is sensed.

Other more specialized feeding techniques are employed by some (e.g., two openbill species are adapted to feed largely on large aquatic snails) with the unusual-shaped bill being adapted to probe into the snail. This is facilitated by saliva containing narcotic agents being dripped into the grasped snail to relax its muscle.

In captivity, it is impossible to replicate the wide range of prey in the stork diet. Whole carcass feeding is normally sufficient, although the following considerations should be noted.

- Avoid excess dietary fat; atherosclerosis is not uncommon.
- Ensure a high-calcium (and high calcium:phosphorus ratio) especially in growing and breeding birds.
- Almost all species are diurnal although the wood stork often feeds at night using tactile fishing techniques.

Anatomy and physiology

Storks are large long-legged birds and may be distinguished from other members of the Ciconiiformes by the large pointed bill and areas of bare skin covering part or all of the head and neck. There are some special adaptations in the bill of some species to facilitate feeding (see above) or to facilitate hunting in water (*Mycteria* spp.). The bill is presumed to be highly sensitive. Both sexes are identical in plumage although females are almost always smaller than males. The neck is similar to the herons in terms of length and vertebral numbers. Feet are like herons and are anisodactyl; however, the middle toe is not pectinated and the toes are basally webbed. Stork feet are adapted for perching and almost all species roost, perch, or nest in trees, the exception being the maguari stork that roosts and nests on the ground because its feet are not adapted for perching. Wings are broad and adapted for soaring flight with some species capable of long-distance migrations. Unlike herons, storks tend to fly with their neck outstretched. The marabou and adjutant storks have specially adapted large airsacs in the ventral neck and dorsal throat. These are used in courtship displays and in body cooling. Storks rarely vocalize except during courtship displays, but will perform beak clacking especially in territorial disputes.

Some species are highly gregarious, nesting and feeding in large numbers. Others are not social even though they may nest or roost in large numbers- they will then disperse to feed alone. This is important in captivity where storks may be extremely aggressive to each other if non-social species are housed together, even in pairs. Thus, storks are rarely suitable for mixed bird species exhibits.

Captive management

Storks are generally kept in pairs, although aggressive territorial species may need to be kept singly and social species in larger groups. Previously they have usually been pinioned or wing-trimmed. While this enables keeping in open-topped aviaries and may reduce injury due to flying into overhead netting, deflighting does prevent normal perching, including arboreal roosting (as above, perching/ trees should be available for stork species rather than traditional methods of stable keeping and ground roosting).

Keeping off the ground may also assist in reducing cold-induced foot problems that may be seen in storks in the United Kingdon. Wing-tip edema-type problems have also been seen by this author (JC) in storks although this may also be related to systemic problems and cold. Otherwise, storks require a lot of space to allow for display of natural foraging behaviors. Feeding practice should also reflect their means of hunting in water. Nesting platforms in trees, or on raised structures with plenty of nesting materials should be provided. Social species can nest in large numbers in the same aviary while non-gregarious should be kept as sole pairs at this time.

Handling

Storks are hard to handle and two handlers should be employed even for the smaller species. The feet are not dangerous but should be restrained to prevent self-injury and fracture. The bill should be restrained. Both stabbing and bite injuries are possible and can be severe. Handlers should wear eye protection.

Anesthesia, therapeutics and analgesia[41-45]

Anesthesia is similar to that described for herons. Isoflurane anesthesia (5% and 2 L oxygen flow rate), administered via specially adapted face masks (2-L drink bottles work well), may be induced using intra-venous alfaxolone or propofol to effect before intubation and isoflurane maintenance. Storks possess a crista ventralis that makes intubation difficult and may necessitate the use of a much smaller endotracheal tube; this will affect the ability to ventilate during anesthesia. Alternatively, the crista may be folded against the glottal wall and a larger ET tube used. This should only be for very short procedures or the crista may necrose, form a granuloma and create respiratory obstruction.

There have been few drug studies in storks, although it does appear that fenbendazole may be toxic in marabou and painted storks. This drug should, therefore, be avoided in storks. Marabou storks have also been found to be sensitive to NSAIDs and these drugs should be used with care in stork species.

Non-infectious diseases

In captivity, fight wounds and trauma are the most common conditions seen in storks. Wild birds suffer commonly from collision injuries especially wind turbines and overhead wires. Heavy metal toxicity is also common in wild birds due to their hunting methods. Non-infectious bacterial disease may also be seen, including impaction and infection of the ventral throat sac in Marabou stork necessitating amputation, and bill necrosis. Neoplasia appears rare though storks are long-lived in captivity. Degenerative cardiac disease has been seen in several species, and this author has seen cases of atherosclerosis in stork species.

Infectious diseases

Infectious disease is not commonly described in storks, although they are susceptible to avian influenza West Nile virus, and a herpesvirus that has been described in storks.[46] Storks may be susceptible to salmonellosis but enteritis appears unusual in these species. Parasitic disease is seen occasionally in storks, including *Paroncherca ciconarum* heartworm. Many parasite species seen in storks require intermediate hosts, which reflects the feeding methods of these species. The diversity of parasites seen indicates a need for regular fecal screening of captive storks with therapy based on the findings. Sensitivity to benzimidazoles (see earlier) should always be considered when treating parasitic infections. Aspergillosis is common in stork chicks, especially in wild social nesting species. This likely indicates the environmental conditions of nesting areas and use of regular sites with potentially rotting organic material.[47,48]

Disclosure: The authors have nothing to disclose regarding this manuscript. If readers know of a nonprofit consortium that would like to support further research and publications on herons and storks, please contact the authors.

References

1. Delacour J. Under-wing fishing of the black heron, *Melanophoyx ardesiaca*. *The Auk*. 1946;63(3):441-442.

2. Meyerriecks AJ. Additional observations on "foot-stirring" feeding behavior in herons. *The Auk*. 1966;83(3):471-472.

3. Dinsmore JJ. Foraging success of cattle egrets, *Bubulcus ibis*. *Am Midl Nat*. 1973:242-246.

4. Coles BH. *Avian Medicine and Surgery*. Hoboken, NJ: Blackwell Scientific Publications; 1986.

5. Oliveira EL, Pacifico Pereira KH, Felippi DA, et al. Blood supply and arteriography of the pelvic limb of the Southern caracara (*Caracara plancus*) and great egret (*Ardea alba*). *Anatomia, Histologia, Embryologia*. 2020;49(6):805-813.

6. Ratliff CM, Hernandez A, Watson CB, et al. Use of margarine for the successful removal of polyisobutylene in an anhinga (*Anhinga anhinga*) and great blue heron (*Ardea herodias*). *J Avian Med Surg*. 2020;34(1): 70-77.

7. Atalgin ŞH, Can M, Bekmezci S. Gross anatomy and mineralization of larynx, trachea and syrinx in gray heron (*Ardea cinerea*). *Turkish J Vet Anim Sci*. 2021;45(1):126-132.

8. Jang JH, Park YS, Kim HJ, et al. Selection of appropriate endotracheal tube in wild birds. *J Vet Clin*. 2018;35(5):184-189.

9. Pereira Junior JJ, Pereira CS, Gering AP, Domingues SF. Use of maropitant citrate in great white egrets (*Ardea alba*). *J LAVECC.* 2017;9.

10. Friend M, Franson JC, Ciganovich EA. Field manual of wildlife diseases: general field procedures and diseases of birds. *US Geological Survey.* 1999.

11. Cotín J, García-Tarrasón M, Jover L, Sanpera C. Are the toxic sediments deposited at Flix reservoir affecting the Ebro river biota? Purple heron eggs and nestlings as indicators. *Ecotoxicology.* 2012;21:1391-1402.

12. Varagiya D, Jethva B, Pandya D. Feather heavy metal contamination in various species of waterbirds from Asia: a review. *Environ Monit Assess.* 2022;194(1):26.

13. Pandiyan J, Poiyamozhi A, Mahboob S, et al. Assessment of the toxic effects of heavy metals on waterbirds and their prey species in freshwater habitats. *Toxics.* 2022;10(11):641.

14. Phalen DN, Drew ML, Contreras C, et al. Naturally occurring secondary nutritional hyperparathyroidism in cattle egrets (*Bubulcus ibis*) from central Texas. *J Wildl Dis.* 2005;41(2):401-415.

15. Bennett DC, Bowes VA, Hughes MR, Hart LE. Suspected sodium toxicity in hand-reared great blue heron (*Ardea herodias*) chicks. *Avian Dis.* 1992;36(3):743-748.

16. Pollock CG, Sleeman JM, Houle CD, Ramsay EC. Vitamin E deficiency and pansteatitis in juvenile boat-billed herons (*Cochlearius cochlearius*). *J Zoo Wildl Med.* 1999;30(2):297-300.

17. Horgan M, Duerr R, Murphy B. Clinical and pathologic findings of an outbreak of Vitamin D3-responsive metabolic bone disease in heron and egret (family Ardeidae) chicks fed capelin (*Mallotus villosus*). *J Zoo Wildl Med.* 2021;51(4):958-969.

18. Pavlat J, Beach H, Platter-Rieger JM, Faulkner M. Steatitis in herons and egrets in Southern California. *Proc Annu Conf Intl Wildl Rehabil Council.* 2004.

19. Neagari Y, Arii S, Udagawa M, et al. Steatitis in egrets and herons from Japan. *J Wildl Dis.* 20111;47(1): 49-55.

20. Yee JK, Liu CM, Platter-Rieger M, et al. Low oleic/stearic desaturation index in great blue herons (*Ardea herodias*) with steatitis in Southern California, USA. *J Wildl Dis.* 2019;55(4):995-999.

21. Liu S, Dolensek EP, Tappe JP. Osteosarcoma with multiple metastases in a Panama boat-billed heron. *J Am Vet Med Assoc.* 1982;181(11):1396-1398.

22. Kubo M, Kobayashi K, Masegi T, et al. A case of chondrosarcoma in a free-flying great egret. *J Wildl Dis.* 2007;43(3):542-544.

23. Rego AA, Matushima ER, Miguel MC. Chondroma in rufescent tiger-heron (*Tigrisoma lineatum*): case report. *Braz J Vet Res Anim Sci.* 1999;36:43-44.

24. Nobel TA. Avian leukosis (lymphoid) in an egret (*Egretta alba*). *Avian Pathol.* 1971;1(1):75-77.

25. Schelling SH. Retrobulbar teratoma in a great blue heron (*Ardea herodias*). *J Vet Diagn Invest.* 1994;6(4): 514-516.

26. Cork SC, Stockdale PH. Carcinoma with concurrent haemosiderosis in an Australian bittern (*Botaurus poidloptilus*). *Avian Pathol.* 1995;24(1):207-213.

27. Malik YS, Milton AA, Ghatak S, Ghosh S. *Role of Birds in Transmitting Zoonotic Pathogens.* Singapore: Springer Nature; 2021.

28. Jubirt MM, Hanson LA, Hanson-Dorr KC, et al. Potential for great egrets (*Ardea alba*) to transmit a virulent strain of *Aeromonas hydrophila* among channel catfish (*Ictalurus punctatus*) culture ponds. *J Wildl Dis.* 2015;51(3):634-639.

29. de Oliveira MC, Camargo BQ, Cunha MP, et al. Free-ranging synanthropic birds (*Ardea alba* and *Columba livia domestica*) as carriers of *Salmonella* spp. and diarrheagenic *Escherichia coli* in the vicinity of an urban zoo. *Vector Borne Zoonotic Dis.* 2018;18(1):65-69.

30. Waterstrat PR, Dorr B, Glahn JF, Tobin ME. Recovery and viability of *Edwardsiella ictaluri* from great blue herons *Ardea herodias* fed *E. ictaluri*-infected channel catfish *Ictalurus punctatus* fingerlings. *J World Aquac Soc.* 1999;30(1):115-122.

31. Locke LM, Ohlendorf HM, Shillinger RB, Jareed T. Salmonellosis in a captive heron colony. *J Wildl Dis.* 1974;10(2):143-145.

32. Phalen DN, Drew ML, Simpson B, et al. *Salmonella enterica* subsp. *enterica* in cattle egret (*Bubulcus ibis*) chicks from central Texas: prevalence, serotypes, pathogenicity, and epizootic potential. *J Wildl Dis.* 2010;46(2):379-389.

33. Erlandsen SL, Bemrick WJ, Wells CL, et al. Axenic culture and characterization of *Giardia ardeae* from the great blue heron (*Ardea herodias*). *J Parasitol.* 1990;76(5):717-724.

34. Ziegler AF, Welte SC, Miller EA, Nolan TJ. Eustrongylidiasis in eastern great blue herons (*Ardea herodias*). *Avian Dis.* 2000;44(2):443-448.

35. Nichols DK, Campbell VL, Montali RJ. Pansteatitis in great blue herons. *J Am Vet Med Assoc.* 1986;189(9):1110-1112.

36. Pollock CG, Sleeman JM, Houle CD, Ramsay EC. Vitamin E deficiency and pansteatitis in juvenile boat-billed herons (*Cochlearius cochlearius*). *J Zoo Wildl Med.* 1999;30(2):297-300.

37. Sönmez E. Plasma T3 (triiodothyronine) and T4 (thyroxine) hormone levels as reference values in rehabilitated healthy gray herons (*Ardea cinerea*). *J Adv VetBio Sci Techn.* 2021;6(3):251-257.

38. Bonar CJ, Lewandowski AH. Use of a liposomal formulation of amphotericin B for treating wound aspergillosis in a Goliath heron (*Ardea goliath*). *J Avian Med Surg.* 2004;18(3)::162-166.

39. del Hoyo J, Elliott A, Sargatal J. Ciconiiformes. In: del Hoyo J, Elliott A, Sargatal J, eds. *Handbook of the Birds of the World.* Vol 1. Barcelona, ES: Lynx Edicions; 1992.

40. Jancock J, Kushlan J, Kahl M. *Storks Ibises and Spoonbills of the World.* San Diego, CA: Academic Press; 1992.

41. Onuk B, Tütüncü Ş, Kabak M. The morphological study of the larynx cranialis in stork (*Ciconia ciconia* L.). *Veteriner Fakültesi Dergisi (Istanbul).* 2014;39(2):148-154.

42. Weber MA, Terrell SP, Neiffer DL, et al. Bone marrow hypoplasia and intestinal crypt cell necrosis associated with fenbendazole administration in five painted storks. *J Am Vet Med Assoc.* 2002;221(3):417-419.

43. Bonar CJ, Lewandowski AH, Schaul J. Suspected fenbendazole toxicosis in 2 vulture species (*Gyps africanus, Torgos tracheliotus*) and marabou storks (*Leptoptilos crumeniferus*). *J Avian Med Surg.* 2003;17(1):16-19.

44. Cuthbert R, Parry-Jones J, Green RE, Pain DE. NSAIDs and scavenging birds: potential impacts beyond Asia's critically endangered vultures. *Biol Lett.* 2007;3(1):91-94.

45. Norton T, Whiteside D. Ciconiiformes. In: Fowler M, Miller E, eds. *Fowler's Zoo & Wildlife Medicine.* Vol 8. St. Louis, MO: Elsevier; 2015.

46. Mack ZE, Bonar CJ, Garner MM, et al. A novel herpesvirus in a white stork associated with splenic and hepatic necrosis. *J Vet Diagn Invest.* 2020;32(3):471-475.

47. Gulcubuk A, Erdogan-Bamac O, Metiner K, et al. A case of pulmonary aspergillosis in white storks. *J Hellenic Vet Med Soc.* 2018;69(2):1004-1009.

48. Olias P, Gruber AD, Winfried B, et al. Fungal pneumonia as a major cause of mortality in white stork (*Ciconia ciconia*) chicks. *Avian Dis.* 2010;54(1):94-98.

Surgical Case Studies from Wild Aquatic Birds

Rebecca S. Duerr, DVM, MPVM, PhD

Session #192

Affiliation: From International Bird Rescue, 4369 Cordelia Rd, Fairfield, CA 94534, USA.

Abstract: Wild marine and freshwater birds commonly present with injuries that defy easy treatment. This masterclass focuses on a discussion of cases where surgical solutions were used to return wild aquatic birds to a releasable state. Some of the procedures that are covered involve mandibular symphysis and bill repair, keel lesion repair, use of drains, grafts, salvage procedures, and tracheal resection/anastomosis procedures.

Injuries

Wild marine and freshwater birds commonly present with injuries that are difficult to treat. Constriction injuries and hook punctures from fishing gear are extremely common, damaging muscles, tendons, and other important structures including bones and joints that absolutely must be functional for the bird to be able to resume its wild lifestyle. Other sources of trauma such as vehicle, window, and miscellaneous stationary object collisions often result in facial or other fractures. Predator-caused injuries also are encountered. At International Bird Rescue's two 100% wild aquatic avian California clinics serving the greater Los Angeles and San Francisco Bay Metropolitan areas, these cases make up a substantial proportion of admissions.

Assessments

When assessing wild birds presenting with these severe injuries, it is helpful to first think about the species' lifestyle and what functional anatomy is required for it to successfully resume its necessary natural history activities. It is also helpful to remember that in the United States, and in many other countries, migratory birds have rules regarding their fate, and the goal of treating these species must always be a return to full function in the wild. The goal of wildlife rehabilitation is not to create disabled animals requiring lifetime placement but to return healthy animals to their wild life. Euthanasia is a reasonable option for individuals too injured to survive or injured such that there is no way for them to be fully functional after healing.

Considerations

Certain groups of avian species rarely exit water and quickly develop additional problems when held out of water for very long (e.g., loons, grebes, diving ducks, alcids, some tubenoses), and have more specific needs for a rapid return to aquatic housing than species that can stand and walk and remain out of water for a limited period of time (e.g., cormorants, coots), although even these species are highly prone to foot problems that require careful management and access to water. Other aquatic avian species can stand and walk well, and can potentially remain dry docked for several weeks, albeit need special attention paid to substrates that avoid foot trauma (gulls, shorebirds, pelicans, herons, dabbling ducks, geese, swans). In all of these cases, whether by injury or anatomic necessity, all of these species require attention to avoid development of lesions caused by abnormal weight-bearing on bony prominences.

Case Studies

Cases are discussed that involve surgical solutions used to return wild aquatic birds to a releasable state. Procedures involving mandibular symphysis and bill repair, keel lesion repair, use of drains, grafts, salvage procedures, and tracheal resection and anastomosis procedures are covered.

Ethical Considerations in Wildlife Medicine

LoraKim Joyner, DVM, MPVM, MDiv

Session #123

Affiliation: From One Earth Conservation, 82-52 211th Street, Hollis Hills, NY 11427, USA.

Note: Adapted and reprinted with permission. Joyner LK. Ethical considerations in wildlife medicine. *Wildl Rehabil Bull.* 2022;39(1):36-46. doi: 10.53607/wrb.v39.248.

Abstract: Component wildlife ethics includes two aspects: an understanding of ethical principles and skills in ethical deliberation. Ethical principles reviewed here include utilitarianism, deontological ethics, environmentalism or respect for nature, virtue ethics, relational ethics, care ethics and reverence for life ethics. Other processes and tools that consider human sociology, behavior and subconscious functioning in moral decision-making include conservation psychology, narrative ethics, socioscience, listening and communication skills, and needs-based ethics. We also consider non-human functioning such as welfare science, conservation behavior and cognitive ethology. Incorporating these tools and instituting ethical practices and programs within our wildlife and conservation management plans and organizations improve our ability to care for ourselves, other humans, wildlife and ecosystems.

Introduction

The stress veterinarians and rehabilitators experience over moral and ethical issues can be intense. This is in part because decisions on levels of care and whether the animal lives or dies may not conform to what is known to be in the best interest of the animal. Furthermore, society's understanding of animals' and humans' relationships has changed dramatically in recent decades, with a plethora of differing viewpoints on animals' agency and value. In a given medical case, no one may agree on how much intervention is appropriate and how much financial resource to expend.

Those working with wildlife may experience even more challenges, given that funding, time and staff are often inadequate. The variety of species and their different needs make for complex treatment and care. Also, wild species often coexist in proximity to humans, which enhances the chance for human-wildlife conflict. The moral complexity of working with wildlife is staggering and not only leads to social conflict but can also cause personal burnout and stress. Finally, pressure on diminishing wildlife populations increases the stakes for a positive outcome, especially when dealing with rare and endangered species. The demand in veterinary medicine can be great, as one professor in ethics once said to the author, "Life is full of tragic choices. There is no correct ethical stance over another, only the presence of one another to support us as we engage in making difficult decisions in our lives" (Anna Peterson, oral communication, October 2008).

Many veterinarians experience the dilemma of how to address the often-competing claims of themselves, other staff members, clients and owners and their families, and the animals. This process of choosing between the interests of self and the interests of others, or between the interests of others, is known as ethical deliberation. The more skilled we can become in ethical deliberation, the greater our ability to resolve conflicts that arise out of our passionate understanding of the best way to treat wildlife, while also caring for ourselves and the people with whom we work. Unfortunately, training in ethical deliberation has not been adequate to the task; people, organizations and animals may languish because of it. For instance, while attending a National Wildlife Rehabilitator Association's symposium, how people got along with one another was mentioned frequently as a pressing concern among those working in wildlife rehabilitation. In one study, the most prevalent cause of conservation project failure was due to interpersonal relationships and conflict.[1] This is also the experience of many in conservation, including Mexican conservationists who reported that up to 50% of conservation projects failed in Mexico due to interpersonal conflict and lack of social capital.

A membership letter in January 2009 from the president of the American Veterinary Medical Association, Dr. James Cook, stated: … "I am worried that differing perspectives on animal welfare have the potential to do what earthquakes and hurricanes couldn't do...divide us. The AVMA has plenty of science-based information to help legislators decide and veterinarians lead on these complex issues, but that isn't enough. I need for you to listen respectfully to your colleagues and engage in a respectful exchange of ideas. We all see things in different ways, but we can't get mired in those differences to the extent that we lose sight of our common goals and veterinary oath." (James Cook, written communication, January 2009).

For the sake of our common goal of animal health, we need to find ways to engage in ethical issues, such as with respectful conversation. How we engage in ethical issues is as important as the ethical principles employed. Using the science of understanding humans and non-humans gives us tools that help us more skillfully and effectively handle ethical situations. This paper will review ethical principles as one framework for handling ethical issues and investigate other frameworks and tools to add to a wildlife medicine ethical toolbox.

Principal Ethical Approaches

Learning ethics happens best when situated in real-life situations, in which the participant is enmeshed. Since this is a written paper, let us all imagine this situation where you have been presented with a young black-crowned night heron that was attacked by a cat. The bird is depressed and may have a broken wing. The reader has never treated this species before and is unfamiliar with how to diagnose, treat, house and release this particular species. In addition, there are no funds to cover costs, and money has been tight around the clinic recently. Finally, today is a very busy day, and there may not be much time to read, go online or call someone for information about caring for the bird. Should you or anyone else accept the bird into your care or, in general, treat wildlife?

Utilitarianism

Approaching the situation of the heron from the viewpoint of utilitarian ethics, the decisions is based on terms of better or worse, basically a cost versus benefit analysis. One seeks to choose a strategy that maximizes good, which, in this case, considers the needs of the heron and the humans involved. Whatever decision is made can be justified because the outcome causes less harm than if no action was taken. The result justifies the means.

In this case, one must take a measure of where the suffering occurs. The heron is wild, and you know little of how to care for the bird. It will suffer in your clinic. Yet, the bird is suffering now. You and your staff will lose time and money to treat the bird. On the other hand, you will gain experience and positive public relations if you accept the bird. You also really like herons, and it is so beautiful. You would like to do the bird some good and learn more about herons by practicing on the bird. You figure that the bird will suffer more if you send it home with the kids who brought it in. You might call up the local rehabilitation clinic, but you know that they too do not have much experience in this species, are swamped with baby birds and do not have many staff or financial resources. All in all, you figure out that it is a greater harm to not accept the bird, so you do.

Disadvantages of the utilitarian approach include trying to determine what suffering is allowable under what conditions. What might be construed as minimal harm by one may be maximum to another. It also purports that individuals can be treated as objects and do not have intrinsic worth or value in and of themselves. In other words, you can do what you wish with animals and people if the result is the least harm and the best for the most beings. This approach can lead to tough circumstances for individuals whose well-being is sacrificed for the benefit of others. Often, animal welfare approaches relating to animals fall under this principle.

Deontological ethics

This approach, on the other hand, elevates the worth and dignity of every individual as the ultimate good. One might know this as Kantian ethics, named for Immanuel Kant. Kant stated that humans have an intrinsic worth and dignity and should therefore be treated always as an end and never merely as a means. The same applies for non-humans; animals are not a means to an end.

In the case of the heron, one might say that under no circumstances should the heron be treated with less than 100% care. This means that considerations of publicity, learning, beauty, one's willingness to contribute or finances do not come into the picture. Knowing that adequate care cannot be provided for the bird, it is not accepted into the clinic. Alternatively, knowing this, the rest of the day's appointments are cancelled, and the bird is driven to a wildlife clinic that specializes in heron care. Alternatively, one might decide that there is not anyone who can care for the bird adequately, so it is euthanized.

The largest criticism to this approach is how hard it is to consistently adhere to absolute statements. For instance, one may say that herons should never suffer, yet citizens allow the presence of a hog production farm nearby whose grounds flooded last year and killed a number of herons with the fecal pollution. One might also have competing rules at stake. For instance, one might say that herons should never suffer, and that humans do not have the right to end the life of another. These two rules can complicate actions if one does not have a way to end the suffering without euthanasia. The stance of animal rights organizations often falls under deontological ethics.

Environmentalist and respect for nature

Sometimes at odds with both deontological and utilitarian ethics is environmentalism or respect for nature. In this approach, humans have duties to a species, not just to individual animals. Our moral concern is not whether a wild animal can live according to its evolved set of behaviors (i.e., deontological ethics says the individual animal has absolute integrity that cannot be violated) or what might cause the greatest harm to individuals or a group of individuals (i.e., utilitarianism). What is held as the ultimate value is the extinction of a species, which is deplorable.

In the case of the heron, one might consider what species of heron it is. Is it common, threatened or endangered? Is it not native to the area? Is it hurting other native wildlife? If the heron is not threatened, one might not be as concerned as if it were a rare species, or was suffering population decline or other environmental threat. One might also elect to not care for the bird because it is just an individual. Resources are directed towards the survival of the whole species, such as donations to conservation and environmental protection.

Criticism here lies in the fact that individuals might suffer because of actions that protect the species or the ecosystem as a whole, such as hunting deer or killing wolves. Who decides which individuals or which species merit less attention than other species or the ecosystem?

Virtue ethics

In virtue ethics, we relate to animals in ways that make us virtuous people. For instance, we say that a virtuous veterinarian cares for all animals. In the case of the heron, one would elect to treat the bird and do everything possible to care for the bird, regardless of other commitments. Alternatively, one could say that a virtuous veterinarian is prudent; takes care of herself or himself, the staff and those s/he supports financially; and is the working and financial success of the clinic. In this case, one might not admit the bird or even spend time with the bird to see that care was provided. Not only might there be competing virtues, but also with the previous two examples, the ethical choice is based on human perspective and not on the animal's.

Relational, care and reverence for life ethics

These are three approaches that are similar in some ways to virtue ethics, in that how an animal is cared for depends on how humans relate to the animal. In relational ethics, if we see our relationship to animals as stewards or as veterinarians, then we are inclined to take care of the heron. Relational ethics does not tell us how to care for the bird and does not take into account the individual bird. It is our relationship to the bird that matters most. With care ethics, we draw on our empathy and say that if an animal suffers, we are obligated to do all we can to care for them. The author believes that the care ethic is strong in wildlife rehabilitation. One possible disadvantage to relying on this as an ethic is that it depends on humans understanding animal physiology and behavior, and correctly interpreting when an animal is suffering. What if our sense of empathy is misplaced or not even triggered? In these cases, we might care for the heron but not realize how the animal suffers due to our intervention techniques. Alternatively, we might not connect with the bird and not be able to properly care for the animal because we see it as not suffering.

The term reverence for life comes from Albert Schweitzer, who said, "In this sense, reverence for life is an absolute ethic. It does not lay down specific rules for each possible situation. It simply tells us that we are responsible for the lives about us. It does not set either maximum or minimum limits to what we must do." Criticism here comes from there being no absolute guide for what we may do. For instance, one might consider the heron the most amazing wonder on Earth. We are responsible not only for this bird but also for all the other amazing species and individuals on the planet. How do we decide? Which animals deserve better care? Humans are inclined to offer greater reverence or compassion to those that look like us. We may wish to refrain from a speciesist stance in which we accord greater worth, respect or care to one species over another. Our subconscious evolved to recognize faces and care for those closely related to us. For instance, the heron with its reserved stance and birdlike ways might be ignored more frequently or given less care or medication than an injured chimpanzee brought to the clinic. If asked, one might not admit to thinking that the heron has less worth than another species; however, the time, money and effort spent on one species say otherwise.

Similar to reverence, Tomas Regan writes of inherent worth. For Regan, every species has a distinctive value that is inherent in their existence. They are a cup that is precious in its own right, no matter how we might fill the cup with our definitions of animal or species, or descriptions of their behavior. No matter how we see the species or imagine their thinking, feeling, behavior and capacity to suffer, all species are valuable and have inherent worth. It is not our thinking, current philosophy or cultural constructs that determine our care, but the existence of the animal himself or herself.

Hybrid ethical views

In all likelihood, most of us would approach the heron with a mixture of ethical approaches, if not all of them! One set of principles alone does not seem satisfactory, and yet, a conglomeration of principles may be no less confusing and leave one straining to resolve the incompatible claims that each ethical approach demands. Combining elements of different ethical views might not yield plausible, logically consistent responsibilities of humans to others.

No matter the approach, one is still faced with the fact that we treat species differently, and in a very real, pragmatic and tragic sense, we consistently compromise our values. In fact, the only consistent approach to ethics is that we are all inconsistent.

The author's rational approach to ethics encompasses the belief that there is no rational, consistent approach to ethics. Decisions are frequently made based on self-interest, past experiences and emotions that do not register in the cognitive realms. Veterinary students at Cornell, those aspiring to work with food animals considered more procedures to be humane for all species than did students aspiring to work with small animals. Both sets of students experienced the same curricula; however, their careers impacted their interpretation of this knowledge.

In the case of the heron, the veterinarian or rehabilitator might be experiencing a difficult day due to an argument with a family member or friend. This results in fewer personal resources to give to a complex and time-consuming situation such as treating the heron. On that day, one might be more inclined to argue that the heron is not suffering, or that it is okay to spend less time on the case than one might otherwise. Consciously, if given time to think or research, one knows that biology, physiology and welfare science indicate that birds do feel stress and pain in therapeutic procedures. At a subconscious level one might take shortcuts in treatment or alternatively compromise self-care. Either way, our subconscious is often the final decision-maker in what we do. Hence, it behooves us to know all we can about how humans think and feel, so we can challenge our assumptions to deliver the best care to ourselves and others. In addition, learning all we can about non-human animals will impact our discernment of what we can do to positively impact their well-being.

Understanding Humans

Although we seek to understand humans as irrational beings, we should not dismiss a rational ethical approach that conforms to ethical principles. Rationality can refine and improve our choices and perhaps ease our own confusion or discomfort. We are largely influenced by the culture around us in ways of which we are not always aware. A greater awareness of how we are influenced helps us understand both ourselves and others whose actions may be at odds with ours or seem inconsistent. Understanding promotes empathy for ourselves and others, which, in turn, promotes a greater ability to discuss ethics. Empathy opens space to increase our understanding of cultural influences, feelings and thoughts, which, in turn, promotes greater empathy.

Humans as feeling animals

Understanding human emotive functioning guides us in discerning how we reason and interact socially. From earlier ape ancestors, humans inherited complex emotional responses hardwired to help us form social attachments and engage in caregiving. The need for social attachments helped us not only raise our young but to offset the biology of earlier apes that leans towards freedom, autonomy, individualism and ego. Whilst the earlier ape biology was successively adaptive to living somewhat individually in a subarboreal niche in the forest, this way of living proved impractical as ape species radiated out in the Africa savannah. There, human ancestors needed to support one another in complex social relationships to maintain social cohesion and reciprocity to combat predation and secure food.

To grow in social complexity, the ability for complex emotions also grew. Since humans are hardwired for complex emotions, we are primed to form attachments in a large variety of forms, including those far from one's base family and community and extending out to other species. In tension with this desire to form attachments is the individualism and ego of ape evolution, which influences any care situation such that humans are also primed to seek benefits for themselves alone and to eschew community, including communities of mixed species.

Along with self-interest, humans also evolved to rely upon each other, seek connection, and appreciate biodiversity. We have developed an innate tendency to focus on life and life-like processes and can respond to non-human animals with a sense of kinship and awe. This appreciation of life and the living world are known as biophilia.

Cognition or rational thinking partners with emotions, the limbic system and subconscious thought-processing to impact our ethical codes and moral actions. In simple terms, a low road uses neural circuitry that runs through the amygdala and other similar automatic nodes without being conscious of it, and the high road sends messages to the prefrontal cortex where one can think about what is happening and intentionally impact our actions. The low road is always operating and, indeed, impacts all our decisions and actions. For this reason, a rational argument alone will not greatly impact human behavior, and indeed, rationality does not exist outside of the emotions that underlie our thinking.

Unfortunately, research in the past has overlooked the role of affect and emotions in moral functioning. Recently, we have learned that care, empathy and other relational-based concerns impact learning and decision-making, as does having a sense of safety and comfort. Girls more than boys can give greater attention to safety and comfort of themselves and others, and less to decision-making.

Overlaying this evolved neural (limbic and cognitive) network to form attachments is the capacity for culture to guide human moral concerns. This is in part because humans have evolved to use rituals to mobilize emotional energy for the benefit of community. In other words, rituals found in community gatherings, such as those in religious traditions, guide behavior for adhering to ethical codes and community taboos and strictures. Furthermore, the use of negative emotions such as guilt, shame and fear developed for social cohesion, as did the use of positive emotions such as pride, satisfaction and happiness.

Humans as learning animals

Knowing how emotions impact reasoning is one factor to consider in social interactions. Intertwined with how we feel is how we learn. Understanding how these two dance together tempers our plan for intentionally growing our capacity for ethical engagement.

A recently discovered brain cell, the mirror neuron, senses both the physical moves another person is about to make and their feelings and prepares us to imitate that movement and feel with them. Mirror neurons exist throughout our lives, ever adapting to social cues around us and how we might care for others. When our body mimics the action of another person, we have a greater sense of what that person or non-human animal felt. We are able to do this not through conceptual reasoning (i.e., high road) but through direct simulation, by feeling (i.e., low road), and not by thinking.

Another recent understanding is how our brains have an incredible capacity to grow and to heal, even as we age and after terrible trauma. According to the theory of neuroplasticity, thinking, learning and acting actually change both the brain's physical structure and functional organization from top to bottom. This means that we heal after emotional and physical trauma to our brains, and the potential to grow the ability to communicate, empathize and think is always present. This is a paradigm shift in our understanding of the brain and brings hope as scientists get closer to designing protocols and strategies to grow and heal brains of all ages. In other words, we can always learn and grow in interpersonal and intrapersonal skills and understandings.

Humans in culture—speciesism

As mentioned earlier, we view species differently, often with a bias towards those animals most closely resembling humans. The result is that we treat different species differently, even though our values are to care for all species equally and accord them respect. The challenge to live according to our values is that we have subconscious and even conscious understandings of species that our culture constructs. For instance, herons and eagles may feel the same amount of pain, yet the author would guess that in most cases, eagles garner much more care and support than herons. This is known as speciesism, which is an attitude or practice that does not grant any nonhuman species equal consideration and respect.

Human dimensions of wildlife and conservation psychology

Besides speciesism, there are many other ways that cultures construct how we view animals, resulting in acting inconsistently. One aspect comes from the work in the social science of human dimensions in wildlife. Here, one sees how values, ideology and value orientations impact behavior. Findings suggest that values related to conformity, tradition, security, and self-enhancement are more utilitarian in concept, while the values of openness to change and self-transcendence support an approach to wildlife that is more protectionist, aesthetic, and mutualistic. Depending on how these collections of values are emphasized, one might lean towards a utilitarian view (i.e., animals can be used as a means to an end) or mutualism (i.e., animals have worth in and of themselves and did not evolve to be manipulated by humans).

People also diverge on ideologic perceptions of wildlife. In hunter-and-gatherer societies, many researchers posit that an egalitarian ideology was present. Wild species were fellow inhabitants of the same world. A domination ideology emerged with pastoral societies as hierarchies formed amongst people and between humans and nonhumans. This domination ideology underlies how humans see themselves as separate from nature in modern times and facilitates the belief that humans' role is to exercise mastery over wildlife.

Value orientations impact behavior as well. For instance, two different people might hold equally important the value of treating all living things humanely. Yet, this value might lead one person to euthanize the heron and the other to attempt to save its life. The difference is due to value orientation. Two value orientations direct a lot of thought about wildlife in North America. One value orientation is domination and the other is egalitarian or mutualism. If one emphasizes domination, this person will likely prioritize human well-being over wildlife, and be accepting of human activities that that result in death or other invasive control of wildlife. This is a utilitarian approach. A mutualism wildlife value orientation, instead, sees wildlife as existing in relationships of trust with humans, as having rights like humans, are part of a human extended family, and merit care and compassion. In the United States, there is an increasing trend towards mutualism orientation and away from domination, though both are prominent within our society.

The field of conservation psychology takes what we know about the science of human behavior and the interdependence between humans and nature and then seeks to promote a healthy and sustainable relationship. Currently, there are books, journals, conferences, websites, departments at universities and careers founded on how we understand humans in relation to other species and how we can use that understanding to impact behavior. Conservation psychology persistently and deeply asks: what is the human place in nature and what is nature's place in the human being? These questions are asked, so that we can sustain care. To care about an issue or a species, people must be informed, people must feel and people should act in ways that allow the expression of both their knowledge and their emotions. Conservation psychologists coach people to care by integrating cognition, emotions and behavior. Behavior change does not happen just at the individual level but involves whole communities. We must also seek to align our behavior with our values in community settings to see if more of our energy, resources and time is spent defending our ideologic stances than in taking concrete steps to improve care for others.

One conservation psychology tool that works at the community level is ethnoornithology. Ethnoornithology explores how people make sense of the lives of birds. It studies the relationships between humans and birds. To gain this understanding, from 2009-2011 the author conducted ethnoornithological research targeting conservationists working in Central America. Meanings that frequently surfaced regarding their efforts included love, conversion, calling, insiders/outsiders, interconnection, death, hope, end times (eschatology and apocalypse), sacrifice, service, suffering, compassion, worth and dignity, awe, wonder, social justice, prophetic voice, resistance and solidarity. Having time for meaning-making activities and sharing allowed the study participants to work together more effectively across differences of class, ethnicity, language, gender, religion, age, values and behavior patterns.

Although it is not possible for everyone involved in wildlife to become proficient with the sociologic aspects of human and wildlife relationships, there is much merit in forming multidisciplinary teams that include social scientists or facilitators to help us navigate the complexity of human thinking and behavior.

Understanding Nonhumans

Cognitive ethology and conservation behavior

Nonhumans experience complex thinking, emotions and behavior. In recent decades, the field of cognitive ethology has emerged to help identify what an animal experiences. Cognitive ethology emphasizes observing animals under more-or-less natural conditions, with the objective of understanding the evolution, adaptation (function), causation, and development of the species-specific behavioral repertoire. In other words, as one studies the subjective lives of animals, one can better understand their levels of stress, suffering or discomfort. This informs how our actions (or inactions) impact the well-being of other species. By studying animal cognition, we develop tools to change attitude and perceptions of non-human animals in society, and hence improve treatment.

Conservation behavior is the application of knowledge of animal behavior to solve wildlife conservation problems. By knowing a species' social, reproductive and antipredator (or predation) behaviors, one can help design conservation strategies that take into account nonhuman cognition and behavior. Such strategies include rescue, rehabilitation, translocations and reintroductions, and hence wildlife medicine.

Welfare science

The scientific approach to animal welfare is one framework that society can use to resolve questions about the proper treatment of animals. It works in conjunction with other frameworks within the broad range of ethical approaches, such as the theories, philosophies and principles of ethics outlined in the beginning of this chapter. Studying animal welfare makes important contributions to discerning animal ethics. Welfare science helps clarify problems, expose compromises, evaluate alternatives, find solutions, and grow one's understanding of how other animals experience their lives. Keeping abreast of recent research in welfare science guides humans in determining the validity of assumptions of animal well-being, while also minimizing the propensity to project one's own subjective experiences onto animals. The Five Domains Model is an excellent tool and includes five categories to assess welfare: nutrition, environment, health, behavior and mental domain.[2]

Compassion and communication tools for engaging ethical complexity

Determining what animals think, how they feel and how they suffer through philosophical arguments, cognitive ethology and welfare science must be part of our ethical decision-making process. Ultimately, one can never know what is best in the morass of ethical vagueness regarding nonhuman life. Instead, one can be compassionate in each moment by considering the needs of all species, which is enforced by having open and sustained discussions with our fellow humans. We might still make tragic choices, but less so. Every deliberation or application of wildlife medicine then becomes a practice ground for the skills of compassion and communication, which impacts our delivery of ethical care. The following are some tools for improving communication and compassion.

Narrative ethics

In narrative ethics, stories are told about ethical choices. While speaking, the teller can clarify their own needs and values, as can the listeners. These stories take the form of case examples that cover more than the medical aspects of a situation. They also highlight moral guides to living the good life, not just in veterinary care, but in all aspects of one's life. These narratives of witness with their experiential truth and passion compel re-examination of accepted medical practices and ethical precepts, which, in turn, allows us, as a community, to develop our ethical abilities. Using narrative ethics that emphasize communication does not preclude the use of principle ethics. Indeed, both contribute to understanding moral life and the process of ethical decision-making in healthcare situations.

In the case of the heron, members of the care team might gather to hear the case report of how decisions were made to care for the heron. There is no right or wrong ethical philosophy or principle here to determine. Instead, the process brings together everyday humans struggling to make the best choice possible in a given situation. In the process of telling, an internal dynamic occurs within both the teller and listeners that stimulates emotions as well as conscious and subconscious thought of past experiences, values and cultural constructs. This dynamic helps us align our behavior with our thoughts and emotions.

Socioscience

"A doctor who only knows Medicine, doesn't even know Medicine."[3]

Socioscience guides the veterinary team member in knowing more than medicine. It is like narrative ethics, in that those in science and medicine take time to examine the ethical implications of their work through intentional periods of presenting and discussing ethical case reports. During these case reports, socioscience stresses morality and ethics as well as the interdependence between science, medicine and society. It does this by considering the psychological and epistemological growth of child or adult individual, and the development of character or virtue. It focuses on growing the individual through relational challenges that focus on complex ethical situations that involve science and human communities.

Relational skills and growth are paramount; habits of mind may suffice for decisions and actions initiated by an individual but do not suffice for real-life complex situations that animal caretakers encounter. A given medical case might entail desires for the flourishing of not just the nonhuman animals but also of self, family, staff, broader communities, global society and habitats full of other species. Relational and communication skills are also important because often the best possible decision or action requires collective decision-making that can be both challenging and uncomfortable. In these decision-making processes, the group constructs meaning and understanding through the pedagogical power of discourse and reasoned argumentation. Humans together, rather than alone, have the power to integrate the emotive, developmental, culture or epistemological connections within the decisions and actions themselves.

Ethical deliberation draws on personal beliefs, individual emotive characteristics and individual identity within a community, such as gender and ethnicity. To engage in discourse that tugs at emotions, core beliefs and identities, mutual respect and tolerance of dissenting views must be supported for the development of more sophisticated learning. Under all levels of discourse, we must examine how power and authority are embedded in scientific and medical enterprises, such as privilege, class, gender and ethnicity. To truly engage in a socioscientific approach to wildlife medicine, means that people will be triggered and uncomfortable as sensibilities are challenged.

Full listening helps us attune to others and their internal states. By stilling the cognitive loops and chatter that go on inside of us, we come to attentive recognition of what another is feeling and, hence, have a greater chance to understand them and offer empathy. When another person feels heard and receives empathy, they, in turn, are in a better place to listen, as well as recognize their own emotional state without it being overridden by concerns of threat from without.

Transformational reasoning occurs when one can clearly internalize and articulate the thoughts, arguments or position of another. This is because one's reasoning becomes integrated with that of another. In socioscience processes, one begins with the presentation of controversial science or medical case studies followed by participants taking turns arguing various viewpoints. It is important to repeat back what one has heard and to argue the case you do not agree with. In this process of pretending to take the other side, one actually gains in empathy for other positions and grows in sophistication with one's newly acquired and more integrated ethical approaches. Participants can also be urged to build consensus regarding the issue to further expand their abilities in discourse.

In practical terms, a group of rehabilitators could hold monthly meetings that discuss the ethical and moral implications of their work. By coming together in this way, the group grows their relational, listening and dialogical skills. This brings greater coherence between their values, subconscious processes and behavior towards one another and the animals in their care. A trained facilitator can help guide the group in both narrative ethics and in socioscience case reports and encourage ever greater active listening and empathy.

Needs-based ethics and compassionate communication

Compassionate communication, based on Marshall Rosenberg's Nonviolent Communication theory, emphasizes honesty and empathy in interpersonal and intrapersonal relationships. Through practice, it leads to shifts in thinking and emotional responses. It is based on the understanding that human beings operate best in social groups when they receive empathy. Greater connection and rapport between individuals, so paramount in social discourse, occurs when communication (e.g., verbal, paraverbal, and body language) used arises naturally from the subconscious's emphasis on the idea of universal needs and not on judgment, blame or domination to get needs met. Instead, empathy through deep listening, authentic sharing of needs and feelings and clear requests suggest the best strategy for people to come up with creative solutions where everyone is heard and everyone, including non-human animals, has their needs considered. By developing concrete ways that people can transact through the medium of needs produces positive emotions and commitment.

In the case of the heron, we seek to empathize with the needs of everyone involved in the case: the heron, herons as a species, veterinarians, veterinary team members, rehabilitators, the people who found the heron or care for the cats that attacked the heron, family members of those working with the heron, the habitat and other species that evolved in balance with the heron (as prey and predators), individuals within local conservation and wildlife groups, such as preserves and the Audubon society, and yourself (as reader).

By equally considering the needs of all involved, one can come up with creative, synergetic solutions that deliver the best care possible to the broadest constituency. This happens because keeping all needs on the table allows one to break free from ideologic stances or cultural constructs that might normally restrain us, such as animal rights versus animal welfare, or domination versus mutualism. Instead, one comes into a spaciousness to hear one another and, indeed, listen to how life is coming through the very worthy lives of the species with which we share our communities. This does not ensure that hard choices will not still need to be made. Even if the ultimate choices one makes are regretful, such as euthanasia or trapping the feral cats living in the preserve, one's work is sustained by connecting fully to the broad diversity of life around us.

Conclusions

Component ethical discourse cannot be achieved by reading this paper, or even the thousands of tomes dedicated to ethics. Ethical processes also differ between cultures, and this article only begins to touch the surface of how multicultural skills are an important part of moral reasoning. It takes practice, hard work and discomfort for our whole lives. We can always improve. We are neither static beings, nor are others, our communities, or our science or medicine. Ethics cannot be achieved then by just one period of focus. One is not alone in this life-long effort because ethical discourse also cannot be done alone. It is a multidisciplinary effort that involves the community, in which our wildlife medicine is embedded. The question of what to do next is not "What will I or you do?" but "What will we do, together?"

References

1. Catalano A, Lyons-White J, Mills M, Knight A. Learning from published project failures in conservation. *Biol Conserv.* 2019;238. doi: 10.1016/j.biocon.2019.108223.

2. Mellor DJ, Beausoleil NJ. The 2020 five domains model: including human-animal interactions in assessments of animal welfare. *Animals.* 2022;10(10):1870.

3. Instituto de Ciéncias Biomedicus Abel Salazar. Available at: https://www.up.pt/portal/en/icbas/. Accessed February 14, 2024.

Section 10

Monday Masterclasses

Susan Horton, DVM;
Crystal Matt, DVM
Kristin Sinclair, DVM, DABVP (Avian Practice), DABVP (Exotic Companion Mammal Practice)
Session Chairs

Management of the Dyspneic Bird

Nico Schoemaker, DVM, PhD, DipECZM (Small Mammal), DipECZM (Avian), and
Yvonne van Zeeland, DVM, MVR, PhD, DipECZM (Avian), DipECZM (Small Mammal), CPBC

Session #174

Affiliation: From the Division of Zoological Medicine, Department of Clinical Sciences, Faculty of Veterinary Medicine, Utrecht University, Yalelaan 108, 3584 CM Utrecht, The Netherlands.

Abstract: One of the most seen clinical abnormalities in birds in practice is dyspnea. Knowledge of the typical avian respiratory anatomy and the specific pathogens seen in birds is extremely useful when treating such birds. During this masterclass several cases are presented, each addressing a different underlying cause. Aside from the cause, attention is also given to the pros and cons of the different diagnostic tools available, as well as the treatment options that can be used in these cases.

Introduction

Respiratory disease is commonly seen in birds. Poor husbandry and nutrition often play a role (e.g., exposure to dust, toxins or allergens, increased susceptibility to respiratory infections such as aspergillosis that can develop secondary to hypovitaminosis A). Respiratory cases will often present with obvious signs of upper and/or lower respiratory disease such as nasal discharge, sneezing, sinus swelling, stridor, cyanosis and/or dyspnea; however, due to the large reserve capacity of the lungs, such clinical signs may not be obvious until disease is advanced. In addition, patients may sometimes present with non-specific clinical signs including anorexia, weight loss, or general malaise.

When a bird is presented with respiratory distress, initial stabilization is often achieved by placing the patient in a warmed, oxygenated incubator. Many patients may furthermore benefit from the administration of an anxiolytic or sedative (e.g., midazolam), analgesic (e.g., butorphanol), and/or bronchodilator (e.g., terbutaline, salbutamol) because these drugs may help them to relax and thereby breathe better. Further work-up and treatment will mainly depend on the location and underlying cause of the respiratory distress.

Diseases of the Nasal Cavity and Sinuses

Presence of nasal discharge, plugging of the opening of the nares, redness and/or swelling of the periorbital area, ocular discharge, sneezing, yawning, shaking the head, rubbing the beak and/or scratching the face are often indicators of rhinitis and/or sinusitis. In these patients, nasal flushes will often be useful, not only because these provide a means to collect material from the nasal cavity and sinuses, but also because it helps to remove mucous and discharge to enable the bird to breathe more easily. Fine needle aspiration from the infraorbital sinuses provides an alternative means of collecting material from the upper airways, which—similar to material from nasal flushes—can subsequently be used for further testing, including cytology, culture and sensitivity testing for fungi or bacteria and/or PCR for specific pathogens (e.g., *Chlamydia psittaci*, *Mycobacterium* spp.). In case of a rhinolith (i.e., concretion of desquamated epithelium, debris, necrotic and inflammatory material), the obstruction can be removed using a small (i.e., dental) curette, after which diagnostic samples may be collected from the mucous membranes which were in direct contact with the rhinolith removed. Depending on the test findings, specific therapies (e.g., topical and/or systemic treatment with, for example antibiotics, antifungals and/or mucolytics) may be initiated. For cases that are unresponsive to initial treatment, endoscopy of the choanal area, including biopsy collection of any abnormalities, as well as imaging (e.g., radiographs, ultrasound, CT or MRI) of the nasal cavity and sinuses may assist in establishing a more definite diagnosis. Of note, nasal flushes should not be performed before diagnostic imaging because this will falsely show soft tissue presence in the sinuses due to the fluid that was flushed. A surgical approach to the sinus may be needed whereby placement of a drain may enable easy flushing of the sinus at the location of the infection.

Tracheal and Bronchial Disease

Obstructions (e.g., tracheal granulomas, strictures, foreign bodies, tracheal parasites such as *Sternostoma tracheacolum* and *Syngamus trachea*) are the most common form of tracheal disease encountered in birds. Often problems are localized at the level of the syrinx which is located at the tracheal bifurcation and represents the location at which the trachea is most narrow. Patients with tracheal disease often present acutely with a typical, loud inspiratory stridor, absence and/or altered vocalizations, increased respiratory rate and effort with open beak breathing, retching and/or gasping for air. In most of these patients, tracheoscopy will provide a definite diagnosis. Radiographs are usually non-diagnostic, while with a CT scan the exact location of the obstruction can be seen, as well as the extent of the obstruction. Initial treatment is aimed at establishing an alternate airway through the placement of an air sac tube in the (left) caudal thoracic air sac. Next, a more extensive endoscopy, or sometimes surgery, can be performed to remove the obstruction and/or collect material for further testing and help guide treatment. For obstructions involving the primary or secondary bronchi, parenteral medication combined with nebulization treatment using mucolytics (e.g., acetylcysteine) and antifungal (e.g., amphotericin B) or antibacterial drugs (e.g., gentamycin) can be attempted. Should medical therapy not be effective, CT-guided surgical intervention during which a bronchotomy is performed to remove the obstruction can also be attempted.

Trauma due to intubation may result in stricture formation with severe respiratory distress as a consequence. Removal of the tracheal rings at the location of the stricture is generally curative in such cases.

Diseases of the Lungs and Air Sacs

In contrast to birds with a tracheal obstruction, birds with primary lung pathology, except those with exposure to airborne toxins such as Teflon, often present with a more chronic history of general malaise, lethargy, weight loss that includes poor body condition and reduced pectoral mass, and anorexia. They also demonstrate an increased respiratory rate and/or effort. Sometimes, harsh sounds may be heard upon auscultation of the coelom, whereas the heart sounds may be louder or bounding due to the increased blood flow resistance in the lung parenchyma. Although radiographs are frequently taken to diagnose abnormalities of lung parenchyma, it has been found that this tool lacks sensitivity and specificity. Therefore, CT imaging and/or endoscopy with subsequent collection of biopsies for culture and sensitivity testing and/or cytologic or histopathologic evaluation will often be warranted to obtain a tentative or definite diagnosis. A CBC and biochemistry may be considered to identify hematologic abnormalities (e.g., polycythemia, leukocytosis) and/or abnormalities in the protein electrophoresis. Many birds with primary lung disease will benefit from the use of parenteral bronchodilators, NSAIDs, antibiotics or antifungals in combination with nebulization therapy, but the use of other drugs may be indicated in specific cases (e.g., furosemide for patients with lung edema due to cardiac failure, sildenafil for patients with signs suggestive of pulmonary hypertension). Endoscopic or surgical removal of fungal granulomas may be beneficial and should be considered as an option within the treatment plan.

Respiratory Disease Due to Extrapulmonary Causes

Birds do not have a diaphragm and they rely on changes in pressure in their air sacs relative to the atmospheric pressure to move air through their non-expandable lungs, so any type of space-occupying process (i.e., egg or other mass, fluid, organomegaly) in the coelomic cavity that compresses the air sacs and prevents them from expanding can result in dyspnea. Typically, these birds will demonstrate shallow breathing with increased respiratory rate. Upon abdominal palpation, a distended, firm abdomen may be palpated. In addition, birds will commonly show a history of lethargy, decreased appetite and general malaise. Depending on the underlying cause for the abdominal distension, diagnostic work-up may include imaging (e.g., radiographs, ultrasound, CT, endoscopy) to identify presence of an abdominal mass or egg, ultrasound-guided coelomocentesis to alleviate congestion and collect fluids for cytology, measurement of total solids and/or culture/sensitivity testing based on which a further therapeutic plan may be initiated aimed at the primary disease process.

Conclusions

Close observation of the patient, in particular the evaluation of its breathing pattern and presence of respiratory sounds, as well as a brief hands-on physical examination will provide important clues to help localize the origin of the bird's respiratory distress. Identifying the origin of respiratory distress will subsequently help further guide the diagnostic and therapeutic process.

References

1. Chitty J, Monks D. Lower respiratory tract disease. In: Chitty J, Monks D, eds. *BSAVA Manual of Avian Practice: A Foundation Manual.* Gloucester, UK: British Small Animal Veterinary Association; 2018: 324-333.

2. van Zeeland YRA. Upper respiratory tract disease. In: Chitty J, Monks D, eds. *BSAVA Manual of Avian Practice: A Foundation Manual.* Gloucester, UK: British Small Animal Veterinary Association; 2018: 299-316.

3. Crosta L. Respiratory diseases of parrots: anatomy, physiology, diagnosis and treatment. *Vet Clin North Am Exot Anim Pract.* 2021;24(2):397-418.

Tips and Tricks for the Avian Practitioner

Anna Osofsky, DVM, DABVP (Avian Practice),
Jennifer Graham, DVM, DABVP (Avian Practice), DABVP (Exotic Companion Mammal Practice), Dipl ACZM,
Anneliese Strunk, DVM, DABVP (Avian Practice),
Marcie Logsdon, DVM,
Brynn McCleery, DVM, DABVP (Avian Practice)

Session #137

Affiliation: From Carrollton West Pet Hospital, Carrollton, TX 75007, USA (Osofsky), Graham Veterinary Consulting, LLC, Madison, AL 35758, USA (Graham), The Center for Bird and Exotic Animal Medicine, Bothell, WA 98011, USA (Strunk), Washington State University Veterinary Teaching Hospital, Pullman, WA 99164, USA (Logsdon), Red Bank Veterinary Hospital, Tinton Falls, NJ 07722, USA (McCleery).

Abstract: This masterclass reviews a variety of tips and tricks for the avian practitioner. Topics that are covered include sedation and anesthesia techniques; wing, beak, and nail trims; surgical tips; tips for acquiring radiographs; incorporating low-stress veterinary visit concepts into practice; and some additional diagnostic and therapeutic techniques.

Sedation and Anesthesia

Humid-Vent®

Avian patients are prone to transtracheal membrane formation post-anesthetic events. Likely predisposing factors include length of the procedure, size and type of the endotracheal (ET) tube, disinfectants inadvertently left on the tube, and whether intermittent positive-pressure ventilation is performed (especially if a mechanical ventilator is used).[1] One trick the authors learned as residents was the use of a Humid-Vent® in any procedure in which the bird is intubated and the procedure is going to last more than 30 min. The Humid-Vent® is a heat and moisture exchanger placed between the anesthesia circuit and the ET tube; it is made of hygroscopic naturally absorbent microwell paper. These can be purchased at JorVet for an approximate cost of $60 for five Humid-Vent® Minis. Due to infectious disease risk, Humid-Vents® should be considered single-use only.

Doppler use in birds

The author (A.O.) uses a Doppler for all exotic pet anesthetic events. It can usually be placed quickly, especially in avian patients, and allows both the clinician and the anesthetist to listen to the patient's heart rate during a procedure. The easiest location for Doppler placement in avian patients is the superficial ulnar artery and vein located on the ventral wing near the cubital joint. The author uses two tongue depressors taped together at one end to hold the Doppler crystal in place. A very small amount of alcohol is used to moisten the skin and part the feathers but not so much that the patient is chilled or that the crystal has contact with the alcohol; doppler gel is placed on the crystal, the crystal is applied to the artery or vein and then held in place with the non-taped end of the tongue depressors (i.e., one tongue depressor on the ventral side of the wing, one on the dorsal). The tightness of the tongue depressors can be adjusted for patient size. For small patients no tape is needed between the two tongue depressors; for larger ones, apply tape to the top of one tongue depressor for three or four revolutions before taping it to the second tongue depressor. The Doppler can also be placed over the palatine artery quite easily in some species, although the patient needs to be anesthetized and intubated for this location to be safely used.

The Doppler can be used to measure blood pressure trends as well, although not when placed on the palatine artery! While studies indicate that there are no known reliable indirect means to measure blood pressure in avian patients,[2-4] measuring trends can be helpful. The blood pressure cuff can be placed on the humerus since there are no large flight feathers located here to interfere with placement.

Face masks

Creative use of face masks is often needed in exotic pet anesthesia. The author (A.O.) finds that avian patients often bite and irreparably damage the rubber diaphragms of face masks; in addition, the diaphragm opening is often not appropriately sized for these patients. The author uses self-adherent stretch bandage material to create custom diaphragms for each avian patient. Multiple layers are needed as this material is much more porous than rubber so exposure to inhalant gasses can occur more readily. Exam gloves can also be used to make a custom diaphragm that will not leak anesthetic gases so readily. Simply stretch the non-finger portion of a glove over the mask opening and tape it to the sides of the mask so the glove material is taut; a small opening is cut into the center of the glove to pass the bird's head through. With this setup, forced ventilation via the mask can often be achieved as long as the seal around the neck is good; to prevent inflation of the cervicocephalic air sacs during ventilation, a self-adherent bandage can be wrapped slightly loosely around the neck of the patient like the neck of a turtleneck shirt. Slight elevation of the patient's head during anesthesia can reduce passive regurgitation of crop contents; the author (A.O.) generally accomplishes this with a folded towel placed underneath the forced air warmer onto which the patient's head, neck, and shoulders can be placed to create a gentle angle.

In addition, even more creative face masks can be made for birds with long beaks. Plastic beverage containers such as 2-L soda bottles can be used for toucans with the bottom cut off to make the diaphragm using a self-adherent stretch bandage, and the pouring spout adapted to fit the anesthetic circuit or two tall, slender water bottles taped together for long-billed birds such as pelicans.

Emergency drugs

Because exotic pets are prone to destabilizing quickly while under anesthesia, the author (A.O.) prefers emergency drug and fluid doses calculated and drawn up before the anesthetic event. The author uses an Excel spreadsheet divided by type of animal (e.g., guinea pig, rat, rabbit, bird, snake, lizard) and drug (e.g., atropine, epinephrine, flumazenil, colloids, crystalloids); the patient's name and weight are entered in the correct spaces and the appropriate doses are calculated. This sheet is also used to calculate the commonly used sedatives (e.g., midazolam and butorphanol). The Association of Avian Veterinarians (AAV) has an even more comprehensive avian anesthesia spreadsheet and calculator available for members; it can be found on the The Golden Egg: Avian Anesthesia and Analgesia web page. If we anticipate the anesthesia will be performed using a face mask, the author always has an appropriately sized ET tube and laryngoscope ready in case of apnea. That said, nothing prevents anesthetic complications in exotic pets more than a vigilant, dedicated anesthetist; for exotic pet procedures, anesthetists should not be acting as concurrent surgical nurses.

Forced-air warming

Hypothermia under anesthesia is a major risk to small exotic patients with their high surface-to-volume ratios. Forced-air warmers have proven superior to infrared heat emitters and circulating water blankets at maintaining body temperature in parrots.[5] The author (A.O.) uses a forced-air warmer system for almost all exotic pet anesthetic events. The blanket is generally placed under rather than over the patient as monitoring respiratory rate and character is essential, even in procedures in which the body cavity does not need to be accessed. A rigid but porous surface (e.g., the top of a perforated sterilization tray) can be used under avian patients if more stability is needed to tape down wings. Reusable fabric blankets are available to keep down cost and allow washing between patients.

Electrocardiogram placement

Monitoring the bird's electrocardiogram can be a useful tool during anesthesia; however, the alligator clips routinely used on mammals can be damaging to fragile avian skin. To make this tool more avian-friendly, pinch the skin where the clip would normally go and then insert a 22- or 25-gauge needle, depending on patient size, through the skin and attach the alligator clip to the end of the needle's tip. To aid in conduction, a small gauze square or cotton ball moistened with alcohol can be placed around the needle tip before placing the alligator clip. Typical sites for application are proximal to the stifles and at the shoulder just before the patagium. Be sure to point the sharp end of the needle away from the patient's body. Add alcohol to the insertion sites judiciously so as not to cool the patient excessively.

Hypafix® adherent bandage

This tip was shared with the author (A.O.) by Lauren Powers, DVM, DABVP (Avian Practice), DABVP (Exotic Companion Mammal Practice). Hypafix® (BSN Medical) is a soft, flexible retention tape that adheres well to bird skin and feathers without causing damage. It can be used effectively for many aspects of avian anesthesia, such as securing an intravenous (IV) catheter in, or a Doppler crystal on, the superficial ulnar vein. It can even be used to tape off feathers around the surgical field.

Protecting the eye

This tip was shared with the author (A.O.) by Anneliese Strunk, DVM, DABVP (Avian Practice). Unlike dogs and cats, many bird species are prey animals and have their eyes on the sides of their heads. This eye position, combined with the common use of forced air warmers during anesthesia, leads to an increased risk of corneal drying. Reapplying artificial tears periodically throughout surgery can help minimize this risk. For added protection, cut a small well in a stack of gauze squares and fill it with ophthalmic lubricant, then place it under the bird's down eye.

AAV's The Golden Egg: Avian Anesthesia and Analgesia

These tips and tricks can all be found on the AAV's The Golden Egg: Avian Anesthesia and Analgesia, only accessible to AAV members. There are many practical resources available in this edition of the Golden Egg such as an anesthetic depth chart, pre-anesthesia fasting recommendations, and videos demonstrating many of the common anesthesia skills. We encourage everyone to give it a look.

Midazolam for sedation

Do you carry midazolam in your practice? If not, you need it! Sedation is useful for many aspects of avian medicine. Sedatives also decrease stress associated with induction but should be combined with analgesics for potentially painful procedures.[6] Midazolam is a commonly used benzodiazepine that can be administered via intramuscular (IM), subcutaneous (SC), or intranasal (IN) routes.[7-13] Sometimes used as a sole sedative for non-painful procedures, including radiographs, midazolam can be reversed with flumazenil. Midazolam is more potent and longer lasting than diazepam, and does not adversely affect mean arterial blood pressure or blood gases in select avian species. Midazolam administration may be occasionally associated with regurgitation in larger macaws and cockatoos. Routine use of flumazenil as a benzodiazepine reversal agent can be rarely associated with adverse reactions in human and canine patients. While the author (J.G.) has witnessed sudden death in a bird following IM injection of flumazenil, no complications have been noted following IN administration by the author and others. Helpful hints: It is ideal for birds to be fasted for 2-3 hours before sedation although not mandatory (if the crop is overly full it is best to aspirate contents before sedation). Midazolam is helpful to reduce anxiety associated with patients in respiratory distress. Midazolam dosage recommendations range from 0.5 to 5 mg/kg depending on species (consult *Carpenter's Exotic Animal Formulary*). The addition of an opioid can provide analgesia and an added sedative effect and anecdotally decreases midazolam-related regurgitation. Flumazenil reversal is generally 0.02-0.1 mg/kg IV or IO but many use this drug IM and IN.

Intubation

With the glottis located at the base of the tongue, endotracheal intubation is relatively easy in most birds. Birds lack an epiglottis, have a longer trachea relative to mammals, and have complete tracheal rings. In birds where visualization of the glottis is difficult, pulling the tongue gently with gauze or atraumatic forceps can improve the field of view. Uncuffed ETs should be used. The use of Cole ETs can help facilitate a better seal to decrease the potential for air leakage. Respiratory secretions can occlude ETs and catheters used as ETs, particularly with smaller patients. Facemasks may be preferable in budgerigar and passerine-size birds for routine procedures. Intubation-induced tracheal stenosis has been reported in multiple species and risk factors may include tracheal trauma, chemical insult (from sterilizing agents), and focal desiccation from airflow. Care should be taken to intubate gently, and shallowly, with caution used when moving the head and neck of intubated birds. The tip of the ET should be lubricated, as for other species. For longer anesthetic procedures in birds prone to mucus plugs, such as Anseriformes, hourly tube replacement can be considered. Birds with an obstructed ET may have difficulties exhaling or capnography machine not recording CO_2. Birds in which the crop is not fully emptied should have their head elevated to prevent passive regurgitation.

Procedures

Ovocentesis update

Chronic egg laying can result in weakness, osteopenia, egg binding, yolk coelomitis, and pathologic fractures. Initial stabilization measures involve the administration of fluids, parenteral calcium, and possibly vitamin D (injectable vitamin A/D/E formulations are most often used). Diagnostics including blood work (including iCa2+), radiographs, ultrasound, and coelomocentesis can be helpful to determine the extent and type of disease. Care should be taken to remove any coelomic fluid before putting the bird in a non-vertical position. Prostaglandin and hormonal therapy may be warranted to induce oviductal contractions once the patient has been stabilized if non-obstructive egg binding is diagnosed. Ovocentesis may be performed to facilitate passage of an egg but runs the risk of inducing coelomitis and is usually reserved for cases in which conservative therapy is unsuccessful. Salpingohysterectomy is recommended for birds with chronic recurring disease or when egg removal is otherwise unsuccessful. A retrospective evaluating 20 birds that underwent percloacal ovocentesis successfully facilitated egg removal in 16 of 20 cases (80%) with no complications observed in 14 of 20 cases (70%).[14] Percloacal ovocentesis is not without potential complications; eggs fragmented in 2 of 20 (10%) cases, with coelomic penetration and coelomitis suspected in a single case (5%). Transcoelomic ovocentesis should be avoided in preference for percloacal approach if ovocentesis is necessary.

Intraosseous catheter

Intraosseous (IO) catheters can be advantageous to use in birds < 60 g, when IV catheterization is difficult, and to prevent hemorrhage if the catheter is removed by the bird. IO catheters should never be placed in pneumatic bones such as the humerus or femur to avoid the risk of air sacculitis, pneumonia, or asphyxiation due to fluid entering the respiratory tract. IO catheters can be placed either in the ulna or the tibiotarsal crest and are generally well tolerated by the patient. Fluid is rapidly absorbed into the systemic circulation when administered intraosseously, with over 50% of the fluid reaching the central circulation within 30 seconds. A major complication of IO catheters is osteomyelitis; therefore, sterile surgical preparation of the area is essential. Placement can be performed in an anesthetized bird or with a local block in severely debilitated birds. After the area is plucked and scrubbed, a local block of 2% lidocaine diluted with saline can be used to numb the area. The dose used can be up to 0.2 mg/100 g bird. Medium- to large-sized birds require a 20- to 22-gauge 1.5- to 2.5-inch spinal needle. A 25-gauge hypodermic needle can be used in smaller birds. A stylet can be made for the hypodermic needle using sterile cerclage wire or 3-0 to 6-0 stainless steel suture, thereby preventing obstruction of the needle lumen with cortical bone. The manus should be pronated while palpating the dorsal tubercle of the distal ulna as a landmark for ulnar IO catheter insertion. If possible, it is best to avoid the extensor digitorum communis, extensor carpi ulnaris and extensor longus digiti. For tibiotarsal placement, the knee should be flexed while the cnemial crest is palpated; the needle should be inserted into the crest in a cranio-medial direction, avoiding the patellar tendon. A padded bandage or lateral splint can be used to stabilize the catheter. Depending on the location, part of a syringe case can be placed over the catheter tip for protection. Compared to IV catheters, more fluid resistance is encountered with IO catheters, making small-volume syringes or continuous rate infusion with a small syringe pump preferable. With appropriate sterility and maintenance, IO catheters can be used for up to three days. IO catheters should not be used in osteopenic birds due to risk of iatrogenic bone fracture.

ET tube to facilitate tracheal wash

The patients should be sedated and anesthetized for intubation.[15-17] A sterile ET tube is placed. A red rubber feeding tube or polypropylene catheter is advanced to the level of the syrinx. Sterile saline is infused into the trachea and immediately aspirated back into the syringe. For safety reasons, keep equipment for an air sac cannula ready; however, if the bird has severe respiratory compromise, first place an air sac cannula. Helpful hints: Volumes of 1-2 ml/kg of nonbacteriostatic 0.9% sterile saline are recommended. Samples can be submitted for cytology, culture (e.g., bacterial and fungal), and PCR analysis. As an alternative, a direct tracheal swab can be performed to sample the proximal trachea. A sterile culturette or mini culturette is used based on the size of the bird. Moisten the swab with sterile saline prior to collecting the sample.

Nasal flush (0.9% saline versus saline and sterile water)

A nasal flush can be helpful to obtain samples for diagnostics as well as therapeutic treatment. Sterile nonbacteriostatic 0.9% saline (preferably warm) should be used if the sample is being submitted for diagnostic purposes. A 50:50 mixture of sterile 0.9% saline and sterile water can be used for therapeutic flushes since this mixture may be a bit less irritating to the bird than straight saline.

The bird should be firmly restrained with the head positioned below the body (i.e., head down, tail up, keel towards the floor). Warm flush is instilled into one naris and a collection device (sterile if obtaining material for culture) is held underneath the bird's head to collect fluid exiting from the choanal slit and out the mouth as well as out of the opposite naris. Note that the nasal passages do not connect in passerine birds so fluid will not exit out the opposite naris in these species. The amount of fluid used can vary but flush volumes can range from 1-10 ml depending on the size of the patient. Fluid obtained in this fashion can be submitted for culture or next-generation DNA sequencing and/or cytologic examination. While a syringe tip can form a seal for flushing in most psittacine species, sterilized gavage needles with a ball tip can be attached to the syringe for larger species such as macaws where a syringe tip is smaller than the naris opening.

Bandage recommendations

The figure-of-eight bandage can be useful to manage fractures, soft tissue injuries, luxations, or when the wing needs to be stabilized to hold a catheter in place.[18] Cast padding is used for the first layer and this is covered by self-adherent bandage material. With the wing in a flexed position, start the bandage at the carpus; wrap the gauze around the carpus creating the top loop of the figure 8. The gauze is now directed toward the elbow and wrapping it around the area just proximal to the elbow helps prevent slippage and creates the lower loop of the figure 8. Care should be taken not to wrap too tightly or create too much bulk and avoid placing pressure over the patagium. A recent recommendation has been made against figure-of-eight bandages for humeral fractures because the bandage worsens the distraction of the humeral fracture. In a captive or companion birds where flight is not critical and surgery is not an option, a body wrap and cage rest, with the hope the fracture does not go to a non-union, can be attempted (personal communication, Julia Ponder, 2017).

A retrospective evaluating tibiotarsal fractures in 86 birds treated with tape splints included birds treated with a tape splint alone (79/86) and tape splint with IM pin (7/86).[19] Median time to fracture stabilization based on palpation was 19 days (range, 7-49 days). In most cases (61/86), the initial splint was maintained until fracture healing was complete. A successful outcome was documented in 92% (79/86) of birds. Increased complications occurred in birds with a predator attack, lack of deep pain, or where the splint was removed prior to 14 days following placement. Overall, findings indicated tape splint as an appropriate means for treatment of tibiotarsal fractures in birds weighing less than 200 g.

Radiographs

Horizontal beam radiographs

Horizontal beam radiographs are routinely used with reptile patients but they can also be quite helpful for avian and small mammal patients in certain situations. The use of this technique requires a mobile x-ray tube head with a moveable imaging detector. Although you cannot evaluate avian films in the same manner as properly positioned films because of the superimposition of wings or legs, you can use this technique in situations where regularly positioned films would be contraindicated. This may be particularly useful if you do not have access to ultrasound in your practice. If you have a bird with coelomic distention, a horizontal beam radiograph can determine if a bird has evidence of a mineralized egg or metallic densities. The major advantage of this approach can be in situations of ascites. For example, chickens are prone to the development of ovarian neoplasia and concurrent ascites is common. The author (J.G.) has used this approach to evaluate hens with ascites for a mass effect in the region of the gonad and kidney before the removal of coelomic fluid. Helpful hints: Never lay a bird flat if you suspect it has coelomic fluid. You can remove fluid via aspiration before positioning for films to reduce the chance of fluid flooding the lungs. Ultrasound is helpful to determine the location and extent of fluid.

Bird board

Many avian positioning devices are commercially available to obtain excellent-quality radiographic images. For example: Bird Board, 8205 Alba Ct., Citrus Heights, CA, USA and VSP Avian Neck Restraint & VSP Avian Restraint Board Magazine, Veterinary Specialty Products, Shawnee, KS, USA. If devices are interposed between the bird and the film sensor, a small increase in kVp (2-4 kVp) may be needed to compensate.

Masking tape for radiographs

Tape can be helpful for positioning avian patients for radiographs but certain tape (e.g., clear tape and cloth medical tape) may damage or remove feathers or scales. The authors find that masking tape can be placed on feathers for radiographic positioning and removed without causing damage to or removing feathers.

Products to Have on the Shelf

Vibravenos

Long-acting injectable doxycycline hyclate (aka Vibravenos; Psittavet) can be administered IM once every 5-7 days making this drug a nice option for bird owners that have difficulty giving oral medications. Vibravenos is not available in the United States so it is generally ordered from the Netherlands. Contact dorresteingm@planet.nl for ordering information. In addition, the volume required for injection is quite large so it should be divided into two sites as appropriate for patient size.

Vinegar for papillomas

Routine avian examinations should include evaluating the vent for symmetry, integrity and hygiene. The cloaca should be everted circumferentially, checking for lacerations, swellings or papillomas. Vinegar (i.e., acetic acid) turns papillomas frosty white. Biospy and histopathology are needed to make a definitive diagnosis of a papilloma.

Grooming

Wing trims utilizing a less severe technique

In some European countries, trimming of avian wings is illegal. While that is not the case in the United States, wing trimming remains a controversial subject. Many owners present their companion birds for wing trims because of a lack of education or to continue what was performed by the breeder or pet store. Owners are also commonly driven to wing trimming by fear of their pet escaping or being injured. Neither of these concerns are alleviated by wing trimming. As avian veterinarians, we must advocate for our patients and educate our clients on the pros and cons of wing trimming, especially since this subject matter is incredibly important when evaluating the welfare of captive avian species.

Education of the client is the first step. The benefits of flight, including the ability to respond normally to noxious stimuli and provide a means of escape, provide choice to the bird, and as a means of exercise, should be discussed. Possible sequelae of wing trimming should also be reviewed including injury (e.g., fractured blood feathers; beak tip trauma or fractures; limb or keel fractures; soft tissue injuries over the keel or tail base; predation injury from inability to get away from cats, dogs, and other birds in the home), feather-damaging behavior from irritation or stress, and psychological damage from lack of choice and subsequent learned helplessness.[20-22] Chronic trauma from falling or crash-landing may result in osteomyelitis and nerve impingement, and unflighted birds may also be at increased risk of obesity, cardiovascular disease (including atherosclerosis), pododermatitis, hepatopathies, and degenerative joint disease and osteoarthritis. Wing trimming alters the aerodynamic shape of the wings, directly affecting a bird's ability to fly. If flight is more difficult or results in falling or trauma, birds can learn over time to not fly—even when the wings are untrimmed—due to these negative experiences. Therefore, even a temporary wing trim could have long-term effects. Studies in mynahs have also shown that flight restriction can reduce pectoralis muscle strength in relatively short periods.

The owner's goals for wing trimming should be explored, with particular focus on whether these goals can be achieved without the removal of flight ability. Alternative options to wing trimming can include recall training, bird-proofing the home, supervising or separating other pets when the bird is out, turning off ceiling fans, and placing stickers on windows to improve visibility. Ultimately, the decision to trim the wings will be the owner's. If the owner is still interested in wing trimming after a thorough discussion, a lesser technique to mitigate some of the negative aspects of wing trimming can be considered. The skinny repeal trim may be an excellent option for these birds.

Skinny or caudal vane wing trim procedure: During a skinny wing trim, only the trailing edge of the feather is trimmed. The length of the vane removed can vary based on the patient's body type, wing load, tail length, body condition score and weight, medical history, and current flight ability, as well as the owner's goals. Therefore, a client-patient-veterinarian relationship should be established, and the bird should undergo a complete physical exam before wing trimming. For birds with reduced flight ability, more of the vane tip (distal portion) can be left intact, thereby improving lift ability. As with all wing trims, feathers should be trimmed symmetrically between the wings and only the primary feathers should be trimmed. The distal most primaries (e.g., typically primaries 9 and 10) should be trimmed as these feathers provide the most thrust during flapping flight.

Effect on flight mechanics: The skinny wing trim reduces compression of air on the downstroke, leading to a gradual loss of elevation. On the upstroke, less resistance is met. This combination leads to the wings moving faster in both directions and increased exercise to the bird in a shorter period. Normal flight posture is maintained, and birds have active control of landing.

Proposed benefits:

- Reduced rachis shredding and feather damaging due to less irritation (i.e., rachis not cut/no sharp edges)
- Reduced risk of breaking blood feathers (i.e., protected by full length of other feathers)
- Reduced risk of trauma from falling and harsh landings
- Normal molt cycle maintained due to normal torque at the germinal disc
- Normal function of the wing providing improved psychological health
- Aesthetically pleasing to the owners (wings fold normally over the back)
- Customizable to the bird's flight abilities (i.e., number of feathers trimmed, top length left intact)

Wing trimming in birds is not meant to prevent flight completely. Rather, trimming serves to prevent upward lift and sustained flight but still allow normal downward gliding and landing. The skinny trim goes beyond traditional trims in its ability to maintain normal flight posture with improved comfort to the patient.

Owners currently control the decision regarding whether to trim a bird's wings. If the veterinarian refuses to trim the wings, this can lead to loss of the client and the bird experiencing a more aggressive, unsafe wing trim elsewhere (e.g., at the pet store, by the breeder, or at home with guidance from the internet). Compromising and offering a skinny trim can improve patient welfare and client-veterinarian trust if leaving the bird flighted is not an option for the owner.

The AAV has an excellent resource on wing trimming options in the Client Education Handouts portion of the website available to members. A wing trim record is also available in the AAV publication library online. Notes about the bird's wing trim can be recorded and scanned into the medical record. Visit www.aav.org.

Beak trims and reducing stress

Beak trims are not required in birds with normally occluding beaks performing normal self-grooming behaviors. If beak elongation is present, a thorough medical work-up is indicated to evaluate for underlying causes. Birds with chronic malocclusion from genetic, congenital, or traumatic causes may require lifelong beak trims. Environmental changes to encourage beak wear and self-grooming should be performed to limit the number of trims required.

The author (B.V.M.) routinely recommends mild sedation for corrective beak trims. Midazolam has been evaluated in a variety of bird species and can provide variable levels of sedation with the added benefits of muscle relaxation, anxiolysis, amnesia, and appetite stimulation that may be helpful for animals with oral pain and minor beak trauma. More information on midazolam usage can be found above (see Sedation and Anesthesia). The author (B.V.M.) commonly uses a dose of 1-2 mg/kg in larger parrots on the low end for cockatoo species and higher doses in smaller parrots and passerine species, up to 4-6 mg/kg reported. IN butorphanol can be utilized in combination with midazolam in patients where pain is expected or to potentiate the sedation effects of the benzodiazepine. The author commonly uses a dose of 1-2 mg/kg IN for butorphanol. The author (B.V.M.) prefers IM, IV, or IO administration of butorphanol for patients with more than moderate pain and may utilize higher doses. Reversal with IN flumazenil can be performed on birds undergoing sedation for beak trimming (see Sedation and Anesthesia for dosing information).

Patients requiring more aggressive corrective beak trims can be anesthetized or potentially have the beak trimmed in stages. The benefits and risks of sedation should be discussed with the owner, doses should be based on patient assessment and the specific procedure (e.g., length of sedation required, any concerns for pain), and patient monitoring should be performed during the sedation.

High-speed rotary tools are commonly used for trimming of the rhamphotheca (horny outer layer) of the beak, and a variety of bits exist in different sizes and shapes to accomplish trims in various bird species. Since rotary tools generate heat, short periods of contact between the bit and keratin are recommended. Care should be taken to avoid continual pressure that could lead to thermal damage to the sensitive dermal tissues below the rhamphotheca. Other handheld tools can also be used, depending on the keratin deformity being addressed. Cuticle nippers are a great option to carefully remove overgrown keratin from the beak.

As mentioned above, beak trimming is not a routine procedure for a bird. If trims are required, development of an enrichment plan is essential for the patient to help increase the use and wear of the beak and to keep the beak in better alignment.

Nail trims

Various methods of nail trimming can be performed in captive birds. For small to medium birds, the author (B.V.M.) commonly uses a Bovie Medical Corporation disposable high-temperature fine-tip cautery pen. These devices are small, battery-operated cautery pens intended for coagulating tissue or arresting bleeding from small vessels by heat conducted through the wire tip. They are commonly used during ophthalmic and general surgeries. The wire loop of the cautery pen can be gently reshaped to fit larger nails. The unit heats up quickly (i.e., approximately 1 second) at the touch of a button. Benefits of this method include a silent and smooth cutting experience compared to a rotary device such as a Dremel, low cost (i.e., approximately $10-15 per pen, sold in boxes of 10 pens), and ability to cauterize if the nail is accidentally cut too short leading to bleeding, although this should always be avoided when possible to prevent pain to the patient. The downside of the cautery pen is that it produces a small amount of smoke and odor. During nail trimming with cautery, it is essential to carefully restrain the bird to prevent any burn injuries, and the cautery unit should always be pre-placed on the nail before being activated. Be careful not to touch the cautery pen to feathers or towels (both of which are flammable), and do not use alcohol around the cautery pens. Do not pull firmly when using the cautery pen, as this can break the wire. In human hospitals, these pens are single-use; however, the pens can be reused for grooming and last for weeks, and they can be gas sterilized between uses if sterility is required.

Another helpful tool for nail trimming is the cylindrical aluminum oxide grinding stone bit (Dremel, bit model 932). The unique shape of this grinding bit allows you to shorten the nail along the side of the bit, followed by rounding and smoothing of the nail tip at the end of the bit which is concave in shape.

Restraint devices

Restraint devices, such as collars and vests, are utilized commonly in avian medicine. In the clinic, restraint devices can help prevent patients from reaching and removing IV or IO catheters, sutures, bandages, air sac cannulas, or other external devices. Birds may be sent home with restraint devices to protect incisions, bandages, or wounds or to prevent self-mutilation.

In humans, physical restraint is infrequently used in modern-day medicine. Instead, medications and behavioral therapy are utilized. A restraint device should also be the last resort for an avian patient in most cases. Alternative options including improved pain management and pharmacotherapy should be considered before placement of a restraint device. When placed, the goal should be temporary use of the device while the patient is healing, a diagnostic work-up is being performed, additional therapeutics are taking effect, or behavioral training with applied behavioral analysis techniques is being instituted.

When a restraint device is used in an avian patient, every effort should be made to choose the least stressful device and to apply the device in a low-stress manner. IN or injectable sedation can be utilized in these patients. Midazolam with or without butorphanol are common drugs that can be administered via a variety of routes in avian patients and can provide sedation, reduced anxiety, and amnesia (see above). No reversal or low-dose reversal with flumazenil is recommended for patients during restraint device applications. This allows the bird to slowly recover from the sedation with the collar or vest in place.

Birds should be monitored closely after application of a restraint device. This should be performed in the veterinary clinic. Some birds will have severe stress reactions to the device and may thrash, fall over, or avoid any movement. Various types of devices may need to be trialed, ideally with mild sedation on board, before determining which is best tolerated. Placing the patient in a well-padded cage with no perches or water bowls is recommended in the immediate period after placement of the restraint device to prevent injury. Food treats can be offered as a distraction and positive reinforcer.

Restraint types:

- Elizabethan collars. These are the most utilized collars. Commercial avian e-collars are often thick and heavy. Although commercial collars come in a variety of circumferences and neck sizes, finding the perfect fit for the patient can be challenging. While the material is more durable against being chewed, the limited size options often result in the patient being able to reach the collar edge or the inside of the neck area. For these reasons, the author (B.V.M.) prefers homemade e-collars made of radiographic film for most medium or small birds. These collars are lighter and can be customized to the patient, depending on the area of the body requiring protection. Although a downward-facing e-collar has been historically recommended for avian patients, the author (B.V.M.) prefers a collar that is parallel with the ground (straight out from the neck) as this reduces irritation to the wings (carpus and propatagium) and prevents the patient from tripping on the collar if a larger restraint device is needed. The neck area can be padded with white medical tape, Elastikon, or moleskin bandaging material. Some birds do well with fleece e-collars which can be purchased online or homemade. These collars are soft and less irritating to the neck and wings. Many of the commercial varieties come with distractor fleece tabs, as well. Hair scrunchies placed around the neck can also be used in some birds.

- Cylindrical neck collars. Polycarbonate plastic spherical or cylindrical collars are available commercially in a variety of sizes. These collars are easy to apply and remove. In medium and larger birds, these collars have limited interference in ambulation and acquisition of food and water; however, in smaller birds, the commercial plastic collars are heavy and can lead to ataxia and other balance issues. These hard plastic collars can also cause pressure wounds on the mandible, and close monitoring is required when utilizing these collars. Pipe insulation or pool noodle foam collars are a low-cost alternative that are lightweight and often well tolerated. This material is also great to have on hand in the clinic to make shoes for birds with pododermatitis. One tip for a proper fit is to ensure the collar is narrow enough to prevent the gnathotheca from hooking into the center of the collar. Like all of the collars mentioned, the collar should be loose enough to spin relatively easily on the neck.

- Vests. Some birds tolerate fleece vests better than collars. Commercial vests exist with reinforced areas over the chest for birds that self-mutilate the pectoral area. Pre-conditioning birds to vests may help improve compliance, similar to training a bird to wear a harness. Operant conditioning with positive reinforcement is recommended.

- Chicken aprons. Hen aprons or chicken saddles are devices that can be placed over the wings to cover the backs and/or shoulders of hens. These devices help protect the hen from trauma during mating inflicted from the spurs and nails of roosters. These devices can also be utilized to protect wounds from conspecific mutilation. Various sizes and styles are available commercially to accommodate different breeds and areas of concern requiring protection. As with vests in parrots, a preconditioning period with plenty of positive reinforcement is recommended to reduce stress associated with application and wearing of the apron. Note: Some companies producing chicken aprons also make neoprene boots for chickens and ducks that can be very helpful in the management of pododermatitis.

Low-Stress Parrot Examinations: The Silver Standard

Decreasing or eliminating the amount of stress a patient experiences in the exam room is beneficial not only for the patient, but also for the owner and veterinary staff as well. Clients are less likely to bring their bird back if they had a negative experience in the exam room, and veterinary team members are more likely to injure, or be injured by, a stressed parrot. Parrots have also been shown to exhibit a stress leukogram in response to handling and restraint.[23] While it can be difficult, or in some cases impossible to completely eliminate the stress experienced by a parrot in a clinical setting, every little bit of stress reduction counts. Below are a few examples of low-input strategies to reduce stress during parrot exams and treatments.

Preparation is key!

Have everything ready first. This seems very straight forward but should never be overlooked. You want to have the bird in hand for as little time as possible. Things should be smooth and swift, but not fast or rushed. Have everything you might need readily available and within easy reach before restraining your bird.

Limit number of restraint times

The fewer times a patient must be restrained the better. Sometimes this may mean delaying the initial hands-on physical exam. If a patient presents for a condition that you know you will want further diagnostics for, it is ok to hold off on doing a physical exam in the exam room. The trick with this is clear communication with the owner. You want to make sure that they understand: 1. Why you are doing this (i.e., to limit the amount of stress on their pet) and 2. Their pet will receive a full physical examination before you perform the diagnostics and any anesthesia, and if you find anything that changes your mind, that you will talk with them again before proceeding. Remember, what happens in the treatment area of any clinic is a complete black box to owners, and you don't want them filling up those blanks with their imagination. Similarly, take advantage of anytime the patient is sedated or anesthetized. If you sedate or anesthetize for radiographs, get owner permission to draw blood at the same time, even if you are unsure as to whether or not you will need to run it.

Tongue depressors

Give your patients something to do during their exam! For medium- to large-size parrots, tongue depressors are the perfect size destruction-distractions. For most birds, chewing on a tongue depressor gives them something to take their mind off the exam. You may need to have them pause for a minute during auscultation, but for the rest of the exam, chew away! Just be careful to ensure that no one is hit by tiny wooden splinters and that the splinters are falling out of the bird's mouth, not into it. Fortunately, most tongue depressors are made of birch wood, which is non-toxic to birds.

Popcorn

Popcorn is the best medication delivery device since the invention of the syringe. There are wonderful methods available for positive-reinforcement training to get a parrot to willingly take medications from a syringe and this is the ideal.[24,25] But some owners are simply not up to that level of training and time commitment. Providing the medication on a treat can be a great alternative to restraining and hand-medicating a parrot on a daily or twice daily basis. The wonderful thing about popcorn is that it is cheap, readily available, easily soaks up medications, and does not crumble or fall apart the way that crackers or bread-type treats do. Of course, this only works for small to moderate volumes of medications that taste good, and you must make sure that if the patient likes to dunk its food that it is given the popcorn well away from the water bowl. This trick has made the difference between treatment or not in many older patients that otherwise would not be receiving their daily arthritis medication.

Weigh-in

For patients that don't step up for new people or won't stand on a perch in an exam room, the easiest way to get a weight during the exam is by zeroing the entire travel crate on a scale and then reading the bird's weight in the negative once it is out of the crate. If your scale doesn't read into the negative, get a new one. If you miss this opportunity, you have a second chance when you go to put the bird back into its crate after the exam is finished.

Handle the stressed bird first

It's time for the annual exam on the Smiths' three birds. You see them once a year and once a year only. The owners mean well enough but don't do anything to desensitize them to their annual visit. As usual, two of the three birds are sitting in their travel enclosures relatively calmly while one is climbing circles in its crate getting more and more worked up. This is the bird that you want to examine first. It is amazing how often the nervous bird will calm down after he is done with his exam, especially if they have been there before and know that after their exam they are done. Whereas if you examine the other two birds first, the nervous one just has more and more time to get worked up.

Initial restraint

Acclimating a parrot to being restrained in a towel is the gold standard and is something that can be accomplished by owners at home or with careful desensitization in the clinic. Not having time to perform this type of desensitization is not an excuse to ignore other little things that can be done to decrease the stress of toweling. If the patient must be grabbed directly out of the crate, move slowly and steadily. Do not pounce on the bird. Similarly, the age-old lights-out trick (i.e., where you move into position, then quickly turn the lights off to prevent the bird from moving away while grabbing it) should only be used when the alternative is going to be more stressful (e.g., a flighted, non-hand tame budgerigar in a cage with multiple others).

Surgery Tips

Magnification and light sources

Visualization of a surgical site is critical. Traditional surgical lights may be adequate in many cases and should be positioned to prevent the development of shadows in the surgical site. A head-mounted light source is invaluable for coelomic procedures since the surgeon tends to be working in a hole. The surgeon may find that a headlamp is beneficial for all procedures.

Another light source to consider is a rigid endoscope, which can be connected to a stand aiming into an incision to provide light when performing procedures deep within the coelom. If connected to a camera, this setup can also be used to provide an even more magnified view of the procedure compared to traditional head-mounted loupes.

There are non-medical headlamps available online, as well as lamps designed specifically for surgery for use alone (e.g., on a headband or mounted on glasses) or in conjunction with loupes. Ideally, lights should have high color fidelity. The scale measuring this is the color rendition index (CRI). A CRI of 100 is equivalent to how colors would appear in direct sunlight. Surgical lights should have a CRI of greater than 95. Lower-quality lights tend to distort colors of tissue, causing them to appear more cyanotic or hyperemic, depending on the light.

There are numerous options for loupes, offering different levels of magnification. Headband and visor magnifiers are available online. They do not require a custom fit, can be used with glasses, and come in a range of magnifications (1.5-3.5x). The downside of these is that the focal distance is not customizable, leading to the need for changes to table height to keep the surgical site in focus. There is also a tendency for the user to change their posture to facilitate this, leading to fatigue and issues with the neck and back (see Ergonomics section below).

The ideal option is a set of custom-designed surgical loupes. These have a personalized focal distance, enhancing the surgeon's ergonomics. They also can be custom-designed with prescription lenses for those who need vision correction.

Ergonomics

Correct posture and ergonomics during surgery will minimize the risk of stress, fatigue, and pain or numbness of the head, neck, back, shoulders, and legs.[26] Long-term poor posture or repetitive motions can lead to the need for medical attention.

There are resources available for best practices to minimize issues associated with surgery, including stretches and processes to minimize risk before and alleviate discomfort after procedures.[27,28]

Key things to consider are listed below.

1. Pre-procedure planning to minimize stress. Perform a walk-through with staff before starting the sedation or anesthesia for the procedure to ensure that all supplies needed for the entire process are ready to go, including lighting and table position.

2. Body position. The surgeon's head should be up with a neutral neck, elbows bent at a 90-degree angle, hands at same level as elbows; set up small breaks for brief range-of-motion (ROM) exercises to prevent prolonged static postures. Monitors being used for endoscopic or ultrasound examination should be at a height that allows this head and neck position to be maintained.

3. Hands and instruments. Keep hands inside the box, do not cross midline, use appropriately sized instruments for the surgeon's hands, utilize ratchets to prevent prolonged squeezing motions, use power grips to hold instruments.

4. Post-procedure. Consider briefing with staff to review the procedure in a non-judgmental way and make adjustments to protocols moving forward.

Transparent plastic drapes

Clear drapes are available in a variety of sizes, with and without fenestrations. The central area often has an area with adhesive, to help isolate the surgical site. One example is the Veterinary Transparent Surgical Drape (Veterinary Specialty Products, Inc., Shawnee, KS, USA).

These drapes allow direct patient monitoring, including the very important tracking of ventilatory depth. In addition, the seal created by the adhesive against the prepared skin helps trap heat under the drape and maintain body temperature, especially when used in conjunction with a warm air thermal heating unit. They also help prevent feather dust and debris from being blown into the surgical site when this heating method is used. Wait until after the drape has been placed and is well adhered before turning on the warm air thermal unit to best prevent surgical site contamination.

Cutting through one side of the drape (from the fenestration to the outer border) allows the surgeon to wrap the drape around a limb.

Tegaderm can be used to create an even smaller fenestration by covering the existing opening and creating a new, customized window. Tegaderm can also be used to reinforce an area where the transparent drape did not achieve an adequate seal, and to cover non-sterile parts of the body (i.e. distal aspect of limbs undergoing amputation).

Glad Press'n Seal Cling Film has been shown to be appropriate as a sterile drape for lab rodents right out of the box.[29] Remove the first 25 cm of the roll before using.

Lone Star® (Ring) Retractor System

The Lone Star® Retractor System (LSR, Cooper Surgical, Trumbull, CT) is a unique customizable ring retractor developed to facilitate visualization in human procedures.[30] The ring can be put together in a variety of configurations of shape and size, and there are sharp and blunted hook stays to hold onto tissue. The hook stays are connected to the ring with an elastic band that can be attached with variable amounts of tension and can be used sequentially to gradually expose more tissue. While they can be used in numerous applications, they are especially useful in birds for visualization during coelomic surgery, and can be used to hold the vent open and evert the oviduct to aid in the egg extraction process, similar to the technique described in guinea pigs for removal of urethral calculi.[31]

References

1. Sykes JM, Neiffer D, Terrell S, et al. Review of 23 cases of postintubation tracheal obstruction in birds. *J Zoo Wildl Med.* 2013;44:700-713.

2. Zehnder AM, Hawkins MG, Pascoe PJ, et al. Evaluation of indirect blood pressure monitoring in awake and anesthetized red-tailed hawks (*Buteo jamaicensis*): effects of cuff size, cuff placement, and monitoring equipment. *Vet Anaesth Analg.* 2009:36:464-479.

3. Acierno MJ, da Cunha A, Smith J, et al. Agreement between direct and indirect blood pressure measurements obtained from anesthetized Hispaniolan Amazon parrots. *J Am Vet Med Assoc.* 2008;233:1587-1590.

4. Johnston MS, Davidowski LA, Rao S, et al. Precision of repeated, Doppler-derived indirect blood pressure measurements in conscious psittacine birds. *J Avian Med Surg.* 2011;25:83-90.

5. Rembert MS, Smith JA, Hosgood G, et al. Comparison of traditional thermal support devices with the forced-air warmer system in anesthetized Hispaniolan Amazon parrots (*Amazona ventralis*). *J Avian Med Surg.* 2001;15:187-193.

6. Mans C. Sedation of pet birds. *J Exot Pet Med.* 2014;23:152-157.

7. Schaffer DPH, Araujo NLLC, Raposo ACS, et al. Sedative effects of intranasal midazolam administration in wild caught blue-fronted Amazon (*Amazona aestiva*) and orange-winged Amazon (*Amazona amazonica*) parrots. *J Avian Med Surg.* 2017;31:213-218.

8. Mans C, Guzman DSM, Lahner LL, et al. Sedation and physiologic response to manual restraint after intranasal administration of midazolam in Hispaniolan parrots (*Amazona ventralis*). *J Avian Med Surg.* 2012;26:130-139.

9. Vasal N, Eskandari MN. Sedative effects of midazolam and xylazine with or without ketamine and detomidine alone following intranasal administration in ring-necked parakeets. *J Am Vet Med Assoc.* 2006;228:383-388.

10. Sadegh AB. Comparison of intranasal administration of xylazine, diazepam, and midazolam in budgerigars (*Melopsittacus undulatus*): clinical evaluation. *J Zoo Wildl Med.* 2013;44:241-244.

11. Vesal N, Zare P. Clinical evaluation of intranasal benzodiazepines, alpha-agonists and their antagonists in canaries. *Vet Anaesth Analg.* 2006;33:143-148.

12. Gilbert DB, Patterson TA, Rose SP. Midazolam induces amnesia in a simple, one-trial, maze-learning task in young chicks. *Pharmacol Biochem Behav.* 1989;34:439-442.

13. Farkas L, Crowe SF. The role of the benzodiazepine-GABA system in the memory processes of the day-old chick. *Pharmacol Biochem Behav.* 2000;65:223-231.

14. Abou-Zahr T, Carrasco DC, Jones SL, Dutton TAG. Percloacal ovocentesis in the treatment of avian egg binding: review of 20 cases. *J Avian Med Surg.* 2019;33(3):251-257.

15. Hawkins MG, Pascoe PJ. Cagebirds. In: West G, Heard D, Caulkett N, eds. *Zoo Animal and Wildlife Immobilization and Anesthesia.* Ames, IA: Blackwell Publishing; 2007:269-297.

16. Plumb DC. *Plumb's Veterinary Drug Handbook.* 8th Ed. Ames. IA: Wiley-Blackwell Publishing; 2015.

17. Beaufrere H. Oxygen therapy. In: Graham J, Doss G, Beaufrere, eds. *Exotic Animal Emergency and Critical Care Medicine.* St. Louis, MO: Elsevier; 2021:450-456.

18. Ponder JB, Redig PT. Avian orthopedics: a game plan for success. *Proc ExoticsCon.* 2016;185-193.

19. Wright L, Mans C, Olsen G, et al. Retrospective evaluation of tibiotarsal fractures treated with tape splints in birds: 86 cases (2006-2015). *J Avian Med Surg.* 2018;32(3):205-209.

20. Driggers T. Flight Mechanics, Parrot Welfare, & Ethical Concerns. 2020. Available at: https://lafeber.com/vet/flight-mechanics-ethical-concerns/. Accessed Jan 23, 2024.

21. Beaufrere H. A review of biomechanic and aerodynamic considerations of the avian thoracic limb. *J Avian Med Surg.* 2009;23(3):173-185.

22. Peng SJL, Chang FC, Sheng-Ting JI, et al. Welfare assessment of flight-restrained captive birds: effects of inhibition of locomotion. *Thai J Vet Med.* 2013:43(2):235.

23. McRee AE, Tully TN, Nevarez JG, et al. Effect of routine handling and transportation on blood leukocyte concentrations and plasma corticosterone in captive Hispaniolan Amazon parrots (*Amazona ventralis*). *J Zoo Wildl Med.* 2018;49(2):396-403.

24. Heidenreich B. Good Bird Inc. Ebook. Train Your Parrot to Accept Oral Medication Video Tutorial 2020. Available at: https://www.youtube.com/watch?v=FwtHWbnZexI. Accessed April 15, 2024.

25. McLaughlin A, Heidenreich B. Giving Medication to Birds Without Fear and Stress – for Birds or Humans! Fear Free Happy Homes. Available at: https://www.fearfreehappyhomes.com/giving-medication-to-birds-without-fear-and-stress-for-birds-or-humans/. Accessed April 15, 2024.

26. American Veterinary Medical Association. Veterinary Ergonomic Guidelines. Available at: https://www.avma.org/resources-tools/avma-policies/veterinary-ergonomic-guidelines. Accessed January 28, 2024.

27. White S. Ergovet. Available at: https://ergovet.com/surgical-ergonomics-movement-posture/. Accessed January 28, 2024.

28. Lee GI, Berguer R, eds. Surgical Ergonomics Recommendations. American College of Surgeons, Division of Education & Surgical Ergonomics Committee. 2022. Accessed January 28, 2024. https://www.facs.org/media/cezf1xvo/surgicalergonomicsrecommendations.pdf.

29. Emmer KM, Celeste NA, Bidot WA, et al. Evaluation of the sterility of Press'n Seal Cling Film for use in rodent surgery. *J Am Assoc Lab Anim Sci*. 2019;58(2):235-239. doi:10.30802/AALAS-JAALAS-18-000096.

30. Doneley B, Black D. Using the Lone Star Retraction System - basic principles and techniques. *Proc Annu Conf Assoc Avian Vet Australasian Cmte*. 2015:1-11. Available at: https://www.aavac.com.au/files/2015-23.pdf. Accessed April 15, 2024.

31. Lewis TT, Lennox AM. Nonsurgical removal of urethral uroliths using a self-retaining retractor with elastic stays in female guinea pigs (*Cavia porcellus*): 16 Cases (2006-2019). *J Exot Pet Med*. 2021;36:11-15. doi: 10.1053/j.jepm.2020.09.006.

Avian Coelomic Surgery Fundamentals: Focus on Reproductive Tract Disease

Angela M. Lennox, DVM, DABVP (Avian Practice), DABVP (Exotic Companion Mammal Practice), DipECZM (Small Mammal)

Session #200

Affiliation: From Avian and Exotic Animal Clinic, 9330 Waldemar Road, Indianapolis, IN 46268, USA.

Abstract: There are many indications for coelomic surgery in birds, most commonly for the treatment of reproductive disease, but also for diagnostic testing and treatment of diseases of the gastrointestinal (GI) tract and spleen. While approaches are well described in textbooks and case reports, an objective large-scale comparison of approaches, diseases and outcomes is not available. Reproductive disease is common in pet avian species and is likely due to alterations of normal cyclical triggers for reproductive activity in the wild, including seasonal light cycles, food availability and the presence of an appropriate mate. Reproductive disease in female birds may involve the ovary, salpinx, and/or cloaca, and an accurate diagnosis is critical for successful resolution, which may include medical and/or surgical management. This masterclass reviews the anatomy of and approaches to the coelomic cavity, and the thorough approach to diagnosis and management of reproductive disease, with emphasis on surgical management.

The Coelomic Approach in Birds

All animal body cavities are divided into compartments. Reptiles are the simplest with two to three cavities. Mammals have three cavities. The avian coelomic cavity has eight compartments. These include the pericardial cavity, the intestinal peritoneal cavity, two pleural cavities, and four hepatic peritoneal cavities. These cavities are of primary importance when considering coelomic surgery. The four hepatic peritoneal cavities (HPC) surround both right and left liver lobes (i.e., left and right ventral and smaller left and right dorsal). The single intestinal peritoneal cavity (IPC) contains most of the GI tract, the spleen and the reproductive tract.[1]

The ventral hepatic cavity(s) are entered when collecting a liver biopsy via the ventral keel approach. Surgeries of the reproductive tract, GI tract and spleen necessitate entering the IPC.

The proventriculus to the terminal GI tract are present within the IPC and are suspended by a dorsal mesentery. The reproductive tract is suspended dorsally by the mesovarium (ovary), the mesosalpinx (oviduct), and mesorchial (testicles). A more thorough description of air sac and coelomic cavity anatomy is available.[1,2]

There is very little fluid normally present within the cavities. In cases of right heart failure or portal hypertension, fluid will accumulate within the HPC first, and may then accumulate within the IPC as well. Fluid production secondary to reproductive disease, will accumulate within the IPC. Ectopic eggs or yolk will also normally be found in the IPC; an exception could be cases of adhesions or erosion of the serosa allowing yolk to enter the lung, air sac, or other neighboring cavities.

The approach to the coelomic cavity depends on the goals of the procedure and varies by species and the exact nature of the intervention, taking into account anatomic variations produced by specific disease conditions. For example, the presence of a fluid or a space-occupying mass may shift the location of the GI tract and the salpinx.

Many more details concerning comparative avian anatomy and modifications for coelomic surgery have been presented.[3] In general, avian skin is very thin, with little subcutaneous fat, and in most cases reduced thickness of muscle layers. Surgical site preparation time can be extensive when feathers must be plucked, or down feathers shaved from larger areas, especially in domestic fowl. Magnification, excellent lighting and microsurgical instruments and techniques are of critical importance. Several retractors have been used in birds, including the Heiss retractor, and others. The author prefers the Lone Star ring retractor, which is a plastic ring with elastic stays and hooks that can be positioned in any configuration desired.

A variety of approaches can be used to access the IPC, including a standard midline or transverse approach with the bird in dorsal recumbency, a left lateral approach with the bird in right lateral recumbency (usually with the left leg extended and taped forward), or with the patient in a slightly oblique position between dorsal and lateral recumbency. A right lateral approach is rarely utilized, especially for reproductive disease, but may be more useful for specific conditions (e.g., right lung, right liver).

For any approach, the author positions all birds in a slightly inclined position (reverse Trendelenburg), which at least subjectively improves ventilation and respiratory parameters. Incisions can be simple (e.g., midline or transverse), or a combination using an “L” or “T/inverted T” shape. The initial incision may need to be expanded or modified depending on the ability to access the area of interest. In general, the gonads tend to be easier to access via a left lateral approach. Excellent images and drawings of incision site options and their relationship to the organs have been published.[2]

The skin incision can be made with fine scissors, or with radiosurgery to reduce bleeding. Regardless of approach, the supraduodenal loop of intestines is located along the mid to caudal ventral midline in psittacine birds and other bird species and can easily be incised or transected during approaches to the IPC. When this loop is manipulated or lifted along with the body wall, it often collapses and is camouflaged by the serosal body wall itself, making identification tricky. A more cranial initial approach can help identify the loop, which is released from attachments to the ventral body wall and safely deflected before extending the incision. Some severe diseases may produce adhesions of organs to organs, and even organs to the body wall, complicating the approach.

Once the IPC is opened, a retractor is placed, and organs are identified. This can be very complicated in cases of gross distension of the reproductive tract, free fluid and egg, and adhesions. In some cases, the amount of fluid is very large. For example, one duck in the author’s practice produced nearly a liter of fluid.

Ventral midline approaches tend to avoid rupture of the air sacs, which is extremely beneficial in cases of marked fluid accumulation. In many cases, air sacs are breeched to access the area of interest, and in some cases to prevent the surgery site from being continually obscured. Anesthesia must be adjusted (e.g., often anesthetic gas concentration increased), as anesthetic gases exit the surgical site.

For closure, the muscle layers are opposed and closed using absorbable monofilament suture in a simple continuous pattern. Skin can be closed using simple interrupted or a continuous pattern.

With careful technique and adequate analgesia, most birds do not disrupt sutures and surgical sites. For this reason, the author and others do not routinely place post-operative collars or bandages. An exception is repair sites when the injury was self-induced by the bird itself as part of self-mutilation syndrome.

Diseases and Management of Reproductive Disease in Companion Birds

Introduction

Reproductive disease is common in pet avian species and is likely due to alterations of normal cyclical triggers for reproductive activity in the wild, including seasonal light cycles, food availability and the presence of an appropriate mate. Reproductive disease in female birds may involve the ovary, salpinx, and/or cloaca, and an accurate diagnosis is critical for successful resolution. More commonly reported ovarian diseases include cystic disease, neoplasia and oophoritis. Diseases of the uterus include salpingitis (infectious or inflammatory), neoplasia, trauma or rupture, and torsion.[4] Egg product (yolk, albumin with or without shell) may be ectopic, or retained anywhere within the salpinx. In some cases, multiple impacted eggs in varying stages (shelled/unshelled) may be found within a greatly distended salpinx. The salpinx may also be fluid-distended, or there may be distention of the IPC which contains the salpinx.

The most difficult decision may be to determine when surgical intervention is the best course of action for resolution. The author finds the following criteria useful:

- Failure of medical management to resolve the condition.
- Failure of manual manipulation to resolve a dystocia.
- Peritonitis with the presence of fluid resulting in clinical signs (e.g., lethargy, increased respiratory effort).
- Chronic oviposition that has resulted in multiple dystocias requiring medical intervention.

Diagnosis of reproductive disease

Confirmation of gender in non-dimorphic species is an important first step in any bird with suspected reproductive disease. In many cases, owners believe they know the gender of the bird, but may be relying on secondhand information and assumptions. Even in cases where owners have found eggs, in multi-bird households, the bird spending time with the egg may not be the bird that produced it. The author once consulted on case where the egg identified radiographically by another veterinarian was a mineral-dense GI foreign body in a male bird. When in doubt, gender should be confirmed objectively.

History may give clues as to the presence of reproductive disease. An important clue is a history of regular egg laying which abruptly stopped; without a directed effort to mitigate hormonal stimulation, these birds have often gone on to develop reproductive disease. Birds with reproductive disease may have a history of reproductive behaviors directed to other birds or their owners. Birds with diseases featuring coelomic distention may show exercise intolerance or increased respiratory effort at home.

Physical examination findings may be unremarkable, but often show a distended coelomic space (i.e., space between the ventral keel and the pelvis), sometimes with a mass or fluid effect. In normal psittacine birds, the coelomic space is narrow and concave. In birds with coelomic distention from any cause, the space may widen and in severe cases take on a more convex appearance. Many birds with coelomic distention show varying degrees of increased respiratory rate and effort, especially after handling. As reproductive disease can produce discomfort or affect general well-being, feather-destructive behavior may be present. Other abnormal physical examination (PE) findings include distention or flaccidity of the vent, and varying degrees of prolapse.

Diagnostic imaging, including radiography, ultrasonography and computed tomography (CT) scanning are very useful and may provide more information on the nature of disease in birds with reproductive disease. In many cases, the exact diagnosis and extent of the disease, including the presence of metastasis in the case of neoplasia, and the presence of adhesions to the body wall and other organs are not fully appreciated until surgery.

Radiographs and CT evidence of reproductive disease can include the presence of mineralized shell of intact or collapsed eggs, round masses around the gonad consistent with ovarian cysts, and a generalized mass effect that may cause displacement of the GI tract. Many birds actively laying or with reproductive disease show hyperostosis, or increased mineralization of the long bones, with loss of the visible medullary cavity. Hyperostosis is transitory in normal laying birds but can persist for months to years in birds with reproductive disease. Hyperostotic changes in abnormal birds can be patchy to severe, even involving bones of the vertebrae. Of interest is the observation that not all birds with severe confirmed reproductive disease show hyperostosis.

Routine bloodwork (i.e., CBC and biochemistry panel) provides general information on overall condition and helps identify the presence of infection and inflammation. Some birds with reproductive disease have elevations in calcium, alterations in lipids (most specifically triglycerides) and elevations of uric acid.

Medical management of reproductive behavior and disease

Reproductive-related behaviors and oviposition are not necessarily diseases. Outside of intentional breeding, incidence should be discouraged and minimized. Behavioral modification to decrease reproductive stimulus for both normal oviposition and reproductive disease is well described and includes techniques such as the removal of nesting sites, establishment of day/night cases, control of light cycles, integration of foraging, and retraining owners to interact with birds in a manner most consistent with social flock members outside of breeding season.

GnRH agonists such as leuprolide acetate and deslorelin have been shown to help reduce both normal reproductive behaviors and treat or manage some forms of reproductive disease, such as salpingitis and certain cancers. GnRH agonists are not useful for resolution of diseases involving egg retention, impaction, and ectopic eggs. Even in these cases, reduction of hormone levels can be a useful adjunct to surgical management for the prevention of continued hormonal stimulation.

Specific reproductive diseases

Retained eggs and dystocia

Eggs (fully shelled or yolk with or without albumin) can be retained in multiple places within the coelomic cavity including the salpinx, the salpinx-cloacal region, partially protruding from the vent and free within the intestinal peritoneal cavity (ectopic). Location, among other factors, determines clinical presentation, severity of disease and options for treatment. Eggs within more distal portions of the reproductive tract may result in compression of the ureters and vasculature, and mechanically interfere with voiding of urine and stool; these birds require urgent care, quick resolution, and are often in poor clinical condition. This is in contrast to birds with ectopic eggs or those retained more proximally. In some cases, these eggs may have been present for many months to years before clinical signs develop.

Urgent care for any sick bird consists of treatment of fluid deficits including hypovolemia and dehydration, warmth, and analgesia if indicated. For birds with distally located eggs causing compression and mechanical blockage of stool and urine also require rapid decision-making for delivery of the egg (discussed in detail below).

For stable birds with formed eggs where there is a reasonable possibility of natural oviposition, initial medical management includes warmth, semi-dark environment, analgesia, nutritional support including calcium (especially in cases of suboptimal diet) and fluids. Birds are monitored for oviposition. Keep in mind that birds are seldom able to deliver collapsed eggs or eggs with a poorly formed or absent shell.

To enhance natural delivery, oxytocin has been suggested and may be effective, but there is scant evidence of efficacy in psittacine birds. DinoprostonePGE2 (Prepadil Pharmacia and Upjohn, Kalamazoo, MI, USA) has been reported to relax the uterovaginal sphincter while inducing uterine contractions. The author has no experience with this drug.

Confirmation of ectopic eggs, especially those without formed shell, is often challenging with any imaging modality. CT may be more useful for identifying the salpinx independent of the egg. Sometimes a markedly abnormal position may give a hint, but a very enlarged and abnormal uterus may contain an egg that appears to be in an unexpected location.

Manual delivery can be considered for formed eggs that appear to be within the distal position in the reproductive tract. While occasionally an egg within the cloaca can be expelled with very little effort in the conscious bird, the safest approach is under general anesthesia after optimizing clinical condition and administering pre-medications. The vent and cloaca are lubricated and gently dilated. Access is greatly facilitated with the Lonestar retractor after the application of a local block (i.e., lidocaine 2 mg/kg) to the vent. Gentle pressure is applied at the proximal aspect of the egg. As the egg begins to present into the cloaca, the small pink raised vaginal fold, which can resemble a cervix, is identified within the mid to left wall of the cloaca. White shell can be seen through the thin wall, and the cervix can be gently identified and dilated with a blunt probe to expose the eggshell. At this point, additional pressure may allow passage of the egg. Alternatively, the shell can be punctured with a hypodermic needle, and the shell and contents carefully removed.

If, despite gentle pressure, the egg cannot be visualized, then there may be adhesions, obstructions, or the egg may be ectopic. Additional pressure may result in rupture of the salpinx and/or delivery of the egg through the thin wall of the cloaca, which results in injury and possible death. Complications of this procedure include uterine trauma and/or rupture, collapse of the egg, hemorrhage, and respiratory compromise secondary to excessive coelomic pressure during delivery.

Collapse of the egg before delivery is an option for eggs that can be directly visualized via the cervix. Aspiration of egg contents through the cloacal wall, or worse the coelomic wall has a high risk of potential harm and is not recommended by the author, especially in the case of an ectopic egg. The author is aware of serious injury in an Amazon parrot where this procedure resulted in laceration of the intestines and peritonitis (Nemetz, personal communication).

After successful delivery, GnRH agonists (if not already given) are recommended, along with steps to mitigate further hormonal stimulation, as discussed above. This can be effective for birds with an otherwise healthy reproductive tract, or for cases of mild reproductive disease such as mild salpingitis. Antimicrobials and anti-inflammatories are beneficial for birds with suspected infectious or inflammatory diseases of the reproductive tract, which may be supported by an abnormal hemogram or inferred. For birds with severe reproductive disease, resolution may ultimately require surgery and salpingectomy with or without partial ovariectomy (discussed below).

If attempts at natural and manual delivery fail, the next options are coelomic surgery or in cases where surgery is not an option, palliative therapy.

Surgical Considerations for Disease of the Salpinx and Ovary

Coelomic surgery of the reproductive tract is well described in avian patients, and approaches are discussed above. While technically challenging, it is facilitated by properly sized surgical instruments, adequate retraction, various-sized hemostatic clips and applicators and magnification.

Surgery of the salpinx

The diseased salpinx may be normal in size and position to grossly enlarged, fluid filled or with adhesions to other coelomic structures. Neoplastic conditions may feature metastasis anywhere in the coelomic cavity. A thorough appreciation of normal anatomy is critical. Of interest are reported numerous cases of persistent right oviduct (PRO) that may enlarge after successful removal of the left oviduct (Nemetz, personal communication).

The salpinx is identified, and a plan developed for ligation of the entire infundibulum, ligation of the vascular supply to the salpinx, and the salpinx as it enters the cloaca. The order of ligation is determined at the time of surgery, and often depends on ease of access. A very enlarged salpinx may be lifted partially out of the incision; smaller organs may require manipulation and ligation within the coelomic cavity.

In some cases, the uterus is so large and filled with debris/eggs or fluid that salpingotomy is performed first to reduce the size for manipulation and ligation. Care is taken to pack the coelomic cavity with gauze to prevent spillage of contents during evacuation.

The cloaca can be identified by pre-placing a lubricated cotton-tipped applicator into the vent and manipulating it through the drape. The surgeon can then readily identify the structure by movement. The colon and ureters are identified and avoided before ligation of the distal salpinx. Ideally, as much salpinx as possible is removed.

Other variations that may become necessary include sectioning, or partial resection of a portion of a very large salpinx to improve visibility of the infundibulum and distal salpinx.

Ligation is performed with appropriately sized suture for larger patients (poultry) and hemostatic clips, depending on the size of the structure to be ligated, which can vary significantly with the size of the patient and of the salpinx itself. For particularly large portions of the salpinx, at the level of the cloaca, a transfixation suture pattern is important to prevent suture from slipping.

Surgery of the ovary

Oophorectomy is considerably more challenging than surgery of the salpinx, and requires excellent exposure, lighting, magnification (often 6X in species < 105 g), knowledge of ovarian anatomy, vasculature (the vascular supply is better termed a plexus rather than discrete vessels), and experience. Considering the risks, practitioners often elect to remove the salpinx without oophorectomy, or remove only grossly abnormal portions. Complete oophorectomy with removal of 100% ovarian stroma is challenging and has not been convincingly documented in psittacine birds in the literature. A study evaluating the use of a vessel-sealing device in immature and adult chickens showed histopathologically confirmed incomplete excision in 40% of birds, and severe hemorrhage in 15% of birds. For this reason, owners should understand that remaining ovary will likely lead to additional reproductive pathology, including hyperplasia of the ectopic right oviduct, if hormonal stimulation is not completely halted.

There are no current retrospective surveys comparing surgeries (i.e., salpingectomy alone and with varying degrees of subtotal ovariectomy) with time to onset of return of confirmed reproductive disease. Nemetz had a case of a cockatiel undergoing simple salpingectomy with no additional therapies that survived another 23 y. Lennox has a case of presumably excised oviductal carcinoma doing well with GnRH agoinists 12 y later.

The ovary is approached after salpingectomy (entire left plus persistent right oviduct if present), and any clots or debris removed. The ventriculus often blocks visualization of the ovary. The ventriculus is connected to the ventral keel and lateral body wall (the IPC serosa and air sac serosa), which holds the ventriculus in position. These connections can be carefully broken down and the ventriculus released which improves visualization. The ureters are identified and checked for patency by observing urine and urate flow, which can be seen about every 15-20 seconds.

The ovary will vary in appearance and size with species and the type of pathologic changes. In addition to ovarian follicles and cysts of variable size, the ovarian artery and the plexus-like vein are located within the short ovarian stalk. The ovary is covered ventrally by serosa, with fascial attachments dorsally. These attachments are visualized with careful manipulation of the ovary and released with gentle careful blunt/sharp dissection. This allows loosening of the ovary for better manipulation.

Partial ovariectomy may be performed by carefully placing hemostatic clips between the area of pathology and the rest of the ovary, and excising. Large ovarian cysts can be twirled along the narrow attachment with a sterile cotton-tipped applicator until released, and then removed. Care should be taken to remove any yolk inadvertently spilled into the IPC.

For subtotal or attempt at complete ovariectomy, the goal is to place a series of hemostatic clips along the stalk parallel to the spine, beginning caudally and progressing cranially. All clips are placed slowly and carefully using controlled pressure to avoid inadvertently puncturing the vasculature or tearing the ovary. The following technique was developed by Nemetz.

A first initial larger (i.e., size medium) hemoclip (Weck® Hemoclip® System, Teleflex Incorporated, Morrisville, NC, USA) is placed caudally to provide additional compression of the stalk. In some cases, this may span the entire ovary (e.g., in smaller birds), but will not in larger birds, or in larger, more irregular ovaries.

The second layer is specifically microclips (Weck® Horizon™ Microclip System, Teleflex Incorporated, Morrisville, NC, USA), which are a different product than hemoclips. Microclips are designed to clamp securely along the entire length of the clip; this is opposed to regular hemoclips, which are designed to ligate specific individual blood vessels and surrounding tissue and are not completely clamped along the entire width of the clip. The use of microclips requires magnification, as these are only about 2 mm in length. Microclips are placed slightly overlapping to seal the entire vascular plexus. Placement is verified by identifying both sides of the stalk. To place subsequent clips, the ovary is cut one-half the length of the microclip to provide room to place the next clip. Placement is tedious and requires significant time and care.

The number of microclips required varies according to the size of the ovary. On average, a conure may require six or seven and a macaw may require twice as many.

Fine scissors are used to detach the ovary slowly in sections, watching carefully for evidence of bleeding. In this case, the stalk has not been completely clamped, and additional clips are required.

Some authors advocate using radiosurgery with a ball tip to destroy any remaining ovarian tissue. Gel foam may help control any mild bleeding.

In all cases, owners must take steps to prevent inadvertent hormonal stimulation due to the presence of remaining ovary, even after apparent complete ovariectomy. This includes behavioral modification, and GnRH agonists. Ongoing monitoring is highly recommended, including periodic radiographs and bloodwork with evaluation of lipids, in particular triglycerides.

Palliative Care

Birds can benefit from palliative care, including periodic ultrasound-guided removal of fluid-producing respiratory impairment. Two cockatiels with ovarian adenocarcinoma confirmed at necropsy were managed with coelomocentesis and GnRH agonists for 9 and 25 mo,[6] Care must be taken to fully assess the quality of life for any bird where palliative care is elected.

Cloacal Prolapse

Prolapse of tissue from the vent has a variety of causes, not always related to the reproductive tract. Except for cases of cloacal cloacitis, papilloma, neoplasia or trauma it is rarely a disease of the cloaca itself. In mild cases where an egg has passed and prolapsed tissue is healthy, simple replacement and stay sutures after carefully cleansing and irrigation of the tissue may allow healing and resolution. Steps are then taken to prevent further oviposition. Severely damaged tissue may require debridement, and damaged salpinx may necessitate hysterectomy. Under no circumstances should the prolapse be replaced and sutured without a thorough investigation of the cause and a plan for resolution.[7,8]

References

1. Taylor WM. Pleura, pericardium and peritoneum: the coelomic cavities of birds and their relationship to the lung-air sac system. In: Speer B, ed. *Current Therapy in Avian Medicine and Surgery.* St. Louis, MO: Elsevier; 2016:345-362.

2. Soft tissue surgical procedures. In: Altmann RB, Clubb SL, Dorrestein GM, Quesenberry K, eds. *Avian Medicine and Surgery.* Philadelphia, PA: WB Saunders; 1997:704-711.

3. Mison M, Mehler S, Echols MS, et al. Approaches to the ceolom and selected procedures. In: Speer B, ed. *Current Therapy in Avian Medicine and Surgery.* St. Louis, MO: Elsevier; 2016:638-665.

4. Petritz OA, Lierz M, Samour J. Advancement in methods for decreasing reproductive success. In: Speer B, ed. *Current Therapy in Avian Medicine and Surgery.* St. Louis, MO: Elsevier; 2016:440-460.

5. Nemetz L. Persistent right oviduct in parrots. *Proc Annu Conf Assoc Avian Vet.* 2010.

6. Keller KA, Beaufrere H, Brandao J, et al. Long-term management of ovarian neoplasia in two cockatiels. *J Avian Med Surg.* 2013;27(1):44-52.

7. Gill KS, Helmer PJ. Cloacal diseases in companion parrots: a retrospective study of 43 cases (2012-2018). *J Avian Med Surg.* 2020;34(4):364-370.

8. Lennox A, Lee A. Cloacal, uterine and rectal prolapse. In: Chitty J, Monks D, eds. *BSAVA Manual of Avian Practice: A Foundation Manual.* Gloucester, UK: British Small Animal Veterinary Association; 2018: 358-365.

The Avian Neurologic Examination and Seizure Management

Natalie Antinoff, DVM, DABVP (Avian Practice)

Session #142

Affiliation: From Texas Avian & Exotic Hospital, 2700 W. State Hwy 114, Grapevine, TX 76051, USA.

Note: Previously presented in Antinoff A. The avian neurologic exam and seizure management. *Proc ICARE.* 2024.

Abstract: Neurologic or seizuring patients require rapid intervention, frequent assessment, and often, very intensive care. Serial neurologic exams are critical in establishing treatment and prognosis, but the neurologic exam itself can be the most challenging aspect of this process. This discussion addresses clinically relevant differences as they apply to the interpretation of the neurologic exam, followed by a discussion of practical therapeutics for the management of the neurologic or seizuring avian patient.

The Avian Neurologic Examination

Perhaps the most challenging aspect of avian medicine is adapting standard tests utilized for mammals and finding a method or technique for performing these tests in birds. Interpretation is also dependent on a thorough understanding of the nerves and their pathways of innervation.

There are six main features to assess in any neurologic examination. These features include the following:

1. Mentation
2. Cranial nerves
3. Posture and gait
4. Postural reactions
5. Spinal reflexes
6. Sensation (cutaneous sensation and deep pain)

A detailed discussion of each of these features is provided below.

Mentation

Mentation includes the level and content of consciousness and appropriateness for surroundings. Abnormalities may include personality changes, changes in awareness, or inappropriate behavioral responses. For example, any bird in an exam room or unfamiliar situation should be alert and aware; a bird that is fluffed, tucks its head, or does not respond to the presence of a stranger has an altered level of consciousness. Delirium or lack of recognition of the owner represents an altered content of consciousness. To further categorize mentation, it is necessary to define basic terms.

Levels of consciousness

Levels of consciousness, from alert to comatose, are defined in Table 1. Depression and lethargy are generally metabolically induced changes, whereas obtunded, stuporous, or comatose indicate changes to the cerebrum or brainstem.

Table 1. Levels of consciousness.

Level	Description
Alert	The bird demonstrates normal awareness of its surroundings.
Lethargic	The bird is aware but demonstrates slightly sluggish responses.
Depressed	The bird exhibits decreased responses to environmental stimuli; sleeping when undisturbed.
Obtunded	The bird exhibits decreased responses to environmental stimuli; sleeping even in strange surroundings, but easily roused.
Stuporous	The bird requires noxious stimuli to elicit awareness of its surroundings.
Comatose	There is no awareness of surroundings; voluntary responses cease; only reflex activity remains.

Content of consciousness

Content of consciousness states are defined in Table 2. These states may indicate structural or metabolic damage to cerebrum. The mental awareness of the patient can be determined, in part, from the historical answers of the owners.

Table 2. Content of consciousness.

State	Description
Confusion	The bird appears confused.
Dementia	The bird demonstrates little awareness or concern for its surroundings (e.g., flying into walls).
Delirium	The bird is over-responsive to stimuli; hyperresponsive or hyperexcitable with the owner or familiar situations.

Cranial nerves

Cranial nerves, their functions, dysfunctions, and descriptions are provided in Table 3.

Table 3. Cranial nerves.

Number	Name	Function
CNI	Olfactory	Sensory for smell. Rarely assessed in birds. The only way to determine function is to provide a noxious olfactory stimulus. May be suggested by changes in appetite.
CNII	Optic	Sensory for vision. Menace response: difficult to interpret in birds; blink is mediated by mandibular branch of CN V (orbicularis oculi) and CN VII. Try to assess vision or perception of an object, or avoidance of obstacles. Pupillary light response: also requires intact CNIII for iris constriction. May be overridden in birds due to presence of striated muscle in iris and voluntary control of papillary constriction and dilation; however, it is possible to assess. Perform early in the exam; use a bright sudden focal light source aimed toward the medial canthus. Birds lack a consensual papillary light response due to complete decussation of the optic nerves at the optic chiasm.
CNIII	Oculomotor	Motor and sensory to upper eyelid muscle. - Dysfunction: drooping of upper eyelid Motor to extrinsic ocular muscle (dorsal) - Dysfunction: ventrolateral deviation of globe Parasympathetic to intrinsic ocular muscle (iris constriction) - Pupillary light response: responsible for iris constriction; suspect dysfunction if patient seems visual but iris does not constrict
CNIV	Trochlear	Motor and sensory to extrinsic ocular muscle (ventromedial). - Dysfunction: dorsolateral deviation of globe
CNV	Trigeminal	Three branches: ophthalmic, maxillary, and mandibular. Mixed sensory (major sensory nerve for the head) and motor (muscles of mastication. Ophthalmic, sensory to facial sensation around eye, upper eyelid, cornea, forehead, nares, upper beak - Dysfunction: lack/loss of sensation to face/beak Maxillary, sensory to muscles of the face/mouth – both eyelids, hard palate, nasal cavity - Dysfunction: lack/loss of sensation to face, beak, mouth/palate Mandibular, motor, lower eyelid, chewing, commissures of beak - Dysfunction: inability to close eye, inability to close beak, diminished beak strength
CNVI	Abducens	Motor and sensory to lateral extrinsic ocular muscles and third eyelid. - Dysfunction: loss of function of third eyelid; medial deviation of globe Note: third eyelid dysfunction or prolapse does not indicate Horner's syndrome or loss of sympathetic innervation
CNVII	Facial	Mixed motor and sensory. Motor to muscles of facial expression - Dysfunction: lack of response to skin touch/facial sensation; facial droop or asymmetry. can be difficult to evaluate. Sensory to taste, also shared with CN IX; may be difficult to evaluate Parasympathetic for many glands of head (tears, mucous) - Dysfunction: decreased secretions

Table 3 Continued.

Number	Name	Function
CNVIII	Vestibulocochlear	Special sensory. Sensory to hearing, balance, and coordination - Dysfunction: difficult to assess hearing in birds; a whistle or loud noise may provide appropriate stimulus Loss of equilibrium-vestibular system Oculocephalic reflex, normal physiologic nystagmus can be induced by moving the head side to side; fast phase is in the direction of movement. - Dysfunction: spontaneous horizontal nystagmus, fast phase away from side of lesion, head tilt and circling towards affected side
CNIX	Glossopharyngeal	Mixed sensory and motor Sensory to tongue, taste, trachea - Dysfunction: decrease/loss of taste (unreliable to test); lack of tongue sensation Motor to pharynx, larynx, crop, syrinx - Dysfunction: dysphagia; loss of voice (?); gag reflex (difficult to assess) Note: anastomoses with with CN X; assess together
CNX	Vagus	Mixed sensory and motor Sensory to larynx, pharynx, viscera - Dysfunction: regurgitation? Motor to larynx, pharynx, esophagus, crop - Dysfunction: regurgitation, voice dysphagia; loss of voice (?); gag reflex (difficult to assess) Parasympathetic to glands, viscera - Dysfunction: lack of rhythmic visceral/gastric contractions; increased heart rate Note: anastomoses with CN IX; assess together
CNXI	Accessory	Motor to superficial neck muscles, cranial scapula muscles - Dysfunction: poor neck mobility; lateral deviation of scapula (not generally clinically detectable)
CNXII	Hypoglossal	Motor to tongue, trachea, syrinx - Dysfunction: deviation of tongue, decreased tone of tongue, change in or absence of song

Clinical notes on cranial nerve examination

The visual response includes CN II, III, IV, and VI. To add the menace test, CN V has been included. Birds have voluntary control of pupillary constriction and dilation. Therefore, there may be no consensual or directed response when shining a light into each eye. Anisocoria may be the result of a lesion in the optic nerve or the oculomotor nerve; however, electrodiagnostic tests may be used in birds to determine whether CN II is intact.

Posture and gait

Postural responses are responsible for maintaining the body in an upright position, and the use of the limbs in maintaining this position. The sensory and motor receptors in the peripheral nervous system (PNS) transmit information to and from the central nervous system (CNS), which interprets input and directs response, translating impulses into actions. Interference in these pathways can occur in either the lower motor neuron (LMN), the upper motor neuron (UMN), or in the brain itself. Generalized weakness or recumbency are non-specific and may involve either metabolic or neurologic pathways.

Observe the patient in its cage. A normal posture is one where the bird can stand on its perch appropriately with its head erect. A head tilt or a twisting of the body are considered abnormal. Postural changes may occur with vestibular or cerebellar disease or asymmetrical lesions of the cervical spinal cord.

Lower motor neuron

Lesions affecting the PNS will lead to loss of muscle power, decreased or absent reflexes, and flaccidity.

Upper motor neuron

Lesions in the CNS will cause increased muscle tone and rigidity in the muscles distal to the lesion. Lesions in the thoracic region of the spinal cord will cause hyperreflexia and rigidity in the hind limbs, while the wings will remain normal. Cervical lesions will cause these signs in all four limbs.

Brain

Opisthotonus: caused by upper motor neuron lesion of the cerebellum or brainstem.

Decerebrate rigidity: extension of all four limbs and trunk, caused by lesion of the rostral brainstem.

Vestibular system

The vestibular system also has central and peripheral components, which function in the same way as the peripheral nerves. Abnormalities manifest as nystagmus, head tilt, rolling, circling, and sometimes leaning to one side. Wide-based stance may also indicate vestibular disease.

Central vestibular system disease: peripheral input is normal but central system is unable to respond or maintain appropriate eye, limb, or body position. Nystagmus is positional. Vertical nystagmus can only be present with central vestibular disease. (Horizontal and rotary nystagmus can exist with either central or peripheral lesions.) Proprioceptive deficits and sometimes changes in mentation are present.

Peripheral vestibular system disease: Horizontal nystagmus, fast phase away from the lesion, but no proprioceptive deficits or changes in mentation. Head tilt and circling may be present.

Evaluation of gait involves assessment of both coordination and strength; both voluntary and involuntary movements require cerebral, cerebellar, vestibular, and proprioceptive pathways.

Cerebral lesion

Voluntary movement: lack of precise/finely tuned movement; missing perches or unable to move in certain directions.

Involuntary movements: seizures, muscle tremors

Cerebellar lesion

Voluntary movement: hypermetric or hypometric movements of limbs.

Involuntary movements: intention tremors

Vestibular lesion

Falling, loss of balance, circling

Proprioceptive lesion

Paralysis or paraplegia of limbs

Postural reactions

Move the bird toward the edge of a horizontal surface and brush the hindlimbs against it. Raptors can be hooded and placed on a low perch. Bring another perch towards them, slightly higher so that the surface touches their feet. With psittacine birds, it can be difficult to blindfold or hood the bird. Another proprioceptive test is the return of the wing to its normal flexed position when pulled away from the body from different positions.

Spinal reflexes

LMN lesions lead to hyporeflexia and diminished or absent responses, and UMN lesions lead to hyperreflexia. LMN lesions involve the reflex arc; UMN lesions involve the CNS proximal to the reflex arc. The standard scale of 0-4 used in companion animals is utilized, where 2 = normal, 3 = hyperreflexic, 4 = clonus, 1 = hyporeflexic, and 0 = absent.

Clinical note: Assess the tone of the thoracic and pelvic limbs. Manipulate all limbs passively and palpate long bone muscle mass to check for atrophy. Flex and extend the limbs to determine differences in muscle tone, and grade as increased, normal, or decreased for each limb.

Increased resistance to manipulation of the limb can occur with lesions of UMN system, increased facilitation of the efferent pathway, and alterations in muscle movement. It may be difficult to separate increased muscle tone from restriction of muscle movement from a mechanical cause. Other causes for increase in tone may be increased muscle irritation or decreased inhibition of the intact LMN pathways. An example of a disease that results in increased muscle rigidity not of UMN origin is tetanus.

Muscle tone can be decreased with both upper and lower motor neuron disease. However, decreased muscle tone is associated more commonly with LMN problems.

The cloacal reflex of birds is similar to the perineal reflex of mammals. A noxious or tactile stimulation of the area around the vent will result in contraction of the external sphincter muscles of the vent and ventral flexion of the tail. This requires that the pudendal nerves be intact.

Sensation

Cutaneous sensation

Cutaneous sensation can aid in assessment of the spinal cord, as well as localization and severity of injury. Birds lack a cutaneous trunci muscle and therefore have no panniculus response. Instead, feather follicles can be used to assess sensation. The location of avian dermatomes has not been described.

Deep pain sensation

Assessment of deep pain is necessary only when there are abnormalities of gait, spinal reflex, or proprioception. Deep pain fibers are located deepest in the spinal cord of all sensory fibers. When evaluating deep pain, *there must be conscious awareness of the stimulus*. Limb withdrawal alone is a reflex arc and does not indicate intact deep pain pathways. However, lack of withdrawal and presence of conscious awareness does indicate that deep pain neurons are intact and probably affords a better prognosis than simple withdrawal response.

Seizures

Seizures are composed of three phases. The aura occurs immediately before the seizure and is typified by a change in behavior. The ictal or seizure phase is often manifested by disorientation with the bird undergoing tonic-clonic convulsions that are associated with extensor rigidity of its wings, legs and neck. The post ictal phase is characterized by some type of abnormal behavior. Birds may act lethargic, disoriented, depressed, and/or restless. This phase may last from seconds to hours, but it often lasts for several minutes.

There are many causes of seizures in birds. These include heavy metal toxicosis, hepatic encephalopathy secondary to liver disease, trauma, and hypocalcemia. Idiopathic epilepsy in red-lored Amazon parrots may be of genetic origin. Seizures also have been reported in greater Indian hill mynahs, but the cause is unknown. Differentials for seizures in cockatiels include chlamydiosis, heavy metal toxicosis, liver disease, and secondary effects of yolk emboli. Raptors may seizure from concussive trauma, hypoglycemia, or systemic diseases, including exposure to neurotoxins.

Nutritional deficiencies have been suggested to cause seizures, particularly in young psittacine birds. Imbalances of calcium, phosphorus, and vitamin D_3, and deficiencies of the B vitamins (B_1 or thiamine, B_2 or riboflavin, B_6 or pryidozine, and B_{12} cyanocobalamine), vitamin E, and selenium have all been implicated as causing seizures.

Other potential etiologies to consider include infectious diseases such as *Chlamydophila* or proventricular dilatation disease, metabolic issues including hepatic or renal failure, or vascular and hypoxic events including atherosclerosis and hyperlipemia.

A diagnostic workup for seizures may include a number of tests based on the history and physical examination, including a complete blood count (CBC); profile, radiographs including skull films; blood concentrations of lead and zinc, acetylcholinesterase activity and/or insecticide residues; inhibition of delta-aminolevulinic acid dehydratase (ALAD) activity and blood protoporphyrins, tests for chlamydiosis, *Aspergillus* titers; electroencephalogram; computer-assisted tomography (CT) and magnetic resonance imaging (MRI).

The treatment plan should be based on the diagnosis and the underlining cause. The most important aspect to remember, regardless of species, is to control the seizures, and maintain normal body temperature. As in any animal, prolonged seizures may result in permanent changes or deficits, but the extent of these deficits cannot be assessed for several days. Supportive care may be required during that time, including fluid therapy, nutritional support, and sometimes even management of the sedated or debilitated patient, including turning the patient, lubricating eyes, and physical therapy.

In any seizing patient, it is wise to check blood glucose, and administer a dextrose bolus if necessary; however, hypoglycemia is very rare in adult birds. If vascular access is not readily available, or if the animal is actively seizing, proceed as you would in any other species. Administer midazolam (or diazepam), 0.5-1.0 mg/kg. Administer intravenously or intraosseously if possible, but this may be a challenge. Midazolam can also be administered intranasally, intramuscularly, or via cloaca with the same or double dose.

Levetiracetam bolus (100 mg/kg) can be administered concurrently with midazolam. Propofol can also be administered to arrest seizures but there is a risk of severe respiratory depression.

If a single dose is not effective, midazolam may be repeated up to three times or can be administered via continuous rate infusion. Start at a rate of 1.0 mg/kg/h. If cerebral edema is suspected, consider mannitol, 0.5-1.0 g/kg IV over 20 min. (Do not administer steroids in birds because they are highly susceptible to the immunosuppressive effects of steroids.) Once seizures have stopped for 12-24 hours, taper the infusion slowly over another 12-24 hours. If the benzodiazepine does not control seizures, consider levetiracetam 50 mg/kg q8h (or 100 mg/kg q12h), or begin a constant rate infusion of phenobarbital, 2-10 mg/kg/h. This may be given in conjunction with the diazepam; monitor for respiratory depression. Once controlled, depending on etiology, phenobarbital can be used for long-term seizure management. Potassium bromide can also be used for seizure control. Adjust dosages based on blood levels. Levetiracetam can be used at 50 mg/kg q8h or 100 mg/kg q12h; zonisamide can be used at 10-30 mg/kg q12h, and gabapentin can be used at 15 mg/kg q8h or 20 mg/kg q12hr. Check blood levels of phenobarbital along with liver values within 2-3 wk of starting therapy; in mammals, potassium bromide levels may not reach a steady state for 60-90 days, but this data is not known for birds.

References

1. Orosz S, Antinoff N. Clinical avian neurology and neuroanatomy. In: Speer B, ed. *Current Therapy in Avian Medicine and Surgery.* 1st Ed. St. Louis, MO: Elsevier; 2016:363-377.

2. Clippinger T, Bennett R, Platt S. The avian neurologic examination and ancillary neurodiagnostic techniques; a review update. *Vet Clin North Am Exot Anim Pract.* 2007;10:803-836.

3. Guzman D. Disorders of the nervous system. In: Samour J, ed. *Avian Medicine.* 3rd Ed. St. Louis, MO: Elsevier; 2016:4211-533.

4. Schnellbacher R, Beaufrere H, Arnold RD, et al. Pharmacokinetics of levetiracetam in healthy Hispaniolan Amazon parrots (*Amazona ventralis*) after oral administration of a single dose. *J Avian Med Surg.* 2014;28(3):193-200.

5. Kabakchiev C, Laniesse D, James F, et al. Diagnosis and long-term management of post-traumatic seizures in a white-crowned pionus (*Pionus senilis*). *J Am Vet Med Assoc.* 2020;256(10):1145-1152.

6. Keller K, Guzman D, Boothe D, et al. Pharmacokinetics and safety of zonisamide after oral administration of single and multiple doses to Hispaniolan Amazon parrots (*Amazona ventralis*). *Am J Vet Res.* 2019;80(2):195-200.

7. De Matos R, Noonan B, Schaefer D, et al. Pharmacokinetics of zonisamide after oral single dosing and multiple-dose escalation administration in domestic chickens (*Gallus gallus*). *Vet Med Sci.* 2021;7(5): 1928-1937.

8. Baine K, Jones M, Cox S, Martin-Jimenez T. Pharmacokinetics of compounded intravenous and oral gabapentin in Hispaniolan Amazon parrots (*Amazona ventralis*). *J Avian Med Surg.* 2015;29(3):165-173.

9. Shaver S, Robinson N, Wright B, et al. A multimodal approach to management of suspected neuropathic pain in a prairie falcon (*Falco mexicanus*). *J Avian Med Surg.* 2009;23(3):209-213.

Index

V

W

Association of Avian Veterinarians
PO Box 9
Teaneck, NJ 07666
USA

ISBN 979-8-3313-0827-8

CURRAN ASSOCIATES INC.
proceedings.com